Neural Bases of Timing and Time Perception

Neural Bases of Timing and Time Perception provides a cutting-edge overview of the main contemporary neuroscientific methods and findings in this burgeoning field.

Featuring an international collection of leading researchers, this text reports the main methodological tools available to address important questions in the field, what discoveries these tools led to, and what avenues remain to be explored. The book provides concise descriptions of the latest neuroscientific developments about time perception and temporal processing (for instance, how to use TMS or tDCS to study time judgments); and signposts avenues for clinicians to develop new insights for understanding pathologies (as in the case of schizophrenia, for instance) from a temporal perspective.

This book will appeal to anyone interested in how we perceive the passing of time, whether from an academic or clinical background.

Giovanna Mioni is an associate professor at the Dipartimento di Psicologia Generale, University of Padova, Italy. Her main research interests include the study of time perception and time processing in healthy and pathological aging using non-invasive brain stimulation techniques. She also investigates the effects of emotional stimuli on subjective time perception.

Simon Grondin is a professor at École de psychologie of Université Laval, Québec. His main research fields are perception and psychophysics, psychological time, and cognitive neuroscience. He is a former editor of the *Canadian Journal of Experimental Psychology* and a former associate editor of *Attention, Perception, & Psychophysics*.

Current Issues in Perception

Series Editors: Simon Grondin, *Université Laval, Canada*
Timothy Hubbard, *Arizona State University, USA*

Current Issues in Perception is a new edited book series conceived to meet the need for brief volumes that reflect the state-of-the-art in research on perception. The series contains cutting-edge research by leading academics and is published by Routledge.

Each volume in the series is tightly focused on a particular topic in perception and consists of 9-10 chapters contributed by international experts in the field. Each volume in the series will contribute to the study of perception by reviewing and synthesizing existing research literature, advancing theory, identifying possible future trends, and shaping future research agendas.

Published titles in this series:

Neural Bases of Timing and Time Perception
Edited by Giovanna Mioni and Simon Grondin

For more information about this series, please visit: www.routledge.com/Current-Issues-in-Perception/book-series/CURRIP

NEURAL BASES OF TIMING AND TIME PERCEPTION

Edited by Giovanna Mioni and Simon Grondin

Routledge
Taylor & Francis Group

NEW YORK AND LONDON

Cover image credit: Shutterstock

First published 2025
by Routledge
605 Third Avenue, New York, NY 10158

and by Routledge
4 Park Square, Milton Park, Abingdon, Oxon, OX14 4RN

Routledge is an imprint of the Taylor & Francis Group, an informa business

Library of Congress Cataloging-in-Publication Data
Names: Mioni, Giovanna, editor. | Grondin, Simon, editor.
Title: Neural bases of timing and time perception / edited by Giovanna Mioni and Simon Grondin.
Description: New York, NY : Routledge, 2025. | Series: Current issues in perception | Includes bibliographical references and index. |
Identifiers: LCCN 2024031044 (print) | LCCN 2024031045 (ebook) | ISBN 9781032583099 (hardback) | ISBN 9781032583082 (paperback) | ISBN 9781003449546 (ebook)
Subjects: LCSH: Time perception. | Neuropsychology.
Classification: LCC QP445 .N45 2025 (print) | LCC QP445 (ebook) | DDC 153.7/53--dc23/eng/20240911
LC record available at https://lccn.loc.gov/2024031044
LC ebook record available at https://lccn.loc.gov/2024031045

ISBN: 978-1-032-58309-9 (hbk)
ISBN: 978-1-032-58308-2 (pbk)
ISBN: 978-1-003-44954-6 (ebk)

DOI: 10.4324/9781003449546

Typeset in Times New Roman
by SPi Technologies India Pvt Ltd (Straive)

CONTENTS

CONTRIBUTORS

Philippe Albouy
École de psychologie and CERVO, Université Laval, Québec Canada

Maria Bianca Amadeo
Unit for Visually Impaired People (U-VIP), Italian Institute of Technology, Genova, Italy

Sonia Betti
Dipartimento di Psicologia Generale, Università di Padova, Padova, Italy

Maddalena Boccia
Department of Psychology, "Sapienza" University of Rome

Mariagrazia Capizzi
Mind, Brain and Behavior Research Center (CIMCYC), University of Granada, Spain
Department of Experimental Psychology, University of Granada, Spain

Nicola Cellini
Dipartimento di Psicologia Generale, Università di Padova, Padova, Italy

Nicola Domenici
Laboratoire des systèmes perceptifs, Département d'études cognitives, École normale supérieure, PSL University, CNRS, 75005 Paris, France

Anne Giersch
French Institute of Health and Medical Research, Strasbourg, France

Joseph Glicksohn
Department of Criminology and The Leslie and Susan Gonda (Goldschmied) Multidisciplinary Brain Research Center
Bar-Ilan University, Israel

Monica Gori
Unit for Visually Impaired People (U-VIP), Italian Institute of Technology, Genova, Italy

Simon Grondin
École de psychologie and CERVO, Université Laval, Québec Canada

Sophie K. Herbst
INSERM, CEA, CNRS, Université Paris-Saclay, France

Luigi Micillo
Dipartimento di Psicologia Generale, Università di Padova, Padova, Italy

Giovanna Mioni
Dipartimento di Psicologia Generale, Università di Padova, Padova, Italy

Martin Riemer
Biological Psychology and Neuroergonomics, Technical University Berlin, Germany

Alice Teghil
Department of Psychology, "Sapienza" University of Rome

Nicola Thibault
École de psychologie and CERVO, Université Laval, Québec Canada

William Vallet
INSERM U1028, CNRS UMR 5292, Université de Lyon 1, Lyon, France

PREFACE

Our ability to perceive the world around us and our sense of self are shaped upon our perception of time. Yet, the way we perceive time is widely debated. Interval timing in the milliseconds-to-minutes range is believed to underlie a variety of complex behaviors in humans and other animals. One of the most challenging issues in the field of timing and time perception is to determine how the brain masters temporal events in different temporal ranges and in a variety of contexts.

There has been an explosion of neuroscientific research in the field of timing and time perception so far in this 21st century to understand how the brain processes temporal information. This research was driven by new technological possibilities as well as by an increased interest in the scientific community toward psychological time. The main purposes of the book are to provide a primer for the main technological tools in neuroscience used to study time perception and to provide the reader with some of the cutting-edge findings in this research field and a description of the research avenues that remain to be explored.

We believe the book will be of interest to both researchers and clinicians. Researchers will find in this book concise descriptions of the latest neuroscientific developments about time perception and temporal processing based on fMRI, EEG, MEG, neurostimulation, and psychophysiological indices, while clinicians will find avenues to develop new insights for understanding pathologies such as schizophrenia, age-related decline, and sensory deficits. In the latter case, understanding how time works in clinical populations could offer an alternative measure of cognitive processing in vulnerable populations;

indeed, impaired timing may underlie problems in specific domains but remains unrecognized as a common denominator across domains.

Psychologists and psychology students, especially those who are drawn to cognitive science, are most likely to take an interest in the book. We hope that, for its useful and informative content about psychological time and brain, students in philosophy and neuroscience will also appreciate reading the book. The question of time leads us humans to a myriad of avenues for response, and the book offers the reader guidance on some of these avenues.

Giovanna Mioni
Simon Grondin

1

TEMPORAL REQUIREMENTS THE BRAIN HAS TO DEAL WITH

Giovanna Mioni
Università di Padova, Padova, Italy

Simon Grondin
Université Laval, Québec, Canada

Introduction

The brain must deal with a variety of temporal demands. It not only processes explicit temporal information but also accommodates various temporal impressions and representations. These time-related processing and representations occur across different levels of a chronometric scale, ranging from microseconds to minutes, hours, days, and even years. This complexity explains why it can be challenging to pinpoint exactly what someone is referring to when discussing time perception (Buhusi & Meck, 2005).

From psychological time to time perception

It is very difficult to establish a definitive taxonomy for the time-related phenomena belonging to psychology (see for instance Merchant & de Lafuente, 2014). For some researchers, psychological time could mean being able to order the chronology of past events (Friedman, 1993), or the propensity to be oriented toward, or attached to, the past, to the present, or to the future (Zimbardo & Boyd, 1999), or could be a series of personal characteristics like being punctual or not (Francis-Smythe & Robertson, 1999). Some people may rather have in mind when hearing about psychological time, that it is the capability to remember that something will need to be done at a given moment in the future (prospective memory; see Mioni et al., 2020), or the capability to manage time or to predict how much time a task will take (Grondin, 2020).

DOI: 10.4324/9781003449546-1

Some avenues of psychological time could take us closer to something referred to as time perception. Time pressure is one such example (Morin & Grondin, 2024). It is the perception of a lack of time, and therefore of some assessment of duration, that is contributing to exert pressure. Even closer to time perception is the impression left by the passage of time (Droit-Volet et al., 2023). This impression could address the present or the past, but voicing it will be based on some speed representation: time passing rapidly or slowly. It is just as if a portion of the time perception world could not be restricted to duration; speed will impose itself. The *present-passage of time* is this impression that time is passing fast or not, for instance, when you are having fun or not. The *past-passage of time* is the impression that a given period (e.g., a week, a year, or the past 10 years) has passed rapidly (Wittmann & Lehnhoff, 2005).

Looking at the past could also be related to other psychological time experiences, and, to some extent, to what could be understood by time perception. Impressions about the past could include this feeling one may have that an event occurred "just as if it was yesterday, or last year" when it occurred several years ago. An assessment of this impression can be made to quantify the distortion between real past duration and the impression from it. Even closer to the contemporary literature on time perception are the judgments associated with what is referred to as retrospective timing. These judgments are those that are made explicitly toward the duration of an activity, or of an event (or a task if it is in the context of an experiment) in cases where the person making this retrospective judgment did not know, at the moment where the activity or event occurred, and that a judgment of time would have to be made (Balcı et al., 2023; Tobin et al., 2010). Therefore, there was no timekeeping activity activated when the event or activity targeted was ongoing; the judgment about time had to be made retrospectively, based on a retrieval process in memory (Block et al., 2018).

Explicit versus implicit timing

Within this book, it is not so much the various forms described above that will be emphasized, although some chapters refer to the passage of time or retrospective timing. The main angle adopted with regard to psychological time is based on the fact that some form of temporal processing is explicitly required, or rather remains incidental. In the former case, referred to as **explicit timing**, the temporal processing either requires a judgment about temporal information per se (duration) or to judge the synchronicity of events. In the latter case, referred to as **implicit timing**, the temporal processing involves tasks that do not require any overt temporal judgment although a failure to take time into account correctly will have an impact on the level of performance reached for the task (Coull & Nobre, 2008).[1]

Explicit timing tasks

Within the domain of explicit timing, there are a lot of methodological nuances to be aware of. A classical distinction is to contrast retrospective and prospective timing. While in the first case, as indicated above, someone is not aware a judgment about time will have to be made, in a prospective timing task, someone knows in advance that such judgment will be required; the duration of a temporal extent will have to be made. This extent could be the duration of a sensory signal (filled interval), or of an event, or of an activity, or the extent between signals (empty interval) or events.

While prospective timing tasks are the most used time perception studies, there is another category of explicit timing tasks where no temporal extent *per se* has to be measured. That is the case for tasks requiring one to judge if signals or events occurred simultaneously or successively, which one of two signals or events occurred first (**temporal order judgment task – TOJ**), or whether there was a gap (an interruption) in a signal. In prospective timing tasks, the temporal extents usually last from a few hundred milliseconds to a few seconds, but in successiveness versus simultaneity or TOJ tasks, the temporal resolution is in the vicinity of less than 100 ms. The reader will note the importance of this latter type of temporal experiences (judging asynchrony of events and their order), within unimodal or multimodal conditions, in Chapter 7 describing the temporal capacities of blind people, in Chapter 8 on Alzheimer's disease (AD) and mild cognitive impairment (MCI), and in Chapter 9 on the temporal experience of schizophrenia.

Another key methodological distinction relies on the fact that explicit and prospective timing tasks may or may not require the use of a motor implementation to delineate a temporal extent. Timing tasks that require a motor response include time **reproduction** and **production tasks** (Chapters 5, 7, and 10). Both tasks require that participants make a motor response in a stopwatch-like fashion, where participants press a key when they perceive a prescribed interval is complete (Mioni, 2018). In a production task, a participant indicates, with chronometric units, the duration of a time interval to produce, while in a reproduction task, a time interval is first presented (e.g., the duration of a sound) to a participant who subsequently has to reproduce it. Note that there exist various ways to reproduce intervals: a continuous press to mark the entire interval, two brief taps to mark the beginning and end of an interval, or a sensory signal marks the beginning of the interval to be reproduced, and it ends with a motor tap by the participant. These variations have an incidence on the results obtained (Mioni et al., 2014). There is also an important motor component in other versions of an interval production task. Instead of a single interval, it is a continuous series of intervals that has to be produced (Wing & Kristofferson, 1973).

These productions occur, for example, once the series of brief sounds marking the interval are presented. Participants may be asked to synchronize (**synchronization-continuation task** – Chapter 5) or not synchronize their taps with the sounds before continuing the productions without the presence of sounds. It is also possible to ask participants to tap at their preferred tempo without a specific instruction to synchronize with a stimulus previously presented (**Spontaneous motor tempo** – see Chapter 5) which is the most natural and easiest pace to follow (Fraisse, 1982). The spontaneous motor tempo is found to average ~2 Hz in adult populations (McAuley et al., 2006). This specific motor signature is slightly faster in children and slower in elderly individuals, but remains close to two movements per second, even across tasks of different levels of complexity (e.g., finger tapping, foot stomping, and hand clapping).

Another classical method for studying time perception, and more specifically explicit timing, is called **verbal estimates**. With this method, a participant is simply asked to estimate the duration of an event or an activity, using chronometric units such as seconds or minutes. This method is of course more useful when relatively long intervals are under investigation and could be used in both retrospective and prospective timing conditions (Grondin & Laflamme, 2015).

An additional method in prospective timing belongs to the general category of "comparison" (Grondin, 2008, 2010). Within this category, we find **duration discrimination**; typically, a participant is placed in a two-forced-choice situation, deciding whether the second of two intervals presented is shorter or longer than the first one (Bausenhart et al., 2018). Another method involving a comparison of intervals is called **bisection task**; in this case, participants are first presented several times with the shortest and longest standard intervals of a set of intervals and then are required to judge whether the new interval is closer to the length of the short or to the long standard previously learned (Penney & Cheng, 2018). Finally, the time **generalization task** requires participants to first learn a reference temporal interval and then decide if each interval presented, out of a set of intervals, is similar to, or different from, the reference temporal interval (Klapproth, 2018).

When there are explicit judgments on the duration of an interval, there are essentially two types of dependent variables of interest: *accuracy* and *precision*. For instance, with a time production task, when several productions are required, it will be relevant to know if the participant is accurate, that is, if the mean productions are close to the targeted intervals to produce, and if this participant is precise, that is, exhibit low variability in the productions. The exact name of the dependent variables depends on the method adopted. In the case of a bisection task, for instance, a psychometric function could be drawn and two indexes of performance, a bisection point (or point of subjective equality in certain conditions) and a difference threshold, could be extracted to quantify the degree of accuracy and precision, respectively.

It is important to keep in mind that several factors (independent variables) are susceptible to influence the estimates of accuracy and precision assessed with these methods. Among these factors, there are the *length of the intervals* under investigation, the *structure* (filled or empty, marked with auditory, visual, or tactile stimuli) of the intervals to be time, and whether presenting single intervals or sequences of intervals (Grondin, 2024).

Considering the *length of the intervals*, most of the studies conducted on prospective and explicit timing processing have focused on intervals ranging from 100 ms to a few seconds. Short intervals, particularly those below 1 s, have garnered significant attention due to their relevance to fundamental adaptive behaviors like speech processing, motor coordination, and music perception. Intervals within the 0.1- to 1-s range are particularly notable for several reasons: the concept of the indifference interval (around 700 ms), the peak sensitivity for tempo discrimination, and the preferred tapping tempo across different age groups. Researchers adopting a neuroscientific perspective highlight a distinction between intervals briefer and longer than 1 s. For instance, this distinction was based on differential pharmacological effects (Rammsayer, 2008). This distinction does not forgo reminding Fraisse's distinction between time perception and time estimation (see Chapter 8). Some researchers claimed that the processing of smaller intervals is sensory-based, or benefits from some automatic processing, whereas the processing of longer intervals requires the support of cognitive resources (Lewis & Miall, 2003). Even if this "1-s" transition period remains somewhat arbitrary, there is certainly some turning point on the time continuum given the benefit one should expect from adopting an explicit counting strategy for processing long temporal intervals (Grondin et al., 1999). Indeed, there are empirical reasons to believe that this transition occurs circa 1.2 s (Grondin et al., 2015); clearly, the Weber fraction for time increases for intervals longer than 1.3–1.5 s (Gibbon et al., 1997; Grondin, 2012, 2014).

Considering the *structure* (filled or empty, marked with auditory, visual, or tactile stimuli), the discrimination of intervals marked with auditory signals is better than the discrimination of intervals marked with visual or tactile stimuli (Azari et al., 2020; Grondin & Rousseau, 1991) and the discrimination is often reported to be better when the intervals are empty rather than filled (Grondin, 1993; Grondin et al., 1998). Also, very brief intervals are much easier to discriminate when the stimuli marking an empty interval are intramodal (delivered within the same sensory modality) than intermodal (Azari et al., 2020; Grondin & Rousseau, 1991; Grondin et al., 2005).

Implicit timing task

Regarding implicit timing tasks, participants may be instructed to produce a motor response to a target stimulus or to provide a perceptual judgment about the stimulus. The key manipulation is that although no explicit temporal estimates are required, the temporal structure of the motor or perceptual task will still influence participant's behavior. Consider, for example, a warned-reaction time (RT) task in which participants simply have to respond to a target stimulus. If the target is separated from the warning signal by different time intervals (or **foreperiods**), the longer the foreperiod, the more likely it is that the target will appear, given that it has not occurred yet (i.e., hazard function; Niemi & Näätänen, 1981). Implicitly processing the foreperiod duration between warning (cue) and target will be beneficial to targets occurring after longer compared to shorter foreperiods (i.e., the *foreperiod effect*; Capizzi & Correa, 2018; Niemi & Näätänen, 1981). The foreperiod effect is contingent upon whether the foreperiod duration fluctuates randomly from trial to trial (*variable foreperiod condition*) or remains constant within a block of trials while varying across blocks (*constant foreperiod condition*). In the *constant condition*, the mean RT typically increases progressively as the foreperiod duration lengthens. Conversely, in the *variable condition*, the mean RT generally decreases as the foreperiod duration increases (Niemi & Näätänen, 1981). In the variable foreperiod condition, however, the target consistently appears at random after the warning signal, hindering the development of reliable response strength. Under these circumstances, individuals have been observed to prepare their responses based on the length of the preceding trial (a phenomenon termed *sequential effect*; Los & Van den Heuvel, 2001). This entails that the reinforced response strength from the previous trial carries over to the subsequent trial, prompting response-related activation at the critical moment experienced in the previous trial. Consequently, shorter RTs are anticipated when the foreperiod of the preceding trial is repeated. This trial-to-trial reinforcement can effectively explain the observed decrease in RT with regard to foreperiod length in a variable foreperiod condition (as demonstrated by Los & van den Heuvel, 2001). Furthermore, this trial-to-trial reinforcement provides insights into the predictions concerning intertrial sequential effects frequently observed in variable foreperiod experiments. It has been frequently documented that when a particular foreperiod is preceded by a longer one in the preceding trial, RT tends to be longer compared to when the preceding foreperiod is equally long or shorter. There are three possible sequences of foreperiods in the *variable foreperiod condition*. First, a foreperiod can be repeated in the subsequent trial, predicting shorter RTs due to reinforced response strength by two consecutive intervals (foreperiods) of the same length. Second, the foreperiod can transition from long to short, leading to longer RTs as the critical moment (target appearance) was not reinforced in the preceding trial. Lastly, the foreperiod can transition from short to long, leading to shorter RTs.

Besides the foreperiod task, implicit timing can also be measured using a **temporal orienting task**, in which participants learn an arbitrary contingency between an informative pre-cue and a specific time interval to predict "when" the target will occur (Nobre et al., 2007; Nobre & Rohenkohl, 2014), or **rhythm timing tasks** in which regular or irregular sequences would implicitly orient (or not, in the case of an irregular rhythm) our attention to a certain point in time corresponding to the temporal pattern of the sequence, resulting in an enhancement of response accuracy and speed (Nobre et al., 2007; Nobre & Rohenkohl, 2014).

As is the case for explicit timing, some factors (independent variables) can also influence RTs in implicit timing tasks. The discussion regarding the length of the intervals (short or long temporal intervals) has been largely addressed and discussed when employing explicit timing tasks, but it is less investigated within the implicit timing literature. The foreperiod effect, indicated by shorter RTs as the length of the intervals between the warning signal and the target increases, is observed with short and long interval ranges. However, it is commonly assumed that a high level of motor preparation can be maintained only for a short period so that, as time uncertainty increases (multiple seconds), participants are less capable of optimizing preparation at the moment of arrival of the target (Näätänen, 1970). The length of the foreperiod is largely investigated when considering the *sequential effect* (see above); indeed, RTs are modulated by the length of the foreperiod in the preceding trial.

Regarding the effect of modality, it has been observed that auditory stimuli induce faster RTs compared to visual stimuli. The facilitatory effect of auditory compared to visual stimuli has been explained considering the immediate arousing effect of auditory compared to visual modality, enhancing the level of motor preparation (Ball et al., 2018; Steinborn et al., 2010). Cases requiring implicit timing are reported in Chapter 4 on the use of magnetoencephalography (MEG) and in Chapter 5 on the use of brain stimulation.

Mechanisms and models

One of the primary debates concerning the perception of time is whether the mechanisms of temporal information are centralized in the brain or distributed in a system (Ivry & Spencer 2004; Teki et al., 2012). The absence of a specific physiological mechanism for representing temporal information has led to controversy regarding the neural basis of time perception. Studies indicate that tasks primarily involving the sensorimotor system can be executed relatively automatically, while those engaging working memory and attention, associated with prefrontal and parietal modules, necessitate more cognitive involvement (Buhusi & Meck, 2005, 2009).

The pacemaker–accumulator model is one theoretical framework suggesting the presence of an internal, central clock mechanism. According to this model, pulses emitted by a pacemaker are temporarily stored in an accumulator (Gibbon et al., 1984). In the information-processing version of this pacemaker–accumulator approach, working memory stores the number of pulses emitted by the pacemaker in a given trial (duration of a given interval), while reference memory retains previously learned numbers of pulses representing a given interval. In a decision phase, estimates are generated by comparing the content in the reference memory with the perceived temporal interval. The pacemaker–accumulator model was later refined by Zakay and Block (1995) to account for the influence of attention and physiological changes, such as arousal, on time estimation. In dual-task situations involving a temporal task or a nontemporal task, directing more attention to a nontemporal task reduces attention toward time, resulting in fewer pulses and a shorter perceived duration (Zakay & Block, 1995). These cognitive models have been mainly used to describe prospective and explicit timing.

The initial examinations into the biological foundations of the pacemaker–accumulator model relied on pharmacological interventions. These studies offered significant support for the notion that there is a separation between the clock phase, influenced by dopaminergic interventions, and the memory phase, influenced by cholinergic interventions (Meck, 2005; Rammsayer, 2008). Some support for this notion came from Parkinson's disease studies indicating modulation of temporal abilities if patients were tested "on" or "off" medication (Jones & Jahanshahi, 2014).

Despite support for the biological plausibility of the pacemaker–accumulator mechanism, alternative biological mechanisms have been considered as potential substrates for interval timing. Since various timing tasks involve diverse activities, such as coordinating complex movements and estimating task durations, it is unlikely that the same brain system is utilized for all timing tasks. Neuroscience-based models aimed at understanding the neural mechanisms of explicit and implicit time processing. According to Coull and Nobre's (2008) review of neuroimaging studies, explicit timing is preferentially associated with basal ganglia, right prefrontal cortex, cerebellar, and supplementary motor area (SMA), whereas implicit timing is more so in the left inferior parietal cortex and premotor areas (see also Wiener et al. 2010a, 2010b). In contrast to these findings, more recent meta-analyses of neuroimaging studies linked the SMA activity to both explicit and implicit temporal processes (Nani et al., 2019; Teghil et al., 2019). For example, Nani et al.

(2019) showed that the SMA was the brain area with the greatest overlap (100%) in a conjunction analysis of all the explicit and implicit timing conditions considered in their meta-analysis (i.e., involving motor and non-motor tasks, as well as intervals below or above 1 s).

An alternative class of theoretical explanations for timing judgments includes intrinsic models (Schwartze & Kotz, 2024), which are specifically dependent on modality or coordination-dependent systems (Jantzen, Steinberg, & Kelso, 2005). A prominent example of this non-dedicated-system view is the state-dependent network, where timing relies on time-dependent changes in the state of neural networks (Buonomano, 2007; Karmarkar & Buonomano, 2007). In this framework, timing does not depend on a clock but on time-dependent alterations in neural network states. An example is the striatal beat-frequency (SBF) model that was developed to explain timing as an emergent activity within thalamo-cortico-striatal loops, where timing is based on the coincidental activation of medium spiny neurons by cortical neural oscillators (Matell & Meck, 2000, 2004). In contrast to traditional pacemaker/accumulator models, where dopamine is assumed to be the neurobiological substrate of pacemaker pulses, in the SBF model, phasic dopamine release acts as a "start gun" signaling the onset of a relevant signal. This synchronizes cortical oscillations and resets the membrane properties of striatal spiny neurons. Tonic dopamine release modulates the speed of the internal clock by altering cortical oscillation frequencies (Matell & Meck, 2004).

Overview of the book

In the first chapter of the book, it is proposed to review the main technological tools used to identify the cerebral structures and networks involved in the processing of various types of temporal information. Each chapter offers an in-depth description of the methodological considerations needed to understand the usefulness and limitations of the tool, and how it has been used to contribute to the corpus of knowledge in the field of timing and time perception.

In Chapter 2, Alice Teghil and Maddalena Boccia describe how functional magnetic resonance imaging (fMRI) can be used to improve our knowledge of brain structures and networks when temporal information, from hundreds of milliseconds to seconds, has to be processed. This chapter provides the reader with a detailed description of the technical foundations of the use of fMRI. After showing what can be done with the application of mass-univariate techniques, they show how the recent application of multivariate techniques led to a more fine-grained characterization of how temporal information is represented in the brain. Finally, after summarizing the most relevant findings in timing and time perception involving the use of fMRI, the authors propose a framework supporting a hierarchical organization of the neural representation of time.

Chapter 3 is dedicated to the description, in timing research, of a frequently used tool in experimental psychology and neuroscience, electroencephalography (EEG). EEG is useful in timing research given its excellent temporal resolution. Nicola Thibault and collaborators report how it is possible to measure the electrical activity from postsynaptic potentials. Both the time (event-related potentials: ERP) and the frequency (spectrum, power, phase of oscillation) domains are useful for studying the processes involved in time perception, emphasizing the use of ERPs in the study of both implicit and explicit time perception tasks, and the primary frequency bands involved in motor synchronization tasks.

In Chapter 4, Sophie Herbst (2018) describes a non-invasive neuroimaging technique, MEG, which offers not only a temporal resolution comparable to one provided by EEG but also an excellent spatial resolution. This chapter is dedicated to the studies based on MEG, leading to the understanding of both implicit and explicit timing tasks. It also describes how the spatial resolution of MEG allows to scrutinize the anatomical sources of both evoked and oscillatory signals often found in sensory and motor areas and how using both MEG and EEG allows to assess time-resolved activities in timing tasks from deeper sources such as the hippocampus and the cerebellum.

Chapter 5 is dedicated to the description of two commonly used methods, transcranial magnetic stimulation (TMS) and transcranial electric stimulation (tES), to non-invasively intervene in brain activity. Sonia Betti and collaborators report the functioning of TMS and tES, and how it is used when studying the brain areas and networks mainly involved in time perception and temporal processing. They emphasize the influence of different factors (the duration of the time intervals investigated, the cognitive set involved in a task, and the stimulus modality used for marking intervals), and also provide criticisms and acknowledge limitations of the studies that used these brain stimulation techniques.

In Chapter 6, it is proposed to consider the key role of bodily signals in the cognitive processing of time. Nicola Cellini and Luigi Micillo provide a primer for the main psychophysiological indices useful to study time, which include body temperature, skin conductance level, muscle reactivity, and cardiac activity. They explain how time perception may be an embodied property of our cognition, which relies on affective and interoceptive states that are dependent on internal bodily signals. The chapter also provides an overview of the current literature on the relationship between psychophysiological indices and temporal processing, explaining why studies addressing this relationship show inconsistent results, and how experimental and methodological challenges in this field could be overcome.

While Chapters 2–6 center on the description of technologies having the potential to serve neuroscientific research for advancing knowledge in the field of time perception, Chapters 7–10 propose other avenues to interrogate the functioning of the brain when dealing with temporal information. The question of the study of the mechanisms underlying timing and time perception is addressed from a pathology and individual differences angle.

In Chapter 7, Maria Bianca Amadeo and collaborators emphasize a sensory perspective. Acknowledging the different levels of efficiency of the sensory modalities for processing temporal information, they address the question of how atypical sensory processing impacts the ability to process time. They highlight the consequences on temporal perception when audition is missing, as is the case in deafness, or when it is vision that is missing, as in blindness.

In Chapter 8, Martin Riemer documents the changes related to time perception in advanced age. After describing the neuroanatomical and neurofunctional changes in patients with AD or with MCI, he summarizes the main findings from studies dedicated to time perception in AD and MCI patients, emphasizing the need to sort out what belongs to deficits in the perception of time and what belongs to deficits in working memory.

Anne Giersch signs Chapter 9 on time perception in individuals with schizophrenia. These individuals are characterized by a fragmentation of the time experience, which led to a series of studies documenting various timing disorders. The main findings of these studies are presented, emphasizing the difficulties of predicting information in time, adapting to changes in delays, either at the millisecond and or second level, and ignoring nonpertinent asynchronies. The chapter also contains information based on fMRI or EEG studies providing some indications about the possible neurobiological bases of the impairments in patients.

Finally, Chapter 10 is dedicated to individual differences in the study of time perception. In this chapter, Joseph Glicksohn proposes different ways of approaching the domain of individual differences in temporal processing and then emphasizes time perception in five major psychobiological traits: extraversion, neuroticism, psychoticism, impulsivity, and sensation seeking. Then, individual differences are approached via their contribution to timing in four major areas of brain activity: the insula, the prefrontal cortex, the SMA, and the cerebellum.

Conclusion remark

Within timing literature, the presence of a variety of timing tasks and pathologies showing time-related deficits, and the use of sophisticated techniques to approach specific timing questions make it difficult to draw simple conclusions about the mechanisms at play in the field of time perception.

Understanding the literature on psychological time and time perception, and eventually the mechanisms, models, or theories, requires paying attention to the specific type of temporal requirement the brain has to deal with, for what is to be explained.

To keep it all straight, one should keep in mind a cascade of questions: Is the requirement a time-related impression or representation (some form of psychological time) or a genuine form of time perception? And if it is time perception, is it implicit or explicit timing? And if it is explicit, is there or not the measurement of a temporal extent? And if there is an assessment of duration, are we referring to prospective or retrospective timing? And if it is prospective, is it with or without a motor component?

Note

1 Note that retrospective timing could be seen to belong to a category where time is incidental, though an explicit judgment about the duration is required.

References

Azari, L., Mioni, G., Rousseau, R., & Grondin, S. (2020). An analysis of the processing of intra- and intermodal time intervals. *Attention, Perception, & Psychophysics*, *82*, 1473–1487. https://doi.org/10.3758/s13414-019-01900-7

Balcı, F., Hüseyin, U., Grondin, S., Gökben, H. S., van Wassenhove, V., & Wittmann, M. (2023). Dynamics of retrospective timing: A big data approach. *Psychonomic Bulletin & Review*, *30*, 1840–1847. https://doi.org/10.3758/s13423-023-02277-3

Ball, F., Michels, L. E., Thiele, C., & Noesselt, T. (2018). The role of multisensory interplay in enabling temporal expectations. *Cognition*, *170*, 130–146. https://doi.org/10.1016/j.cognition.2017.09.015

Bausenhart, K. M., Di Luca, M., & Ulrich, R. (2018). Assessing duration discrimination: Psychophysical methods and psychometric function analysis. In A. Vatakis, F. Balcı, M. Di Luca, & Á. Correa (Eds.), *Timing and time perception: Procedures, measures, & applications* (pp. 52–78). Brill. https://doi.org/10.1163/9789004280205_004

Block, R. A., Grondin, S., & Zakay, D. (2018). Prospective and retrospective timing processes: Theories, methods, and findings. In A. Vatakis, F. Balcı, M. Di Luca, & A. Correa. (Eds.), *Timing and time perception: Procedures, measures, and applications* (pp. 32–51). Brill. https://doi.org/10.1163/9789004280205_003

Buhusi, C. V., & Meck, W. H. (2005). What makes us tick? Functional and neural mechanisms of interval timing. *Nature Reviews Neuroscience*, *6*(10), 755–765. https://doi.org/10.1038/nrn1764

Buhusi, C. V., & Meck, W. H. (2009). Relativity theory and time perception: Single or multiple clocks? *PLoS ONE*, *4*(7), e6268.

Buonomano, D. V. (2007). The biology of time across different scales. *Nature Chemical Biology*, *3*, 594–597.

Capizzi, M., & Correa, Á. (2018). Measuring temporal preparation. In A. Vatakis, F. Balcı, M. Di Luca, & Á. Correa (Eds.), *Timing and time perception: Procedures, measures, and applications* (pp. 216–232). Brill. https://doi.org/10.1163/9789004280205_011

Coull, J., & Nobre, A. (2008). Dissociating explicit timing from temporal expectation with fMRI. *Current Opinion in Neurobiology*, *18*(2), 137–144. https://doi.org/10.1016/j.conb.2008.07.011

Droit-Volet, S., Monier, F., & Martinelli, N. N. (2023). The feeling of the passage of time against the time of the external clock. *Consciousness and Cognition*, *113*, 103535 https://doi.org/10.1016/j.concog.2023.103535

Fraisse, P. (1982). Rhythm and tempo. In D. Deutsch (Ed.), *The Psychology of Music* (pp. 149–180). Academy Press, Inc.

Francis-Smythe, J., & Robertson, I. (1999). Time-related individual differences. *Time & Society*, *8*, 273–292. https://doi.org/10.1177/0961463X99008002004

Friedman, W. J. (1993). Memory for the time of past events. *Psychological Bulletin*, *113*, 44–66. https://doi.org/10.1037/0033-2909.113.1.44

Gibbon, J., Church, R. M., & Meck, W. H. (1984). Scalar timing in memory. *Annals of the New York Academy of Sciences*, *423*(1), 52–77.

Gibbon, J., Malapani, C., Dale, C. L., & Gallistel, C. (1997). Toward a neurobiology of temporal cognition: Advances and challenges. *Current Opinion in Neurobiology*, *7*(2), 170–184. https://doi.org/10.1016/S0959-4388(97)80005-0

Grondin, S. (1993). Duration discrimination of empty and filled intervals marked by auditory and visual signals. *Perception & Psychophysics*, *54*, 383–394. https://doi.org/10.3758/BF03205274

Grondin, S. (2008). Methods for studying psychological time. In S. Grondin (Ed.), *Psychology of time* (pp. 51–74). Emerald Group Publishing.

Grondin, S. (2010). Timing and time perception: A review of recent behavioral and neuroscience findings and theoretical directions. *Attention, Perception & Psychophysics*, *72*, 561–582. https://doi.org/10.3758/APP.72.3.561

Grondin, S. (2012). Violation of the scalar property for time perception between 1 and 2 seconds: Evidence from interval discrimination, reproduction, and categorization. *Journal of Experimental Psychology: Human Perception and Performance*, *38*, 880–890. https://doi.org/10.1037/a0027188

Grondin, S. (2014). About the (non)scalar property for time perception. In H. Merchant, & V. de Lafuente (Eds), *Neurobiology of interval timing* (pp. 17–32). Springer.

Grondin, S. (2020). *The perception of time – Your questions answered*. Routledge.

Grondin, S. (2024). The processing of short-time intervals: Some critical issues. In H. Merchant, & V. de Lafuente (Eds.), *Neurobiology of interval timing* (Vol. 2). Springer.

Grondin, S., & Laflamme, V. (2015). Stevens's law for time: A direct comparison of prospective and retrospective judgments. *Attention, Perception, & Psychophysics*, *77*, 1044–1051. https://doi.org/10.3758/s13414-015-0914-5

Grondin, S., Laflamme, V., & Mioni, G. (2015). Do not count too slowly: Evidence for a temporal limitation in short-term memory. *Psychonomic Bulletin & Review*, *22*, 863–868. https://doi.org/10.3758/s13423-014-0740-0

Grondin, S., Meilleur-Wells, G., & Lachance, R. (1999). *When to start explicit counting in a time-intervals discrimination task: A critical point in the timing process of humans. Journal of Experimental Psychology: Human Perception and Performance*, *25*(4), 993–1004. https://doi.org/10.1037/0096-1523.25.4.993

Grondin, S., Meilleur-Wells, G., Ouellette, C., & Macar, F. (1998). Sensory effects on judgments of short-time intervals. *Psychological Research*, *61*, 261–268. https://doi.org/10.1007/s004260050030

Grondin, S., & Rousseau, R. (1991). Judging the relative duration of multimodal short empty time intervals. *Perception & Psychophysics*, *49*, 245–256. https://doi.org/10.3758/BF03214309

Grondin, S., Roussel, M.-E., Gamache, P.-L., Roy, M., & Ouellet, B. (2005). The structure of sensory events and the accuracy of judgments about time. *Perception, 34*, 45–58. https://doi.org/10.1068/p5369

Herbst, S. K., Fiedler, L., & Obleser, J. (2018). Tracking temporal hazard in the human electroencephalogram using a forward encoding model. *Eneuro, 5*(2). https://doi.org/10.1523/ENEURO.0017-18.2018

Ivry, R. B., & Spencer, R. M. (2004). The neural representation of time. *Current Opinion in Neurobiology, 14*, 225–232. https://doi.org/10.1016/j.conb.2004.03.013

Jantzen, K. J., Steinberg, F. L., & Kelso, J. S. (2005). Functional MRI reveals the existence of modality and coordination-dependent timing networks. *Neuroimage, 25*(4), 1031–1042.

Jones, C. R. G., & Jahanshahi, M. (2014). Motor and perceptual timing in Parkinson's disease. In H. Merchant, & V. de Lafuente (Eds.), *Neurobiology of interval timing. Advances in experimental medicine and biology* (pp. 265–290). Springer-Verlag.

Karmarkar, U. R., & Buonomano, D. V. (2007). Timing in the absence of clocks: Encoding time in neural network states. *Neuron, 53*(3), 427–438.

Klapproth, F. (2018). Towards a process model of temporal generalization. In A. Vatakis, F. Balcı, M. Di Luca, & Á Correa. (Eds.), *Timing and time perception: Procedures, measures, & applications* (pp. 149–164). Brill. https://doi.org/10.1163/9789004280205_008

Lewis, P. A., & Miall, R. C. (2003). Distinct system for automatic and cognitively controlled time measurement: Evidence from neuroimaging. *Current Opinion in Neurobiology, 13*, 1–6. https://doi.org/10.1016/S0959-4388(03)00036-9

Los, S. A., & van den Heuvel, C. E. (2001). Intentional and unintentional contributions to nonspecific preparation during reaction time foreperiods. *Journal of Experimental Psychology: Human Perception and Performance, 27*(2), 370–386. https://doi.org/10.1037/0096-1523.27.2.370

Matell, M. S., & Meck, W. H. (2000). Neuropsychological mechanisms of interval timing behavior. *Bioessays, 22*(1), 94–103. https://doi.org/10.1002/(SICI)1521-1878(200001)22:1<94::AID-BIES14>3.0.CO;2-E

Matell, M. S., & Meck, W. H. (2004). Cortico-striatal circuits and interval timing: Coincidence detection of oscillatory processes. *Cognitive Brain Research, 21*(2), 139–170. https://doi.org/10.1016/j.cogbrainres.2004.06.012

McAuley, J. D., Jones, M. R., Holub, S., Johnston, H. M., & Miller, N. S. (2006). The time of our lives: Life span development of timing and event tracking. *Journal of Experimental Psychology: General, 135*, 348–367. https://doi.org/10.1037/0096-3445.135.3.348

Meck, W. H. (2005). Neuropsychology of timing and time perception. *Brain and Cognition C, 58*(1), 1–8. https://doi.org/10.1016/j.bandc.2004.09.004

Merchant, H., & de Lafuente, V. (2014). Introduction to the neurobiology of interval timing. In H. Merchant, & V. de Lafuente (Eds), *Neurobiology of interval timing. Advances in experimental medicine and biology* (pp. 1–13). Springer. https://doi.org/10.1007/978-1-4939-1782-2_1

Mioni, G. (2018). Methodological issues in the study of prospective timing. In A. Vatakis, F. Balcı, M. Di Luca, & Á Correa. (Eds.), *Timing and time perception: Procedures, measures, & applications* (pp. 79–97). Brill.

Mioni, G., Grondin, S., McLennan, S. N., & Stablum, F. (2020). The role of time-monitoring behaviour in time-based prospective memory performance in younger and older adults. *Memory, 28*, 34–48. https://doi.org/10.1080/09658211.2019.1675711

Mioni, G., Stablum, F., McClintock, S. M., & Grondin, S. (2014). Different methods for reproducing time, different results. *Attention, Perception & Psychophysics, 76*, 675–681. https://doi.org/10.3758/s13414-014-0625-3

Morin, A., & Grondin, S. (2024). Mindfulness and time perception: A systematic integrative review. *Neuroscience & Biobehavioral Reviews*, *162*, 105657. https://doi.org/10.1016/j.neubiorev.2024.105657

Näätänen, R. (1970). The diminishing time-uncertainty with the lapse of time after the warning signal in reaction-time experiments with varying fore-periods. *Acta Psychologica*, *34*, 399–419.

Nani, A., Manuello, J., Liloia, D., Duca, S., Costa, T., & Cauda, F. (2019). The neural correlates of time: A meta-analysis of neuroimaging studies. *Journal of Cognitive Neuroscience*, *31*(12), 1796–1826. https://doi.org/10.1162/jocn_a_01459

Niemi, P., & Näätänen, R. (1981). Foreperiod and simple reaction time. *Psychological Bulletin*, *89*(1), 133–162. https://psycnet.apa.org/doi/10.1037/0033-2909.89.1.133

Nobre, A. C., & Rohenkohl, G. (2014). Time for the fourth dimension in attention. In A. C. Nobre, & S. Kastner (Eds), *The Oxford handbook of attention*. Oxford Academic. https://doi.org/10.1093/oxfordhb/9780199675111.013.036

Nobre, A. C. K., Correa, A., & Coull, J. (2007). The hazards of time. *Current Opinion in Neurobiology*, *17*, 465–470. https://doi.org/10.1016/j.conb.2007.07.006

Penney, T. B., & Cheng, X. (2018). Duration bisection: A user's guide In A. Vatakis, F. Balcı, M. Di Luca, & Á Correa. (Eds.), *Timing and time perception: Procedures, measures, & applications* (pp. 98–127). Brill. https://doi.org/10.1163/9789004280205_006

Rammsayer, T. H. (2008). Neuropharmacological approaches to human timing. In S. Grondin (Ed.), *Psychology of time* (pp. 295–320). Emerald Group.

Schwartze, M., & Kotz, S. A. (2024). Timing patterns in the extended basal ganglia system. In *Neurobiology of Interval Timing*, 275–282. Springer.

Steinborn, M. B., Rolke, B., Bratzke, D., & Ulrich, R. (2010). The effect of a cross-trial shift of auditory warning signals on the sequential foreperiod effect. *Acta Psychologica*, *134*(1), 94–104. https://doi.org/10.1016/j.actpsy.2009.12.011

Teghil, A., Boccia, M., D'Antonio, F., Di Vita, A., de Lena, C., & Guariglia, C. (2019). Neural substrates of internally-based and externally-cued timing: An activation likelihood estimation (ALE) meta-analysis of fMRI studies. *Neuroscience & Biobehavioral Reviews*, *96*, 197–209. https://doi.org/10.1016/j.neubiorev.2018.10.003

Teki, S., Grube, M., & Griffiths, T. D. (2012). A unified model of time perception accounts for duration-based and beat-based timing mechanisms. *Frontiers in Integrative Neuroscience*, *5*, 90. https://doi.org/10.3389/fnint.2011.00090

Tobin, S., Bisson, N., & Grondin, S. (2010). An ecological approach to prospective and retrospective timing of long durations: A study involving gamers. *PLoS ONE*, *5*(2), e9271. https://doi.org/10.1371/journal.pone.0009271

Wiener, M., Turkeltaub, P., & Coslett, H. B. (2010a). The image of time: A voxel-wise meta-analysis. *Neuroimage*, *49*(2), 1728–1740. https://doi.org/10.1016/j.neuroimage.2009.09.064

Wiener, M., Turkeltaub, P. E., & Coslett, H. B. (2010b). Implicit timing activates the left inferior parietal cortex. *Neuropsychologia*, *48*(13), 3967–3971. https://doi.org/10.1016/j.neuropsychologia.2010.09.014

Wing, A. M., & Kristofferson, A. B. (1973). Response delays and the timing of discrete motor responses. *Perception & Psychophysics*, *14*(1), 5–12. https://doi.org/10.3758/BF03198607

Wittmann, M., & Lehnhoff, S. (2005). Age effects in perception of time. *Psychological Reports*, *97*, 921–935. https://doi.org/10.2466/pr0.97.3.921-935

Zakay, D., & Block, R. A. (1995). An attentional-gate model of prospective time estimation. *Time and the Dynamic Control of Behavior*, *5*, 167–178.

Zimbardo, P., & Boyd, J. (1999). Putting time in perspective: A valid, reliable individual-differences metric. *Journal of Personality and Social Psychology*, *77*, 1271–1288. https://doi.org/10.1037/0022-3514.77.6.1271

2

STUDYING TIME PERCEPTION WITH fMRI

Methodological considerations and neural networks for processing time intervals

Alice Teghil and Maddalena Boccia

Sapienza University of Rome, Rome, Italy

Introduction

This chapter focuses on the contribution of fMRI to the study of time perception, and on brain networks supporting duration processing. Starting with the main experimental designs, we will present the standard analysis pipelines (mass-univariate approach) and the most advanced neuroimaging strategies (information-based approach) devoted to the study of both regional and network contributions (effective and resting-state connectivity), along with their evidence for neural mechanisms supporting time perception. This is not intended to be an exhaustive presentation of physical principles and techniques of fMRI, for which the reader is referred, for example, to Poldrack et al. (2011). Rather, it is intended to be a guide for those who are new to neuroimaging of time perception and wish to understand, plan, or analyze fMRI designs and data in this area of cognitive neuroscience. The final part of the chapter will summarize findings that emerged from this line of research, providing an overview of brain networks supporting the perception and representation of time intervals across different contexts and timescales.

Some issues in studying timing with fMRI

The advent of task-based fMRI marked a turning point in research on brain correlates of time perception. In the last 20 years, this line of research has tremendously increased our knowledge of which brain regions and neural networks support time processing; moreover, traditional approaches involving

DOI: 10.4324/9781003449546-2

mass-univariate analyses have been recently complemented by studies using multivariate and connectivity approaches, allowing us to characterize not only where but also how time is represented in the brain.

A general issue in studying timing with fMRI, particularly when using mass-univariate approaches, concerns the choice of a control task allowing to rule out factors associated with basic sensorimotor and cognitive processes of non-interest (Coull, 2014). The approach of subtracting activations observed during a "control" condition from those detected during a condition of interest dates back to the early age of neuroimaging research. According to this subtraction principle, statistically significant blood-oxygenation-level-dependent (BOLD) signal differences observed between the condition of interest and the control condition should point to brain regions involved in the task of interest (Amaro & Barker, 2006). Although more sophisticated comparison strategies have been developed since that early age of fMRI, controlling for potential confounding factors associated with basic sensorimotor processes is still considered essential in fMRI designs.

While this statement virtually applies to every domain of cognition, the control of processes of non-interest has proven particularly tricky when it comes to studying duration processing with fMRI. This is because performing a timing task requires not only to "keep track of time" per se but – besides involving perceptual processing and motor preparation – relies on a set of additional cognitive functions, including attention, working memory, long-term memory, and decision-making.

The issue is further complicated considering that different timing tasks, and their methodological variants, involve separate cognitive processes to a variable degree: consider the difference between a classical reminder task, in which a standard and a comparison duration are presented in each trial, and a duration bisection task, in which the long and the short anchor are presented at the beginning of the experiment, but only comparison durations are presented in experimental trials (Bausenhart et al., 2018; Grondin, 2010). Although both paradigms involve the estimation and comparison of durations, the second one relies more strongly than the first one on long-term memory (Levy et al., 2015; see also Chapter 1).

In a recent meta-analysis, Naghibi et al. (2024) identified three levels of control that can be achieved in relation to non-timing processes in fMRI studies. The minimal level of control requires to use the same task structure and present the same stimulus across the timing and the comparison condition, allowing control for confounds related to sensorimotor, attentional, memory, and decision demands. A more stringent control involves also matching conditions for stimulus dynamics or task difficulty. Finally, the higher level of control involves all the conditions mentioned above. As it will be apparent going through this chapter, researchers investigating time processing with fMRI have developed increasingly careful procedures to control for confounding factors.

Studying time perception with fMRI

From BOLD signal to experimental design

fMRI assumes that brain regions consume more oxygen when actively engaged in a task compared to when they are not and exploits the different magnetic properties of oxyhemoglobin (Hb, diamagnetic) and deoxyhemoglobin (paramagnetic, dHb). In short, local neuronal activity results in a higher metabolic demand and increased blood flow. As a result, the ratio between Hb and dHb changes, and this results in a change in the local MR signal. This difference is known as the BOLD signal. Thus, the magnitude of the BOLD signal is an indirect measure of neuronal activity. Due to the properties of the local vascular system, there is an intrinsic delay between neuronal activity and the increase in blood flow; this mechanism is known as the hemodynamic response function (HRF) and its shape characterizes the BOLD signal. After an initial dip (Figure 2.1a and b), a peak in the BOLD signal is observed ~4–6 s after stimulus onset; a return to baseline is observed ~12 s after stimulus onset (Figure 2.1b), followed by an undershoot. In the case of prolonged stimulation, a plateau is observed (Figure 2.1b). Despite the presence of a delay, the BOLD response is stereotyped. Thus, we can interpret the BOLD signal as a linear function of local neural activity, based

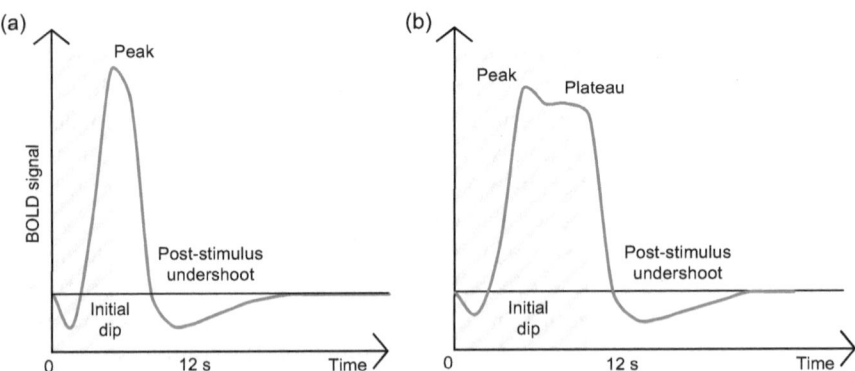

FIGURE 2.1 Temporal dynamic of BOLD signal, depending on the stimuli presentation as an event (a) or prolonged within a block (b). Boxes represent stimulation.

on the canonical models of the HRF. In fact, the BOLD signal exhibits linear time invariant properties.

fMRI can be used to measure task-related activation (task-based fMRI) or spontaneous signal fluctuations in the absence of external stimuli or task demands (resting-state fMRI [rsfMRI]). Based on the temporal organization of stimuli/conditions, task-based fMRI studies can be defined as block or event-related designs (for a review of fMRI paradigms, see Amaro & Barker 2006; Clark, 2012; Petersen & Dubis, 2012). In block designs, stimuli used to test different conditions are presented sequentially within different blocks (Figure 2.2a). Different blocks of similar stimuli are typically 12 s long and are arranged in a pseudorandomized order during the fMRI scan (e.g., ABBA), with an inter-block interval of ~12 s. By repeatedly presenting stimuli belonging to the same conditions, block designs result in sustained and relatively larger BOLD signal changes (Figure 2.1b), increasing the signal-to-noise ratio and statistical power. Block designs are thus well suited

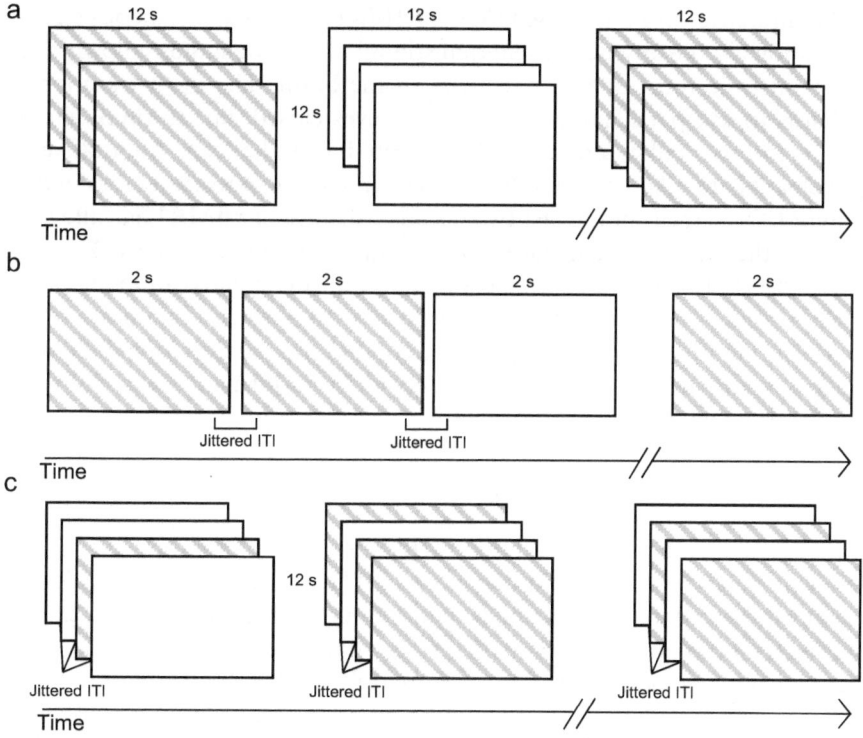

FIGURE 2.2 Examples of block design (a), event-related design (b), and mixed design (c).

to detect regions of interest (ROIs) and, more generally, for subtraction analyses. However, using this type of experiment does not allow a more fine-grained assessment of trial-by-trial effects. Event-related designs (Figure 2.2b) allow to delineate the relation between brain functions and discrete events, testing specific responses to trials, and to model factors such as errors, response times, and subjective judgments. Initially, the implementation of event-related designs was time-consuming. In fact, the temporal presentation of experimental stimuli, especially the inter-stimulus interval (ISI), was regulated by the temporal evolution of the HRF, with ISIs as long as the canonical HRF (~12 s). Thus, experiments were longer and presented fewer experimental trials than blocked designs. In fast event-related fMRI designs, trials are efficiently spaced by jittering the ISI duration to allow deconvolution of the overlapping HRFs. This type of design allows a more detailed assessment of brain correlates of many psychophysical experiments and conditions, and the increased number of stimuli increases statistical power. Mixed fMRI designs allow to test the interaction between block- and event-related factors (Figure 2.2c).

At difference with task-based fMRI, rsfMRI exploits spontaneous signal fluctuations in the absence of external stimuli or task demands, allowing to assess intrinsic functional connectivity in the brain. The most traditional design is to ask participants to keep their eyes closed, not to think about anything in particular, and not to fall asleep during the scan. Alternatively, participants are asked to keep their eyes open and/or to fixate a cross in the center of the screen. In the last two decades, the use of rsfMRI has allowed to define the intrinsic organization of the brain in large-scale networks (Yeo et al., 2011). Furthermore, many studies have investigated differences in intrinsic functional connectivity associated with specific conditions (e.g., neurodegenerative disorders, psychiatric conditions, and developmental disorders) or with individual differences in cognitive performance (e.g., Sulpizio et al., 2016; Teghil et al., 2020a, 2023).

Preprocessing and first-level GLM

Before entering the main analysis, fMRI images are preprocessed (for a detailed explanation, see Poldrack et al., 2011). First, images are typically visually inspected to detect artifacts that prevent further analysis (e.g., excessive head motion and scanner artifacts). Spatial distortions associated with the gradient-echo echoplanar imaging used for fMRI acquisition are typically corrected using images at two different echo times, with the phase difference between the two images used to calculate the local field inhomogeneity and to create a map quantifying the distance that each voxel has shifted (field map). After removing the first (non-steady-state) volumes, the images are temporally and spatially preprocessed. First, a slice-timing correction is applied, since the data are acquired one slice at a time. Thus, images are realigned to reduce the misalignment between images in an fMRI time series occurring due to head motion. Functional volumes are then typically coregistered with the high-resolution T1-weighted image from the same subject to improve the resolution of the functional images. Coregistered images from different individuals are then aligned in a common spatial framework, normalizing them to a standard template (usually MNI152), in order to combine data for group analysis. Finally, individual functional images are smoothed to reduce noise (this last step applies only to the univariate analysis pipeline, since multivariate pattern analysis (MVPA) is performed on unsmoothed images; see below).

The preprocessed functional images are then entered into the first-level analyses that estimate subject-specific parameters and variance. For each voxel in the brain, an estimate of the expected BOLD signal for each condition of interest is computed by convolving stimulus onsets and durations with a canonical HRF (see Figure 2.1 for canonical HRF for block and event-related designs). In the general linear model (GLM) for first-level analyses, each voxel time series is set as the dependent variable, while conditions and intrinsic confounders (e.g., motion parameters) are the independent variables. In standard univariate analyses (see below), conditions or task-related parametric modulators (e.g., reaction times) are usually included. Instead, GLM preceding multivariate pattern analyses usually includes stimuli as independent variables (see below). The output of the GLM is a subject-specific parametric map.

Concerning rsfMRI, potential confounding effects are estimated and removed from the preprocessed functional run/session. Confounding effects include noise components from cerebral white matter and cerebrospinal areas, estimated subject-motion parameters and identified outlier scans. Temporal frequencies below 0.008 Hz or above 0.09 Hz are removed to focus on slow-frequency fluctuations while minimizing the influence of physiological, head-motion, and other noise sources.

Second-level analysis and related evidence for time perception in task-based fMRI

Second-level analyses are aimed to make inferences at the group level and take as inputs the subject-specific parameter estimates and variance estimates from the first-level model. Mass-univariate analysis and MVPA are the main categories of second-level models. The former is mainly aimed at revealing which brain areas underlie the experimental condition (activation-based analysis; Kriegeskorte & Bandettini, 2007). The latter aims to identify a perceptual representation or cognitive state based on multivoxel regional fMRI signals (information-based analysis; Kriegeskorte & Bandettini, 2007). Thus, mass-univariate analyses aim to demonstrate the involvement of a region or a voxel in a task, whereas MVPA aims to reveal the representational content of regions or voxels (Mur et al., 2009). In addition to these approaches, population receptive field (pRF) mapping has been developed to test spatial tuning properties of topographically organized areas (e.g., retinotopic properties in the visual cortex; Dumoulin et al., 2008). All of these approaches are aimed at investigating local cerebral function. Two main approaches have been developed to test neural networks underlying specific experimental conditions, namely psychophysiological interaction analysis (PPI) and dynamic causal modeling (DCM).[1]

Mass-univariate analyses

Classical subtraction analysis, in which activation maps result from simply subtracting the average activation during one task from activation during another task, is an example of mass-univariate analyses. Univariate data are typically analyzed using a *t*-test or analysis of variance for each voxel in the brain (voxel-wise analysis) or locally in specific ROIs, defined based on previous literature or meta-analyses or based on a functional localizer or contrast that is independent from the main analysis to be performed in the ROI (Poldrack, 2007). Overall, mass-univariate analyses aim to understand the neurofunctional anatomy of cognitive processes. Within mass-univariate analyses, one way to enrich the design is to parameterize conditions, with a larger number of conditions allowing to estimate the correlation between condition parameters and brain activity.

Early fMRI investigation of time processing – mainly assessing motor timing and relying on paced finger-tapping or reproduction paradigms – largely used block designs that are especially well-suited for subtraction analyses. A common approach in fMRI mass-univariate studies of timing using finger-tapping involves subtracting brain activation during the synchronization phase from that during the continuation phase, in order to isolate neural activations related to internally driven movement timing. Following this approach, Jäncke et al. (2000) asked participants to tap their right index finger in synchrony with isochronous tones (auditory condition) or visual stimuli presented at the same frequency (visual condition), and then to maintain the same tapping rate without the pacing stimulus; the continuation phase was then contrasted with the synchronization phase (Jäncke et al., 2000). Lewis et al. (2004) adopted a similar design, presenting not only isochronous auditory patterns but also more complex multi-interval rhythms; the authors also included a "random" condition, requiring a button press in response to tones presented at unpredictable times that was modeled as a covariate of non-interest at the second-level analyses. Contrasting the continuation with the synchronization phase, this study reported activation in a network involving the bilateral supplementary motor area (SMA), the putamen, the medial superior frontal and post-central gyrus, and the inferior parietal and occipital cortex (Lewis et al., 2004). An even more controlled design was adopted by Garraux et al. (2005). Their paradigm involved four conditions: timing, order, self-initiated (SI), and visually triggered (VT), allowing not only to match motor and working memory demands between the two conditions of interest (timing vs. order), but, through the inclusion of two other conditions differing only in how movements were initiated (SI vs. VT), also to control for effects related to variations in movement initiation between the timing and order conditions. Using the contrast [TIMING – ORDER] – [SI – VT], the authors showed that the right putamen was differentially activated by timing compared with order of movements.

Duration production and reproduction paradigms have also been widely used in fMRI timing studies with block designs. For example, Lewis and Miall (2002) had participants either produce time intervals around 3 s indicating the end of their estimate with a button press (time condition), press the same button with a specific force (pressure condition), or press the button without attending to force and time (motor condition). Contrasting the BOLD signal change during the time with that in the pressure condition, the authors showed activation of the pre-SMA and SMA proper, dorsolateral prefrontal cortex (DLPFC), premotor cortex, insula and inferior parietal cortex; thus, besides showing that the pre-SMA/SMA were involved in timing not only during the production of overlearned sub-second motor patterns (e.g., finger tapping), the authors showed that the DLPFC was involved in timing of durations spanning multiple seconds also during motor tasks, and its activation was thus not specific to perceptual tasks. Similarly, Bueti et al. (2008) were able to probe brain regions specifically associated with the use of temporal information for action (motor condition – control motor condition) and for perception (perceptual condition – control perceptual condition), showing a common involvement of the putamen; moreover, they showed that the motor timing task, compared with its control, was associated with higher activity in the right pre-SMA, left premotor cortex, left middle frontal gyrus (midFG), bilateral intraparietal cortex, as well as left fusiform gyrus and bilateral cerebellum, compared with the perceptual timing task minus its control [(motor condition – control motor condition) – (perceptual condition – control perceptual condition)].

Duration discrimination studies have also employed different control tasks, with the general aim of controlling for demands associated with sustained attention and working memory updating. In the auditory domain, Schubotz et al. (2000) presented sequences of three tones and asked participants to detect either rhythm (timing task) or pitch deviants (Schubotz et al., 2000; see also Nenadic et al., 2003 for a similar design). Similarly, in the visual domain, Lewis and Miall (2003a) presented lines varying in length over time and asked volunteers to discriminate whether test lines were shorter or longer either in duration (time condition) or in length (control condition) compared to a learned standard.

Another aspect which consideration is crucial to interpret the results of subtraction analyses in timing studies is the relative difficulty of the compared conditions: indeed, if the timing task is more difficult than the control task, differential activation cannot be univocally related to timing (Coull, 2014; Henry et al., 2015; Livesey et al., 2007). Whereas some studies have attempted to match the difficulty of the timing and the control task adjusting difficulty based on participants' performance (Lewis & Miall, 2003a; Nenadic et al., 2003), others specifically tested the effect of task difficulty on brain correlates of temporal processing. Tregellas et al. (2006) asked participants to discriminate the duration of short tones, with a standard duration of 200 ms. The task entailed an "easy" and a "difficult" condition, in which test durations were, respectively, more or less similar to the standard duration; in a control condition, two tones were presented with an equal duration. The difficult > easy contrast revealed activation of the DLPFC, the SMA, putamen, and insular-opercular cortex, suggesting that the involvement of these regions in timing in the subsecond range was at least in part modulated by task difficulty. Similarly, Livesey et al. (2007) compared activation during a duration discrimination task with that during two control tasks, respectively, easier or more difficult than the timing task. They reasoned that regions showing a reverse pattern of activation when timing was contrasted to an easier or more difficult task should be related to task demands, rather than with timing per se. In order to control for confounds associated with working memory updating and sensorimotor processing, both the time and control conditions involved the presentation of two consecutive disks varying both in duration and color; in the timing task, participants had to indicate whether the first or second stimulus had been presented for longer, whereas in the color task, they had to decide which of the two flashed red the more. The timing versus easy control contrast activated the pre-SMA, right inferior parietal lobe, the putamen, the inferior frontal gyrus (IFG) and insula, the prefrontal cortex, and the cerebellum; the timing versus hard control contrast, however, only activated a region in the ventral part of the left inferior parietal cortex, the bilateral insula/IFG, and the left putamen. Also, different regions, including some dorsal prefrontal regions, appeared to be more activated by the more difficult task. This study thus provided evidence in favor of a key role for regions such as the left inferior parietal cortex, putamen, and anterior insula in timing of durations in the second range.

While the studies mentioned above provided an overall view of brain networks involved across different tasks and paradigms, studies adopting event-related designs allowed a more fine-grained characterization of neural substrates supporting different components of timing processes. Rao et al. (2001) were the first adopting this approach to investigate activation associated with different stages of duration discrimination (encoding vs. decision-making/response preparation). In this study, durations to be discriminated were defined by two pairs of tones; a pitch comparison as well as a sensorimotor control condition (involving the same stimuli but requiring only a keypress after the presentation of the second pair of tones) was included. Subtraction analyses (time > control and time > pitch) were performed at four time points after trial onset: 2.5, 5.0, 7.5, and 10.0 s (Figure 2.3).

Considering that the peak of the positive signal change component of the BOLD response occurs approximately 4–6 s after stimulus onset, Rao et al. (2001) took the BOLD response having a maximal signal change at 2.5 and 5.0 s after trial onset as a proxy of the response occurring during the encoding of temporal information. Also, since the comparison duration was presented approximately 3.5 s after trial onset, activation occurring at the 7.5 and 10 s time points was assumed to reflect decision and motor preparation processes. Results showed that activation of the right putamen specifically

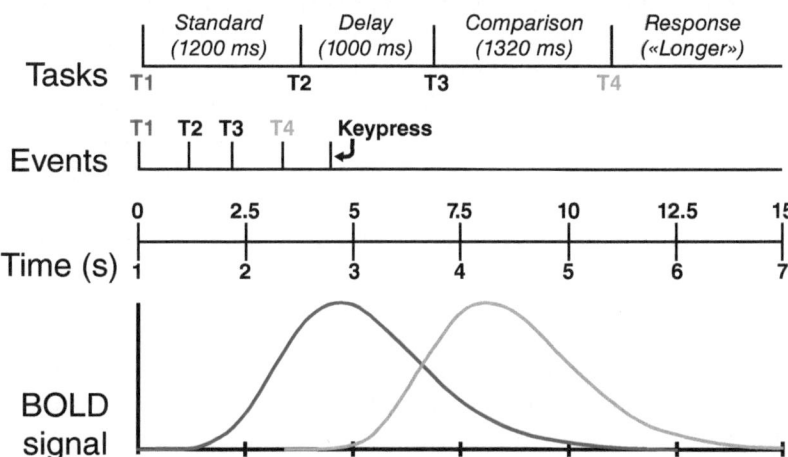

FIGURE 2.3   Schematic representation of task events and their correspondence with the acquisition of images and predicted hemodynamic response functions. Adapted from Rao et al. (2001).

occurred at earlier time points (2.5 s after trial onset), suggesting a role of this region in encoding time intervals. Activation of the DLPFC only occurred at later epochs, whereas that of the inferior parietal cortex was apparent during both early and late epochs. However, none of the latter was specific to the timing task, in line with the possibility that – at least at the short timescale assessed in this study – such activations reflected, respectively, comparison and decision-making and attentional processes, rather than time processing per se.

Even greater precision in isolating brain activation patterns associated with different stages of temporal processing has been achieved by following event-related studies, adopting the approach of time-locking BOLD signal to task events separated by a variable delay (jittering) (Coull et al., 2008, 2015; Bueti & Macaluso, 2011; Harrington et al., 2010; Wencil et al., 2010; Wittmann et al., 2010, 2011). A widely used paradigm in event-related studies investigating duration discrimination (Coull et al., 2008; Morillon et al., 2009; see also Livesey et al., 2007, for a similar paradigm) has been developed by Coull et al. (2004). Each trial started with the presentation of a cue directing participants to estimate either the duration or the color (control condition) of task stimuli. After a delay, two flickering colored disks were presented, both having one of three possible durations, and an overall percept of one of three possible shades of purple (the disks rapidly changed color); participants were instructed to compare either the duration or color of the second stimulus with that of the first one. This control task allowed to control for attention and working memory demands associated with duration discrimination, since, in order to perform the color task, participants had to attend to the stimulus for its whole duration, as for the timing task (in which duration cannot be estimated until the end of stimulus presentation). The pre-SMA, the left putamen, the right frontal operculum, premotor cortex, DLPFC, as well as the superior, inferior, and middle temporal gyri and the bilateral inferior parietal cortex increased activation with increasing attention to time. In a later study from the same group (Coull et al., 2008), task parameters were further modulated in order to distinguish brain activity associated with different task components. In this study, the sample and probe stimuli were separated by an ISI ranging between 2 and 4 s, thus allowing the temporal deconvolution of BOLD signal corresponding to the two stimuli. Also, to reduce confounds associated with preparation of the motor response, fingers associated with different discrimination responses were varied on a trial-by-trial basis, and this mapping was only specified in the response phase, allowing to distinguish the cognitive discrimination from the preparation of the motor response. The SMA, DLPFC, and right superior temporal gyrus (STG) were more activated during the time compared to the color condition across the whole task; however, activation of the SMA was observed during both the encoding of the first stimulus and that of the second stimulus (with activation during the comparison stage being more rostral, toward the pre-SMA), whereas activation of the left putamen was associated with the encoding of the first stimulus, and activation of DLPFC and STG with that of the second one; these data thus provided strong evidence for a key role of the SMA in timing the duration of events.

Adopting a partially different approach, Bueti and Macaluso (2011) used a mixed design to test whether changes in brain activation occurred proportionally with the perceived duration of subsecond visual and auditory stimuli. To this aim, they developed a highly controlled duration reproduction paradigm, in which physical properties of the to-be-timed stimuli were manipulated in order to bias their perceived duration, and separately modeled the encoding and reproduction phases. Contrasting duration encoding in the visual and auditory modality with the corresponding control tasks (responding to color and position, respectively), they showed that the SMA, putamen, right insula, right middle and superior temporal gyri, right cerebellum, and left temporo-parietal junction were activated during duration encoding independently from stimulus modality. During this phase, specifically in the visual condition, the right putamen, insula, and middle/superior temporal cortex showed also an activation profile that reflected perceived time, suggesting that a distributed network may encode perceived stimulus duration.

At difference with the abovementioned studies testing timing of subsecond and peri-second durations, Wittmann et al. (2010) investigated brain activation during timing of auditory tones in the range of multiple seconds (3–18 s). Their task involved an encoding phase, a reproduction phase in which participants had to press a key when they judged that a second tone had lasted the same duration as the encoded one, and a control condition requiring a keypress as fast as possible after the end of tones. Considering the length of the intervals, a secondary memory task was used in order to prevent chronometric counting, requiring to remember digits presented at the beginning of the trial. During encoding, activation was found in the SMA, as well as in the bilateral posterior insula and superior temporal regions; this activation shifted more rostrally in the reproduction phase. Notably, an examination of the time course of brain activation during encoding showed that the right posterior insula and regions of the right STG displayed a climbing pattern of activity with a peak at the end of the stimulus, suggesting an ongoing accumulation of temporal information; the SMA, instead, showed an inverted U-shaped function (Wittmann et al., 2010).

A more fine-grained characterization of neurofunctional correlates of cognitive processes is possible by means of fMRI adaptation, which is based on evidence that repeated exposure to a specific experimental condition results in the neural adaptation effect, a particular form of the carryover effect (Aguirre, 2007). fMRI adaptation aims to demonstrate the presence of a neural representation, based on the assumption that a reduction in the event-related BOLD signal amplitude occurs due to the repetition of the same "condition" over consecutive trials (see Malach, 2012, for a review of adaptation methods and approaches). Adaptation procedures have been successfully used by Hayashi et al. (2015; 2020) to assess the presence and features of duration-tuned neural populations, hypothesizing that such neural populations would have shown weaker BOLD responses following the repeated presentation of the same and similar durations. In one study, participants performed a duration discrimination and a control shape discrimination task. In the duration discrimination task, an adaptor (reference) stimulus of either 400 or 600 ms was presented, and participants had to decide whether a second stimulus with a variable duration was different from the reference stimulus. Analyzing the offset response to the test stimuli, different temporal and parietal regions showed adaptation; however, only the right SMG showed adaptation to the range of tested durations. Duration adaptation occurred not only during the timing task but also during the control task, in which duration information was not task-relevant (Hayashi et al., 2015). In a later study, neural adaptation in the right SMG was found to correlate with behavioral measures of perceived time, suggesting that duration tuning in this region may contribute to subjective duration experience (Hayashi & Ivry, 2020). Interestingly, the region highlighted in these studies was different from the more dorsal part of the right IPL associated with suprasecond duration processing (Hayashi et al., 2014), suggesting that different regions of the IPS may be involved in processing subsecond and suprasecond durations (Hayashi et al., 2015).

Multivariate pattern analyses

Unlike standard univariate analyses, MVPA of fMRI data aims to identify the distributed patterns of neural activity associated with specific cognitive or perceptual processes. Specifically, MVPA focuses on patterns of activity across multiple voxels in the brain (e.g., searchlight approach) or ROIs. Activity patterns are extracted for each exemplar in the experimental condition (such as different durations in a timing task). These patterns are then used to train a machine learning classifier (e.g., support vector machine) to discriminate between the experimental conditions of interest (training dataset). The classifier is then applied to new, independent fMRI data (test dataset) to test its ability to accurately predict the experimental conditions. If the classifier is indeed able to discriminate between experimental conditions (i.e., it decodes the experimental condition), then information is present in the activity patterns. Unlike the decoding approach mentioned above, which focuses on decoding or classifying neural activity patterns, representation similarity analysis aims to quantify the similarity (or dissimilarity, representation dissimilarity analysis) between neural activity patterns (extracted as in the decoding approach) and an expected similarity matrix.

Using multivariate methods, Henry et al. (2015) investigated attention to duration at multiple timescales, using a task requiring to discriminate either the duration or the modulation rate of brief (~500 ms) auditory stimuli, that were varied orthogonally and which discrimination difficulty had been matched in a pilot study. Using a searchlight-based analysis, the authors correlated participants' behavioral patterns in the different stimulus conditions with betas estimated from design matrices comprising the same conditions for the two tasks, thus highlighting regions generally associated with task performance. A second analysis further revealed a bilateral fronto-parietal network in which activation scaled as a function of attention to duration features. Hayashi et al. (2018) also used both ROI-based and searchlight MVPA to decode duration information. Participants performed both functional localizer scans, in which they had to discriminate either the duration or orientation of stimuli between 240 and 983 ms, and task runs, in which they only performed the duration discrimination task. Localizer scans were then used to identify brain regions activated during the time and orientation tasks that served as ROIs for the ROI-based MVPA. The left IPL, right SPL/IPL, and right IFG ROIs all showed above-chance classification accuracy, allowing to decode stimulus duration; these findings were replicated in the searchlight analysis, providing overall further evidence for the role of parietal regions in representing subsecond intervals.

Population receptive fields

Population receptive field is a technique used to estimate the spatial tuning properties of cortical neurons in the human brain. The goal of pRF mapping is to identify the receptive field properties of individual neurons by measuring the fMRI response to sensory stimuli with different length scales. Originally developed to estimate the visual field map (Dumoulin et al., 2008), pRF has been successfully used to investigate a wide range of processes that can be characterized in terms of spatial tuning, including numerosity (Harvey & Dumoulin, 2017) and timing. Protopapa et al. (2019) showed the existence of chronotropic maps in the human SMA combining mass-univariate analyses with pRF analyses. In a duration discrimination task with durations from 0.2 to 3 s, they found that different durations were represented in a spatially organized manner within the SMA, with shorter and longer durations being associated, respectively, with neural activity in the anterior and posterior SMA. A cluster in the left IPS also showed duration-selective responses, although without a clear topographic organization. Also using pRF, Harvey et al. (2020) identified a network of regions exhibiting timing-selective responses. Participants were trained in a visual duration discrimination task before scanning. Then, during fMRI, they were presented with repetitive visual events – a circle appearing and disappearing – that gradually varied in duration (periods gradually increasing from 50 to 1,000 ms in 50 ms steps); participants were asked to press a button when a white circle was presented instead of the usual black circle. A widespread network, spanning from the occipital visual areas to the parietal multisensory to frontal action planning regions, showed neural responses to specific ranges of visual event timing (Figure 2.4a). In the same sample of participants, Hendrikx et al. (2022) showed that the early visual cortex encodes the duration of visual stimuli in a monotonic fashion, with response decreasing with increasing retinotopic distance from stimulus location, whereas timing-tuned neural responses begin in the medial temporal area (MT+) and appear to be

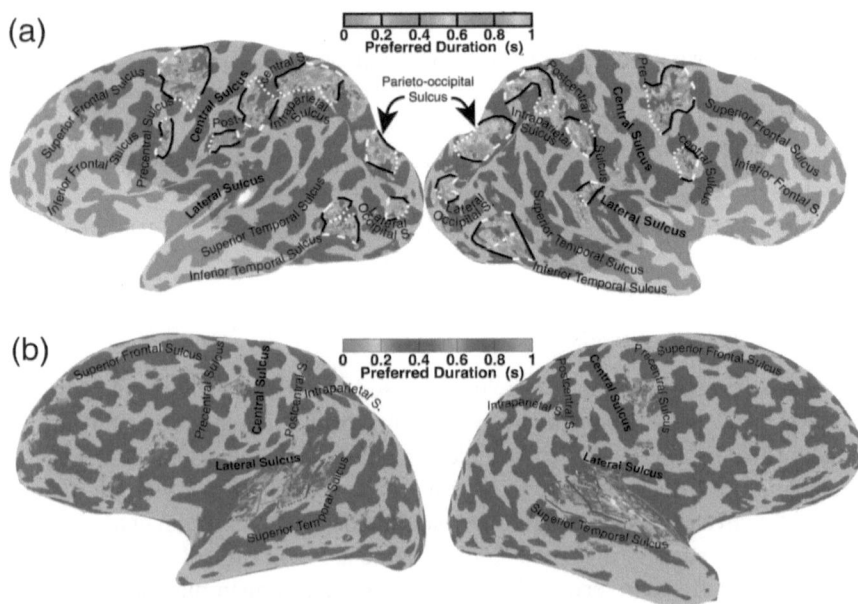

FIGURE 2.4 Distribution of duration tuned neural responses in two individual participants for visual (a) and auditory (b) stimuli. Adapted from Harvey et al. (2020) and van Ackooij et al. (2022).

independent of stimulus location. Timing tuned responses have also been observed for auditory stimuli: van Ackooij et al. (2022) presented participants with repetitive white noise bursts varying in duration and period and observed timing tuned response in the bilateral Hesch's gyrus and surrounding regions, extending to the STG, and in a region of the premotor cortex (Figure 2.4b). When asking participants to perform an explicit timing task (duration reproduction) on auditory stimuli, Bueti et al. (n.d.) also observed timing tuned neural responses in the auditory parabelt, SMA and IPS.

Psychophysiological interaction analysis

Task-based fMRI also allows for testing of whether and how functional connectivity between brain regions changes as a function of task (or condition) using PPI (McLaren et al., 2012). PPI allows to test context-dependent connectivity and provides the opportunity to understand how brain regions interact in a task-dependent manner, by modeling BOLD responses either in a target brain region (seed-to-seed) or in each voxel in the brain (seed-to-voxel) in terms of the interaction between a psychological process and the neural signal from a source region. In other words, PPI allows testing whether experimental conditions modulate the functional connectivity between a source region and a target region or voxel. If the seed–target connectivity is more affected by condition A than B, then one can assume that these regions are more activated during condition A. PPI is crucial to understand functional integration in the brain and to characterize the psychological or behavioral meaning of such integration (Friston et al., 1997).

In a study investigating brain correlates of movement order and timing (see the *Mass-univariate analyses* section), Garraux et al. (2005) used PPI to assess the functional connectivity of the right putamen during the timing condition, showing that a set of brain regions, including the bilateral midFG, superior frontal sulcus and IPL, as well as the right pre- and postcentral gyrus, were preferentially coupled with the right putamen during the timing condition, and with the right cerebellum during the order condition. Using a block design together with PPI analysis, Davranche et al. (2011) showed that the left IPS was not only activated during temporal orienting when temporal cues were aimed to speed a motor response but also when they allowed speeding perceptual discrimination. The functional connectivity of left IPS was also affected by task condition, as it showed connectivity with the bilateral inferior frontal and precentral gyrus during the motor condition, and with the extrastriate visual cortex during the perceptual condition, suggesting that the left IPS may drive temporal orienting affecting the activity in sensory and motor regions according to task demands. Liu et al. (2023) also used PPI to examine the connectivity pattern underlying duration reproduction, finding increased connectivity of the left caudate with the bilateral STG, post-central gyrus, left precuneus and IFG, and of the left cerebellum with the right pallidum, anterior cingulate cortex, middle and superior frontal gyrus, and left medial superior and middle frontal gyrus compared with a sensorimotor control task.

Dynamic causal modeling

Dynamic causal modeling (Friston et al., 2003) is a type of effective connectivity analysis that aims to infer the causal relations between different brain regions and to model neural processes underlying task-related changes in these relations. It is based on an a priori selection of interconnected brain regions, usually based on the existence of functional or structural connections between regions of the model and previous knowledge and hypotheses on neural circuits involved in the experimental conditions. Models in DCM are specified a priori to describe how neural activity in each region changes over time and how this activity is affected by activity in other regions and experimental conditions. Bayesian statistics are used to fit the models to fMRI data and to estimate the model parameters and their uncertainty. DCM can be used to investigate how the strength of connectivity between different brain regions changes as a function of experimental tasks or manipulation. Protopapa et al. (2023) used DCM to investigate effective connectivity within a network including the SMA, IPS, IFG, cerebellum, and primary visual areas, previously found to be activated by the encoding of subsecond durations (Protopapa et al., 2019). In this study, the best model was found to be that in which the cerebellum had feedback and feedforward connections with all other regions of the network, whereas the SMA only had feedforward connections with the cerebellum, IPS, and V1, but did not affect activity in any brain region; these findings thus suggested that the SMA may be involved in the last stage of explicit duration recognition.

Resting-state functional connectivity

Resting-state functional connectivity (rsFC) has been widely used to test intrinsic functional connectivity between pairs of regions (ROI-to-ROI analyses), between a seed region and the rest of the brain (ROI-to-voxel analysis), and between voxels (voxel-to-voxel analysis). In all of these analyses, the Fisher-transformed bivariate correlation coefficient is the most commonly used index of functional coupling. ROI-to-ROI analyses produce a single statistical matrix of T- or F-values characterizing the effect of interest between all possible pairs of ROIs. Similarly, ROI-to-voxel analyses yield a statistical map representing the functional connectivity between the ROI and each voxel in the brain. Voxel-to-voxel analyses test all functional connections between each pair of voxels in the brain. Different strategies have been developed to provide interpretable measures at each individual voxel.

Intrinsic connectivity (IC), global correlation (GCOR), and local correlation (LCOR) are the most used a priori approaches. IC and GCOR are both measures of network centrality, respectively, based on the strength of connectivity and both the strength and sign of connectivity between a given voxel and the rest of the brain. LCOR, instead, is a measure of local coherence characterized by the strength and sign of connectivity between a given voxel and its neighboring areas.

Principal component analysis (PCA) and independent component analysis (ICA) are the most widely used data-driven approaches for rsFC. The goal of PCA is to find the number of orthogonal components that maximize the explained variance of data and separate relevant information from noise. This approach identifies patterns of brain activity that are common to many regions and uses these patterns to reduce the dimensionality of the data and identify different brain networks. PCA does not assume that the components represent independent sources of activity. Unlike PCA, ICA separates individual elements into their underlying components and models the dataset as a constant number of spatially or temporally independent components.

A growing body of research over the last 20 years has shown that studying patterns of resting-state functional connectivity provides relevant information on brain correlates of cognitive functions (Stevens & Spreng, 2014). Measures of resting-state connectivity have been proposed to reflect patterns of synaptic efficacy in cortical networks that are affected by previous experience: the repeated co-activation of specific brain regions during a task would affect their functional connectivity, affecting the efficiency of their coupling and in turn task performance (Harmelech & Malach, 2013; Sadaghiani & Kleinschmidt, 2013). Studying the relation between variations in performance in a behavioral task and differences in resting-state functional connectivity thus provides an indirect indication of brain networks involved in performing such a task. In a study from our group (Teghil et al., 2020a), in which we investigated the role of interoceptive processing and the insular cortex in duration processing in the suprasecond range, participants performed outside of the scanner a multi-second duration reproduction paradigm, previously tested in a behavioral study (Teghil et al., 2020b). Volunteers had to reproduce time intervals in a condition in which a regular auditory pattern provided an external cue on elapsing time, and in one in which this pattern was irregular, preventing participants from referring to environmental information to estimate elapsed time. A first seed-to-seed analysis was performed to assess the relation between individual differences in accuracy in the two conditions of the timing task, and the strength of the resting-state functional connectivity between a set of a-priori defined ROIs, corresponding to the brain network involved in duration processing based on previous meta-analyses (Nani et al., 2019; Teghil et al., 2019; Wiener et al., 2010a). Within this network, a selective association was found between accuracy in the irregular condition of the timing task, and the strength of Intrinsic connectivity (IC) of the right precentral gyrus with the ipsilateral putamen and insular cortex ROIs. A second set of analyses assessed the relationship between connectivity within networks associated with individual variations in self-reported awareness toward bodily sensations (interoceptive sensibility), and performance in the two conditions of the timing task. Within a network modulated by individual variations in interoceptive sensibility, the strength of the resting-state connectivity of the right posterior insula with the precentral gyrus and posterior STG in the same hemisphere was significantly and specifically correlated with performance in the irregular, but not in the irregular, condition of the timing task.

Mitchell et al. (2018), instead, exploited rsfMRI to investigate the relation between Intrinsic connectivity (IC) patterns and differences in behavioral performance, assessing the effect of tolcapone – a COMT inhibitor preferentially acting on dopamine tone in frontal regions – on suprasecond duration production. They found that the Intrinsic connectivity (IC) of the right IFG with the ipsilateral putamen was decreased in strength on tolcapone compared with placebo, and this decrease in connectivity correlated inversely with behavioral measures of duration production.

Finally, a study by Northoff et al. (2018) provides an example of the use of rsfMRI to study time perception in psychiatric conditions. This study assessed Intrinsic connectivity (IC) patterns associated with an altered subjective experience of time in depressed and manic bipolar patients, who, respectively, report an accelerated and decelerated time experience. The authors hypothesized that the inner reference for subjective time could be associated with intrinsic activity in the somatomotor network, whereas neural activity in sensory networks would determine the speed of "outer" time (i.e., time of environmental events). Results showed that the ratio of neuronal variability (an index of the speed of neural activity) between the somatomotor and sensory networks displayed an opposite pattern in depressed and manic patients, in line with their alterations in perceived time. Interestingly, these results fit well with findings reported above that individual variations in the strength of the functional coupling between different regions part of the sensorimotor network are associated with duration reproduction in irregular contexts that rely on the self-development of temporal representations independent from external cues (Teghil et al., 2020a; Teghil et al., 2019).

Coordinate-based meta-analyses of fMRI studies

From the beginning of 2000s, several attempts have been made to integrate the results from all the disparate fMRI studies investigating time perception using different paradigms, experimental designs, and analytic approaches. Whereas label-based reviews have proven useful in highlighting regions commonly activated across fMRI timing studies and factors modulating their involvement (e.g., Lewis & Miall, 2003b), the development of coordinate-based meta-analytic methods has allowed to quantitatively synthesize results from fMRI studies, going beyond the limitations of individual experiments, including low sample size and the use of different experimental and analysis procedures (Müller et al., 2018). Starting from 2010, different quantitative meta-analyses have been performed on fMRI studies on time perception in order to answer to different questions, mainly using activation likelihood estimation (ALE) (Eickhoff et al., 2009, 2016; Turkeltaub et al., 2012).

ALE assesses whether activation foci from different experiments have a significantly higher clustering compared to what is expected under the null distribution of a casual spatial association of results from these same experiments; thus, it allows to assess the overlap between activation foci from individual experiments modeling foci as probability distributions (Eickhoff et al., 2009), and provides a quantitative measure of the probability of activation of a given brain region across different neuroimaging studies. ALE can thus be used to test hypotheses concerning the convergence or divergence of activation in fMRI studies according to methodological or theoretical dimensions of interest.

In the first ALE meta-analysis on timing, Wiener et al. (2010a) categorized fMRI experiments according to whether timed durations were sub- or suprasecond, and task demands were mainly motor or perceptual. The same approach was later adopted by Nani et al. (2019). These meta-analyses showed that the SMA was consistently activated across the four conditions; moreover, subsecond and suprasecond tasks, respectively, involved in a more consistent manner subcortical (e.g., caudate, putamen, and cerebellum) and cortical (IFG, precentral gyrus, IPL, SMG, midFG, and insula) structures.

Evidence from ALE further suggests that additional factors modulate brain networks involved in time perception. Beyond the predominance of motor versus perceptual demands, these factors include the nature of the stimuli (sequential vs. discrete), the goal of the task (overt time estimation vs. using time for prediction) (Naghibi et al., 2024; Wiener et al., 2010b), as well as the nature of the temporal representation required by the task (i.e., whether it is externally-cued, when an external sensory signal is available to guide timing, or whether it is internally-based, when a temporal representation has to be generated independently from external cues) (Teghil et al., 2019).

Whereas the pre-SMA and left insula show common and timing-specific activation across conditions (Mondok & Wiener, 2023; Naghibi et al., 2024), timing sequential, auditory stimuli more strongly entails activation of the SMA-proper and of the dorsal striatum; timing discrete visual stimuli, instead, recruits the frontoparietal network (Naghibi et al., 2024). Also, using time for prediction entails activation of the left IPS (Naghibi et al., 2024; Wiener et al., 2010b). Finally, internally-based timing mainly entails activation of the SMA, anterior insula, and the right basal ganglia and IPS, whereas externally-cued timing involves stronger activation of a circuit involving the right IFG and left SMA, precentral gyrus, and anterior insula (Teghil et al., 2019).

Neural networks for processing time intervals

Going through this chapter, it will be clear that several brain regions have been implicated in time processing. It is now generally accepted that different mechanisms and brain networks may support duration and temporal processing depending on multiple factors, including the timescale, the sensory modality of the stimuli, and the requirements of the task (Bueti & Macaluso, 2011; Merchant et al., 2013; Paton & Buonomano, 2018; Wiener et al., 2011). This section provides a brief overview of brain regions supporting time perception summarizing relevant findings from fMRI studies.

Evidence from mass-univariate studies, as well as from ALE meta-analyses (Nani et al., 2019; Naghibi et al., 2024; Teghil et al., 2019; Wiener et al., 2010a), has strongly implicated in timing a striato-thalamo-cortical network, including the SMA and right IFG; this network, mainly including motor structures, has been shown to be involved not only in motor but also in perceptual timing tasks (Coull et al., 2016; Merchant & Yarrow, 2016; Naghibi et al., 2024). It has been thus proposed that duration processing relies on a distributed network, including cortical and subcortical structures, whereas the SMA and basal ganglia would form a "core" timing network, generally involved in explicit timing, and other regions would be involved in timing in a context-dependent fashion (Merchant et al., 2013).

Concerning the role of subcortical structures, one of the most influential theories on neurobiological mechanisms of interval timing (striatal beat frequency model) suggested, indeed, a key role of the basal ganglia in duration processing (Matell & Meck, 2004). As reviewed in previous sections, event-related fMRI studies have shown specific activation of the putamen during the encoding of interval duration (Bueti & Macaluso, 2011; Coull et al., 2008; Harrington et al., 2010), suggesting that activation of this region during timing may reflect the storage of stimulus duration in working memory (Coull, 2014; Coull & Droit-Volet, 2018). Results of meta-analytic studies, however, have suggested a more consistent involvement of the basal ganglia in the subsecond, rather than suprasecond range (Lewis & Miall, 2003b; Nani et al., 2019; Wiener et al., 2010a).

The cerebellum has also been largely implicated in time processing, especially in timing of subsecond durations (Nani et al., 2019; Wiener et al., 2010a). It has been highlighted that the specific neuronal features of the granular layer of the cerebellum may indeed be optimal for supporting timing processes with millisecond precision (Bareš et al., 2019). Whereas the cerebellum could be thus well-suited to time intervals between 100 and 2,000 ms, longer intervals appear to additionally recruit the striato-thalamo-cortical circuit (Petter et al., 2016). Activation of the cerebellum in fMRI studies of timing has been further associated with the processing of sequential time intervals in weakly rhythmic contexts, suggesting that this structure may play a specific role in adjusting behavior in accordance with the temporal correspondence between sensory input and motor output (Naghibi et al., 2024).

Different cortical areas have also been implicated in timing. As mentioned above, the SMA is consistently activated in fMRI timing studies (Cona et al., 2021; Nani et al., 2019; Naghibi et al., 2024; Wiener et al., 2010a). Activation of this region has been associated with accumulation of temporal information and coding of temporal magnitude: among different findings, the SMA has been shown to be the only region activated during both the encoding of the reference and the comparison stimulus in duration discrimination, suggesting that its role can be specific for duration processing (Coull et al., 2008). Accordingly, the SMA has been found to increase parametrically its activation with increasing duration, but not with increasing traveled distance (Coull et al., 2015). Activity in the SMA further increased together with the duration of the first stimulus in an interval discrimination task (Wencil et al., 2010), and its activation correlated positively with perceived stimulus duration (Tipples et al., 2015). Based on the evidence of a general involvement of the SMA in duration processing across different tasks and contexts, it has been proposed that this region may work as a hub in the time processing network, flexibly recruiting different brain regions according to task demands (Mondok & Wiener, 2023). In line with this possibility, different portions of the SMA have been shown to be associated with different task features, with more anterior regions (including the pre-SMA) preferentially recruited, respectively, during perceptual and motor timing tasks (Naghibi et al., 2024; Schwartze et al., 2012; Wiener et al., 2011).

Several fMRI studies using different paradigms (e.g., Cerasa et al., 2005; Harrington et al., 2010; Wencil et al., 2010) also revealed the involvement of frontal regions in time perception. Whereas activation of the left IFG has been sometimes interpreted as reflecting the use of verbal or more general auditory rehearsal strategies (Hinton et al., 2004; Wiener et al., 2010a), activation of the right IFG and of the DLPFC during timing has been mainly associated with attentional, working memory, and decision-making demands (Coull et al., 2008; Henry et al., 2015; Lewis & Miall, 2003a; Rao et al., 2001).

A role for the parietal cortex in timing has also been suggested by different fMRI studies (Bueti et al., 2008; Coull et al., 2000; Hayashi et al., 2015, 2018; Rao et al., 2001). The role of right parietal regions in timing has been often interpreted in terms of attentional demands involved in duration processing (Harrington et al., 2004; Lewis & Miall, 2003a; Rao et al., 2001). However, it has also been proposed that the right inferior parietal cortex represents a common neural substrate for the representation of magnitudes, including time, space, and numbers (Bueti & Walsh, 2009; Walsh, 2003). According to this account, the representation of space and time would be learned in parallel through action, and this anatomical organization would allow the integration of spatiotemporal information during coordination and motor control (Bueti & Walsh, 2009). In line with this possibility, the bilateral IPS has been shown to represent both time and space along a dorsoventral gradient of activation, suggesting that this organization may support the integration of time and space for action preparation (Cona et al., 2021). As reviewed in previous sections, neural responses in the right SMG show specific adaptation to the duration of presented stimuli (Hayashi et al., 2015), and duration selectivity has been observed in the IPS (Hayashi et al., 2018; Protopapa et al. 2019). These findings suggest that, at least in the subsecond range, the posterior parietal cortex may be involved in time representation per se, possibly through the presence of duration tuned neural populations (Hayashi et al., 2018; Protopapa et al. 2019) (see below). Conversely, the involvement of parietal regions in the suprasecond range could more strongly reflect a role of attentional modulation (Lewis & Miall, 2003a, 2003b).

Several fMRI studies reported activation of the insula during timing tasks (Ferrandez et al., 2003; Lewis & Miall, 2003a; Livesey et al., 2007; Rao et al., 2001; Tregellas et al., 2006). Compelling evidence has been further provided by the results from the fMRI study by Wittmann et al. (2010) mentioned earlier in this chapter (later replicated in Wittmann et al., 2011), in which a climbing pattern of activation, peaking at the end of the stimulus to be timed, was specifically observed in the posterior insula, suggesting the building of a representation of duration in this region through the accumulation of bodily changes. Moreover, the insula appears to be more likely involved in timing when the surrounding environment does not provide reliable cues allowing to estimate the duration of events since individual variations in resting-state connectivity of the right insula are specifically associated with timing performance in this condition (Teghil et al., 2020a). Evidence for a consistent involvement of both the SMA and the insula in timing is overall in line with embodied accounts of time perception, suggesting that the representation of duration is intrinsically rooted in bodily processing, including the motor system and the interoceptive network (Naghibi et al., 2024; Mondok & Wiener, 2023).

Finally, a body of evidence suggests that the duration of events may be represented locally by modality-specific sensory regions. A mechanism that is likely to operate at the level of primary sensory regions is duration tuning: evidence from studies reviewed in previous sections, using pRFs (Protopapa et al., 2019; Harvey et al., 2020; Hendrikx et al., 2022; van Ackooij et al., 2022), points indeed to the possibility that a first neural representation of event timing in the subsecond range may be derived from the activity of early sensory regions. In this vein, timing-tuned responses in the primary visual (Harvey et al., 2020; Hendrikx et al., 2022) and auditory cortex (van Ackooij et al., 2022) could translate implicit information resulting from response dynamics of these regions into a neurally explicit representation of temporal information (Hendrikx et al., 2022; van Ackooij et al., 2022). Notably, responses from different sensory modalities appear to converge into supra-modal and more abstract representations of time at later processing stages: time-tuned responses have been indeed reported to be hierarchically organized in a progression toward associative brain regions, such as the parietal cortex (Harvey et al., 2020), in line with evidence from studies showing duration selectivity in the inferior parietal lobule (Hayashi et al., 2018; Protopapa et al., 2019). It has been thus suggested that duration tuning in the IPS may represent an intermediate stage of duration processing, in which information coming from primary sensory regions is automatically decoded (Protopapa et al., 2019; 2023). Finally, at the top of the processing hierarchy, the SMA has been proposed to act as a temporal read-out, allowing the explicit recognition of durations, in line with evidence that topographically organized representations of time intervals have been observed in this region using explicit timing tasks (Protopapa et al., 2019).[2]

Conclusions

The use fMRI applied to the study of duration processing has provided invaluable insights on how our brain perceives, produces, and reproduces time. Whereas the use of carefully controlled designs, together with the application of mass-univariate techniques, has allowed to identify individual process components of different timing tasks and to characterize brain networks involved in the processing of durations in the range of hundreds of milliseconds and seconds, the more recent application of multivariate techniques has allowed to go beyond simple localization, starting to disclose how time is represented in the brain. This body of research suggests that several brain regions are involved in the processing of time depending on multiple factors, including the task, the sensory modality of the stimuli, and the range of durations under investigation. Considering how our experience of duration appears to arise from a set of complex interactions between endogenous and environmental signals, an important future challenge for fMRI will be to understand where and how these sources of information can be integrated in the brain, and how this process is shaped by individual differences in cognitive functioning and allows the development of our sense of time.

Notes

1 Due to the high-dimensional nature of the data, the probability of making a type I error (family-wise error rate) increases dramatically in all the analyses mentioned above. Bonferroni correction for multiple comparisons and false discovery rate are the most used strategies to overcome this issue. Although these general approaches can be used when considering multiple independent regions of interest, when computing a statistical parametric map from a second-level voxel-wise analysis, it is important to note that neighboring voxels are not independent. In fact, signal changes are not localized in a single voxel but are usually spread across neighboring voxels. One strategy is to use Gaussian random field theory (Chumbley et al., 2010; Worsley et al., 1992, 1996), which is at the basis of cluster-level inference.

2 Other mechanisms for duration coding that have been proposed to operate at the level of primary sensory regions include climbing neural activity and population clock models (see Paton & Buonomano, 2018; Tsao et al., 2022, for reviews); however, since they describe neuronal-level rather than brain networks dynamics, they will not be discussed in this chapter.

References

Aguirre, G. K. (2007). Continuous carry-over designs for fMRI. *NeuroImage*, *35*(4), 1480–1494. https://doi.org/10.1016/j.neuroimage.2007.02.005

Amaro, E., Jr, & Barker, G. J. (2006). Study design in fMRI: Basic principles. *Brain and Cognition*, *60*(3), 220–232. https://doi.org/10.1016/j.bandc.2005.11.009

Bareš, M., Apps, R., Avanzino, L., Breska, A., D'Angelo, E., Filip, P., Gerwig, M., Ivry, R. B., Lawrenson, C. L., Louis, E. D., Lusk, N. A., Manto, M., Meck, W. H., Mitoma, H., & Petter, E. A. (2019). Consensus paper: Decoding the contributions of the cerebellum as a time machine. From neurons to clinical applications. *Cerebellum (London, England)*, *18*(2), 266–286. https://doi.org/10.1007/s12311-018-0979-5

Bausenhart, K. M., Di Luca, M., & Ulrich, R. (2018). Assessing duration discrimination: Psychophysical methods and psychometric function analysis. In *Timing and Time Perception: Procedures, Measures, & Applications* (pp. 52–78). Brill.

Bueti, D., Kulashekhar, S., Maass, S. C., & van Rijn, H. (n.d.). The topographic representation of time and its link with temporal context and perception. https://doi.org/10.2139/ssrn.3902136

Bueti, D., & Macaluso, E. (2011). Physiological correlates of subjective time: Evidence for the temporal accumulator hypothesis. *NeuroImage, 57*(3), 1251–1263. https://doi.org/10.1016/j.neuroimage.2011.05.014

Bueti, D., & Walsh, V. (2009). The parietal cortex and the representation of time, space, number and other magnitudes. *Philosophical Transactions of the Royal Society B: Biological Sciences, 364,* 1831–1840. https://doi.org/10.1098/rstb.2009.0028

Bueti, D., Walsh, V., Frith, C., & Rees, G. (2008). Different brain circuits underlie motor and perceptual representations of temporal intervals. *Journal of Cognitive Neuroscience, 20*(2), 204–214. https://doi.org/10.1162/jocn.2008.20017

Cerasa, A., Hagberg, G. E., Bianciardi, M., & Sabatini, U. (2005). Visually cued motor synchronization: Modulation of fMRI activation patterns by baseline condition. *Neuroscience Letters 373,* 32–37. https://doi.org/10.1016/j.neulet.2004.09.076

Chumbley, J., Worsley, K., Flandin, G., & Friston, K. (2010). Topological FDR for neuroimaging. *NeuroImage, 49*(4), 3057–3064. https://doi.org/10.1016/j.neuroimage.2009.10.090

Clark V. P. (2012). A history of randomized task designs in fMRI. *NeuroImage, 62*(2), 1190–1194. https://doi.org/10.1016/j.neuroimage.2012.01.010

Cona, G., Wiener, M., & Scarpazza, C. (2021). From ATOM to GradiATOM: Cortical gradients support time and space processing as revealed by a meta-analysis of neuroimaging studies. *NeuroImage, 224,* 117407. https://doi.org/10.1016/j.neuroimage.2020.117407

Coull, J. T. (2014). Getting the timing right: Experimental protocols for investigating time with functional neuroimaging and psychopharmacology. *Advances in Experimental Medicine and Biology, 829,* 237–264. https://doi.org/10.1007/978-1-4939-1782-2_13

Coull, J. T., Charras, P., Donadieu, M., Droit-Volet, S., & Vidal, F. (2015). SMA slectively codes the active accumulation of temporal, not spatial, magnitude. *Journal of Cognitive Neuroscience, 27*(11), 2281–2298. https://doi.org/10.1162/jocn_a_00854

Coull, J. T., & Droit-Volet, S. (2018). Explicit understanding of duration develops implicitly through action. *Trends in Cognitive Sciences, 22*(10), 923–937. https://doi.org/10.1016/j.tics.2018.07.011

Coull, J. T., Frith, C. D., Büchel, C., & Nobre, A. C. (2000). Orienting attention in time: behavioural and neuroanatomical distinction between exogenous and endogenous shifts. *Neuropsychologia, 38*(6), 808–819. https://doi.org/10.1016/s0028-3932(99)00132-3

Coull, J. T., Nazarian, B., & Vidal, F. (2008). Timing, storage, and comparison of stimulus duration engage discrete anatomical components of a perceptual timing network. *Journal of Cognitive Neuroscience, 20*(12), 2185–2197. https://doi.org/10.1162/jocn.2008.20153

Coull, J. T., Vidal, F., & Burle, B. (2016). When to act, or not to act: That's the SMA's question. *Current Opinion in Behavioral Sciences, 8,* 14–21. https://doi.org/10.1016/j.cobeha.2016.01.003

Coull, J. T., Vidal, F., Nazarian, B., & Macar, F. (2004). Functional anatomy of the attentional modulation of time estimation. *Science 303*(5663), 1506–1508. https://doi.org/10.1126/science.1091573

Davranche, K., Nazarian, B., Vidal, F., & Coull, J. (2011). Orienting attention in time activates left intraparietal sulcus for both perceptual and motor task goals. *Journal of Cognitive Neuroscience, 23*(11), 3318–3330. https://doi.org/10.1162/jocn_a_00030

Dumoulin, S. O., & Wandell, B. A. (2008). Population receptive field estimates in human visual cortex. *NeuroImage, 39*(2), 647–660. https://doi.org/10.1016/j.neuroimage.2007.09.034

Eickhoff, S. B., Laird, A. R., Grefkes, C., Wang, L. E., Zilles, K., & Fox, P. T. (2009). Coordinate- based activation likelihood estimation meta-analysis of neuroimaging data: A random-effects approach based on empirical estimates of spatial uncertainty. *Human Brain Mapping 30*, 2907–2926. https://doi.org/10.1002/hbm.20718

Eickhoff, S. B., Nichols, T. E., Laird, A. R., Hoffstaedter, F., Amunts, K., Fox, P. T., Bzdok, D., & Eickhoff, C.R. (2016). Behavior, sensitivity, and power of activation likelihood estimation characterized by massive empirical simulation. *NeuroImage 137*, 70–85. https://doi.org/10.1016/j.neuroimage.2016.04.072

Ferrandez, A. M., Hugueville, L., Lehéricy, S., Poline, J. B., Marsault, C., & Pouthas, V. (2003). Basal ganglia and supplementary motor area subtend duration perception: An fMRI study. *NeuroImage 9*, 1532–1544. https://doi.org/10.1016/S1053-8119(03)

Friston, K. J., Buechel, C., Fink, G. R., Morris, J., Rolls, E., & Dolan, R. J. (1997). Psychophysiological and modulatory interactions in neuroimaging. *NeuroImage, 6*(3), 218–229. https://doi.org/10.1006/nimg.1997.0291

Friston, K. J., Harrison, L., & Penny, W. (2003). Dynamic causal modelling. *NeuroImage, 19*(4), 1273–1302. https://doi.org/10.1016/s1053-8119(03)00202-7

Garraux, G., McKinney, C., Wu, T., Kansaku, K., Nolte, G., & Hallett, M. (2005). Shared brain areas but not functional connections controlling movement timing and order. *The Journal of Neuroscience, 22*, 5290–5297. https://doi.org/10.1523/JNEUROSCI.0340-05.2005

Grondin S. (2010). Timing and time perception: A review of recent behavioral and neuroscience findings and theoretical directions. *Attention, Perception & Psychophysics, 72*(3), 561–582. https://doi.org/10.3758/APP.72.3.561

Harmelech, T., & Malach, R. (2013). Neurocognitive biases and the patterns of spontaneous correlations in the human cortex. *Trends in Cognitive Sciences, 17*, 606e615. https://doi.org/10.1016/j.tics.2013.09.014

Harrington, D. L., Boyd, L. A., Mayer, A. R., Sheltraw, D. M., Lee, R. R., Huang, M., & Rao, S. M. (2004). Neural representation of interval encoding and decision making. *Cognitive Brain Research 21*, 193–205. https://doi.org/10.1016/j.cogbrainres.2004.01.010

Harrington, D. L., Zimbelman, J. L., Hinton, S. C., & Rao, S. M. (2010). Neural modulation of temporal encoding, maintenance, and decision processes. *Cerebral Cortex (New York, 1991), 20*(6), 1274–1285. https://doi.org/10.1093/cercor/bhp194

Harvey, B. M., & Dumoulin, S. O. (2017). Can responses to basic non-numerical visual features explain neural numerosity responses? *NeuroImage, 149*, 200–209. https://doi.org/10.1016/j.neuroimage.2017.02.012

Harvey, B. M., Dumoulin, S. O., Fracasso, A., & Paul, J. M. (2020). A network of topographic maps in human association cortex hierarchically transforms visual timing-selective responses. *Current Biology, 30*(8), 1424–1434.e6. https://doi.org/10.1016/j.cub.2020.01.090

Hayashi, M. J., Ditye, T., Harada, T., Hashiguchi, M., Sadato, N., Carlson, S., Walsh, V., & Kanai, R. (2015). Time adaptation shows duration selectivity in the human parietal cortex. *PLoS Biology, 13*(9), e1002262. https://doi.org/10.1371/journal.pbio.1002262

Hayashi, M. J., & Ivry, R. B. (2020). Duration selectivity in right parietal cortex reflects the subjective experience of time. *The Journal of Neuroscience, 40*(40), 7749–7758. https://doi.org/10.1523/JNEUROSCI.0078-20.2020

Hayashi, M. J., Kantele, M., Walsh, V., Carlson, S., & Kanai, R. (2014). Dissociable neuroanatomical correlates of subsecond and suprasecond time perception. *Journal of Cognitive Neuroscience, 26*(8), 1685–1693. https://doi.org/10.1162/jocn_a_00580

Hayashi, M. J., van der Zwaag, W., Bueti, D., & Kanai, R. (2018). Representations of time in human frontoparietal cortex. *Communications Biology, 1,* 233. https://doi.org/10.1038/s42003-018-0243-z

Hendrikx, E., Paul, J. M., van Ackooij, M., van der Stoep, N., & Harvey, B. M. (2022). Visual timing-tuned responses in human association cortices and response dynamics in early visual cortex. *Nature Communications, 13*(1), 3952. https://doi.org/10.1038/s41467-022-31675-9

Henry, M. J., Herrmann, B., & Obleser, J. (2015). Selective attention to temporal features on nested time scales. *Cerebral Cortex (New York, N.Y.: 1991), 25*(2), 450–459. https://doi.org/10.1093/cercor/bht240

Hinton, S. C., Harrington, D. L., Binder, J. R., Durgerian, S., & Rao, S. M. (2004). Neural systems supporting timing and chronometric counting: an FMRI study. *Brain research. Cognitive brain research, 21*(2), 183–192. https://doi.org/10.1016/j.cogbrainres.2004.04.009

Jäncke, L., Loose, R., Lutz, K., Specht, K., & Shah, N. J. (2000). Cortical activations during paced finger-tapping applying visual and auditory pacing stimuli. *Brain Research, 10*(1-2), 51–66. https://doi.org/10.1016/s0926-6410(00)00022-7

Kriegeskorte, N., & Bandettini, P. (2007). Analyzing for information, not activation, to exploit high-resolution fMRI. *NeuroImage, 38*(4), 649–662. https://doi.org/10.1016/j.neuroimage.2007.02.022

Levy, J. M., Namboodiri, V. M., & Hussain Shuler, M. G. (2015). Memory bias in the temporal bisection point. *Frontiers in Integrative Neuroscience, 9,* 44. https://doi.org/10.3389/fnint.2015.00044

Lewis, P. A., & Miall, R. C. (2002). Brain activity during non-automatic motor production of discrete multi-second intervals. *Neuroreport, 13*(14), 1731–1735. https://doi.org/10.1097/00001756-200210070-00008

Lewis, P. A., & Miall, R. C. (2003a). Brain activation patterns during measurement of sub- and supra-second intervals. *Neuropsychologia, 41*(12), 1583–1592. https://doi.org/10.1016/s0028-3932(03)00118-0

Lewis, P. A., & Miall, R. C. (2003b). Distinct systems for automatic and cognitively controlled time measurement: Evidence from neuroimaging. *Current Opinion in Neurobiology, 13*(2), 250–255. https://doi.org/10.1016/s0959-4388(03)00036-9

Lewis, P. A., Wing, A. M., Pope, P. A., Praamstra, P., & Miall, R. C. (2004). Brain activity correlates differentially with increasing temporal complexity of rhythms during initialisation, synchronisation, and continuation phases of paced finger tapping. *Neuropsychologia, 42*(10), 1301–1312. https://doi.org/10.1016/j.neuropsychologia.2004.03.001

Liu, M., Huang, G., Zhao, K., & Fu, X. (2023). The functional brain network of subcortical and cortical regions underlying time estimation: An functional MRI study. *Neuroscience, 519,* 23–30. https://doi.org/10.1016/j.neuroscience.2023.02.019

Livesey, A. C., Wall, M. B., & Smith, A. T. (2007). Time perception: Manipulation of task difficulty dissociates clock functions from other cognitive demands. *Neuropsychologia, 45*(2), 321–331. https://doi.org/10.1016/j.neuropsychologia.2006.06.033

Malach R. (2012). Targeting the functional properties of cortical neurons using fMR-adaptation. *NeuroImage, 62*(2), 1163–1169. https://doi.org/10.1016/j.neuroimage.2012.01.002

Matell, M. S., & Meck, W. H. (2004). Cortico-striatal circuits and interval timing: Coincidence detection of oscillatory processes. *Cognitive Brain Research, 21*(2), 139–170. https://doi.org/10.1016/j.cogbrainres.2004.06.012

McLaren, D. G., Ries, M. L., Xu, G., & Johnson, S. C. (2012). A generalized form of context-dependent psychophysiological interactions (gPPI): A comparison to standard approaches. *NeuroImage, 61*(4), 1277–1286. https://doi.org/10.1016/j.neuroimage.2012.03.068

Merchant, H., Harrington, D. L., & Meck, W. H. (2013). Neural basis of the perception and estimation of time. *Annual Review of Neuroscience, 36*, 313–336. https://doi.org/10.1146/annurev-neuro-062012-170349

Merchant, H. & Yarrow, K. (2016). How the motor system both encodes and influences our sense of time. *Current Opinion in Behavioral Sciences, 8*, 22–27. https://doi.org/10.1016/j.cobeha.2016.01.006

Mitchell, J. M., Weinstein, D., Vega, T., & Kayser, A. S. (2018). Dopamine, time perception, and future time perspective. *Psychopharmacology, 235*, 2783e2793. https://doi.org/10.1007/ s00213-018-4971-z

Mondok, C., & Wiener, M. (2023). Selectivity of timing: A meta-analysis of temporal processing in neuroimaging studies using activation likelihood estimation and reverse inference. *Frontiers in Human Neuroscience, 16*, 1000995. https://doi.org/10.3389/fnhum.2022.1000995

Morillon, B., Kell, C. A., & Giraud, A. L. (2009). Three stages and four neural systems in time estimation. *The Journal of Neuroscience 29*(47), 14803–14811. https://doi.org/10.1523/JNEUROSCI.3222-09.2009

Müller, V. I., Cieslik, E. C., Laird, A. R., Fox, P. T., Radua, J., Mataix-Cols, D., Tench, C. R., Yarkoni, T., Nichols, T. E., Turkeltaub, P. E., Wager, T. D., & Eickhoff, S. B. (2018). Ten simple rules for neuroimaging meta-analysis. *Neuroscience and Biobehavioral Reviews, 84*, 151–161. https://doi.org/10.1016/j.neubiorev.2017.11.012

Mur, M., Bandettini, P. A., & Kriegeskorte, N. (2009). Revealing representational content with pattern-information fMRI--an introductory guide. *Social Cognitive and Affective Neuroscience, 4*(1), 101–109. https://doi.org/10.1093/scan/nsn044

Naghibi, N., Jahangiri, N., Khosrowabadi, R., Eickhoff, C. R., Eickhoff, S. B., Coull, J. T., & Tahmasian, M. (2024). Embodying time in the brain: A multi-dimensional neuroimaging meta-analysis of 95 duration processing studies. *Neuropsychology Review*, https://doi.org/10.1007/s11065-023-09588-1

Nani, A., Manuello, J., Liloia, D., Duca, S., Costa, T., & Cauda, F. (2019). The neural correlates of time: A meta-analysis of neuroimaging studies. *Journal of Cognitive Neuroscience, 31*(12), 1796–1826. https://doi.org/10.1162/jocn_a_01459

Nenadic, I., Gaser, C., Volz, H. P., Rammsayer, T., Häger, F., & Sauer, H. (2003). Processing of temporal information and the basal ganglia: New evidence from fMRI. *Experimental Brain Research, 148*(2), 238–246. https://doi.org/10.1007/s00221-002-1188-4

Northoff, G., Magioncalda, P., Martino, M., Lee, H. C., Tseng, Y. C., & Lane, T. (2018). Too fast or too slow? Time and neuronal variability in bipolar disorder-A combined theoretical and empirical investigation. *Schizophrenia Bulletin, 44*(1), 54–64. https://doi.org/10.1093/schbul/sbx050

Paton, J. J., & Buonomano, D. V. (2018). The neural basis of timing: Distributed mechanisms for diverse functions. *Neuron, 98*(4), 687–705. https://doi.org/10.1016/j.neuron.2018.03.045

Petersen, S. E., & Dubis, J. W. (2012). The mixed block/event-related design. *NeuroImage, 62*(2), 1177–1184. https://doi.org/10.1016/j.neuroimage.2011.09.084

Petter, E. A., Lusk, N. A., Hesslow, G., & Meck, W. H. (2016). Interactive roles of the cerebellum and striatum in sub-second and supra-second timing: Support for an initiation, continuation, adjustment, and termination (ICAT) model of temporal processing. *Neuroscience and Biobehavioral Reviews, 71*, 739–755. https://doi.org/10.1016/j.neubiorev.2016.10.015

Poldrack R. A. (2007). Region of interest analysis for fMRI. *Social Cognitive and Affective Neuroscience, 2*(1), 67–70. https://doi.org/10.1093/scan/nsm006

Poldrack, R.A., Mumford, J.A., & Nichols, T.E. (2011). *Handbook of functional MRI data analysis*. Cambridge University Press.

Protopapa, F., Hayashi, M. J., Kanai, R., & Bueti, D. (2023). Topographic connectivity in a duration selective cortico-cerebellar network. *Scientific Reports, 13*(1), 20674. https://doi.org/10.1038/s41598-023-47954-4

Protopapa, F., Hayashi, M. J., Kulashekhar, S., van der Zwaag, W., Battistella, G., Murray, M. M., Kanai, R., & Bueti, D. (2019). Chronotopic maps in human supplementary motor area. *PLoS Biology, 17*(3), e3000026. https://doi.org/10.1371/journal.pbio.3000026

Rao, S. M., Mayer, A. R., & Harrington, D. L. (2001). The evolution of brain activation during temporal processing. *Nature Neuroscience, 4*(3), 317–323. https://doi.org/10.1038/85191

Sadaghiani, S., & Kleinschmidt, A. (2013). Functional interactions between intrinsic brain activity and behavior. *NeuroImage, 80*, 379e386. https://doi.org/10.1016/j.neuroimage.2013.04.100

Schubotz, R. I., Friederici, A. D., & von Cramon, D. Y. (2000). Time perception and motor timing: A common cortical and subcortical basis revealed by fMRI. *NeuroImage, 11*(1), 1–12. https://doi.org/10.1006/nimg.1999.0514

Schwartze, M., Rothermich, K., & Kotz, S. A. (2012). Functional dissociation of pre-SMA and SMA-proper in temporal processing. *NeuroImage, 60*(1), 290–298. https://doi.org/10.1016/j.neuroimage.2011.11.089

Stevens, W. D., & Spreng, R. N. (2014). Resting-state functional connectivity MRI reveals active processes central to cognition. *Wiley Interdisciplinary Reviews. Cognitive Science, 5*(2), 233–245. https://doi.org/10.1002/wcs.1275

Sulpizio, V., Boccia, M., Guariglia, C., & Galati, G. (2016). Functional connectivity between posterior hippocampus and retrosplenial complex predicts individual differences in navigational ability. *Hippocampus, 26*, 841e847. https://doi.org/10.1002/hipo.22592

Teghil, A., Boccia, M., D'Antonio, F., Di Vita, A., de Lena, C., & Guariglia, C. (2019). Neural substrates of internally-based and externally-cued timing: An activation likelihood estimation (ALE) meta-analysis of fMRI studies. *Neuroscience and Biobehavioral Reviews, 96*, 197–209. https://doi.org/10.1016/j.neubiorev.2018.10.003

Teghil, A., Boccia, M., Nocera, L., Pietranelli, V., & Guariglia, C. (2020b). Interoceptive awareness selectively predicts timing accuracy in irregular contexts. *Behavioural Brain Research, 377*, 112242. https://doi.org/10.1016/j.bbr.2019.112242

Teghil, A., Bonavita, A., Procida, F., Giove, F., & Boccia, M. (2023). Intrinsic hippocampal connectivity is associated with individual differences in retrospective duration processing. *Brain Structure & Function, 228*(2), 687–695. https://doi.org/10.1007/s00429-023-02612-3

Teghil, A., Di Vita, A., D'Antonio, F., & Boccia, M. (2020a). Inter-individual differences in resting-state functional connectivity are linked to interval timing in irregular contexts. *Cortex, 128*, 254–269. https://doi.org/10.1016/j.cortex.2020.03.021

Tipples, J., Brattan, V., & Johnston, P. (2015). Facial emotion modulates the neural mechanisms responsible for short interval time perception. *Brain Topography*, *28*(1), 104–112. https://doi.org/10.1007/s10548-013-0350-6

Tregellas, J. R., Davalos, D. B., & Rojas, D. C. (2006). Effect of task difficulty on the functional anatomy of temporal processing. *NeuroImage*, *32*(1), 307–315. https://doi.org/10.1016/j.neuroimage.2006.02.036

Tsao, A., Yousefzadeh, S. A., Meck, W. H., Moser, M. B., & Moser, E. I. (2022). The neural bases for timing of durations. *Nature Reviews. Neuroscience*, *23*(11), 646–665. https://doi.org/10.1038/s41583-022-00623-3

Turkeltaub, P. E., Eickhoff, S. B., Laird, A. R., Fox, M., Wiener, M., & Fox, P. (2012). Minimizing within-experiment and within-group effects in activation likelihood estimation meta-analyses. *Human Brain Mapping 33*(1), 1–13. https://doi.org/10.1002/hbm.21186

van Ackooij, M., Paul, J. M., van der Zwaag, W., van der Stoep, N., & Harvey, B. M. (2022). Auditory timing-tuned neural responses in the human auditory cortices. *NeuroImage*, *258*, 119366. https://doi.org/10.1016/j.neuroimage.2022.119366

Walsh, D. (2003). A theory of magnitude: Common cortical metrics of time, space and quantity. *Trends in Cognitive Science 7*, 483–488. https://doi.org/10.1016/j.tics.2003.09.002

Wencil, E. B., Coslett, H. B., Aguirre, G. K., & Chatterjee, A. (2010). Carving the clock at its component joints: Neural bases for interval timing. *Journal of Neurophysiology*, *104*(1), 160–168. https://doi.org/10.1152/jn.00029.2009

Wiener, M., Matell, M. S., & Coslett, H. B. (2011). Multiple mechanisms for temporal processing. *Frontiers in Integrative Neuroscience*, *5*, 31. https://doi.org/10.3389/fnint.2011.00031

Wiener, M., Turkeltaub, P., & Coslett, H. B. (2010a). The image of time: A voxel-wise meta-analysis. *NeuroImage*, *49*(2), 1728–1740. https://doi.org/10.1016/j.neuroimage.2009.09.064

Wiener, M., Turkeltaub, P. E., & Coslett, H. B. (2010b). Implicit timing activates the left inferior parietal cortex. *Neuropsychologia*, *48*(13), 3967–3971. https://doi.org/10.1016/j.neuropsychologia.2010.09.014

Wittmann, M., Simmons, A. N., Aron, J. L., & Paulus, M. P. (2010). Accumulation of neural activity in the posterior insula encodes the passage of time. *Neuropsychologia*, *48*(10), 3110–3120. https://doi.org/10.1016/j.neuropsychologia.2010.06.023

Wittmann, M., Simmons, A. N., Flagan, T., Lane, S. D., Wackermann, J., & Paulus, M. P. (2011). Neural substrates of time perception and impulsivity. *Brain Research*, *1406*, 43–58. https://doi.org/10.1016/j.brainres.2011.06.048

Worsley, K., Evans, A., Marrett, S., & Neelin, P. (1992). A three-dimensional statistical analysis for rCBF activation studies in human brain. *Journal of Cerebral Blood Flow and Metabolism*, *12*, 900–918. https://doi.org/10.1038/jcbfm.1992.127

Worsley, K. J., Marrett, S., Neelin, P., Vandal, A. C., Friston, K. J., & Evans, A. C. (1996). A unified statistical approach for determining significant signals in images of cerebral activation. *Human Brain Mapping*, *4*(1), 58–73. https://doi.org/10.1002/(SICI)1097-0193(1996)4:1<58::AID-HBM4>3.0.CO;2-O

Yeo, B. T., Krienen, F. M., Sepulcre, J., Sabuncu, M. R., Lashkari, D., Hollinshead, M., Roffman, J. L., Smoller, J. W., Zöllei, L., Polimeni, J. R., Fischl, B., Liu, H., & Buckner, R. L. (2011). The organization of the human cerebral cortex estimated by intrinsic functional connectivity. *Journal of Neurophysiology*, *106*(3), 1125–1165. https://doi.org/10.1152/jn.00338.2011

3

STUDYING TIME PERCEPTION WITH ELECTROENCEPHALOGRAPHY

Nicola Thibault
Université Laval, Québec, Canada

William Vallet
Université de Lyon 1, Lyon, France

Philippe Albouy and Simon Grondin
Université Laval, Québec, Canada

Introduction

The literature on the neuroimaging of timing and time perception is now quite abundant. However, our understanding of how the human brain perceives and reproduces time intervals is mostly founded on functional magnetic resonance imaging (fMRI) and positron emission tomography (PET) (Albouy et al., 2020; Nani et al., 2019). These neuroimaging tools allow for great spatial localization (Gosseries et al., 2008), but are limited in terms of temporal resolution (see Chapter 2 for fMRI). Electroencephalography (EEG) is a tool that alleviates this limitation by offering another perspective on brain activity with considerably better time resolution.

Electroencephalography (EEG)

The brain activity recorded with EEG reflects the summation of inhibitory and excitatory post-synaptic potentials which are recorded by electrodes placed on the scalp (Nunez & Srinivasan, 2006). Hans Berger was the first to record brain activity in 1924 (İnce et al., 2021). The EEG served as a diagnostic tool for neurologic and psychiatric conditions. One can imagine how useful this non-invasive tool was for clinicians in an era where the only ways of identifying "diseased" areas of the brain were lumbar puncture, pneumoencephalography, and ventriculography (Tudor et al., 2005). However, the localization of the recorded signal was far from perfect. With EEG

DOI: 10.4324/9781003449546-3

recordings, the raw signal is typically the summation of task-related cognitive signals, ongoing fluctuation of electrical activity, and physiological artifacts such as eye blinks, heart rate, other motor movements, and electrical line noise (Fatourechi et al., 2007; Ge et al., 2017; Radüntz et al., 2017). Today, the artifacts can somewhat be attenuated using denoising procedures such as independent components analysis (ICA) and filtering techniques (Gorjan et al., 2022). The ICA is based on a wide class of supervised machine learning algorithms. It assumes that the signals have independent spatial sources that can be associated with physiological activities, such as blinking and eye saccades. It separates signal patterns into independent components and allows the suppression of unwanted components of the signal which includes the artifacts (Jain & Rai, 2012). Filtering techniques for the signal include *notch filtering* (a filter that attenuates specific frequencies, such as electrical outlets) and *bandpass filtering* (a filter that keeps certain frequencies of interest). After the ICA and the filters have been applied, the preprocessed signal is considered cleaner and freed from a considerable portion of artifacts.

Localizing the brain generators of signal recording on the scalp has shown to be more challenging than was initially anticipated by researchers and clinicians. Developers of the EEG analysis tools described two distinct challenges. The first is called the forward EEG problem. It states that the conductivity of different tissues below the electrodes is not homogeneous (scalp, skull, and brain). Also, with EEG, the recorded electrical activity stems mostly from the pyramidal neurons. While the dendrites of these neurons are oriented orthogonal to the brain's surface, this is not the case for other types of neurons that are also recorded by the EEG, in which case, the electrical signal can cancel out (Hallez et al., 2007). Researchers now use the boundary element method (BEM) to solve the forward EEG problem (He et al., 1987; He et al., 1999). The BEM takes into consideration anatomic information of the head as well as conductivity characteristics of the scalp, the skull, and brain tissues to increase spatial resolution. The second challenge, the EEG inverse problem, refers to the ability to identify brain sources that shape the brain signal into the sensors. The inverse problem requires the resolution of the forward problem by modeling head tissues and other anatomical characteristics. Once the BEM matrix is created, one may inverse this matrix to solve the inverse problem (Awan et al., 2019). The current chapter does not aim to elaborate on the mathematical modeling and algorithms employed for solving these specific problems. It is, however, important to mention both forward and inverse problems as they are commonly employed in the EEG literature. The localization of brain signal sources is now well established with EEG, so much so that some authors consider it as a neuroimaging technique and no longer simply as an electrophysiological technique (de Peralta Menendez et al., 2004; Michel & He, 2019; Michel & Murray, 2012).

TABLE 3.1 Oscillatory rhythms and their involvement in cognitive functions

Oscillatory rhythms	Function
Delta (1–4 Hz)	Sleep and attentional processes (Amzica & Steriade, 1998; Harmony et al., 1996)
Theta (4–8 Hz)	Executive functions and memory (Albouy et al., 2017; Schacter, 1977)
Alpha (8–12 Hz)	Cortical excitability, attentional, memory, motor, and sensory processes (Başar et al., 1997; Hanslmayr et al., 2011; Klimesch et al., 1999)
Beta (12–30 Hz)	Motor, somatosensory processes, and predictive temporal encoding (Jenkinson & Brown, 2011; Morillon & Baillet, 2017)
Gamma (>30 Hz)	Attentional processes and sensory integration (Wang, 2010)

The classical approach to EEG is to observe irregularities or patterns in the brain signal. Another way to look at the EEG signal is through spectral analysis. This type of analysis is applied to continuous EEG recording to calculate the power of oscillatory rhythms. Commonly observed oscillatory rhythms are delta (1–4 Hz), theta (4–8 Hz), alpha (8–12 Hz), beta (12–30 Hz), and gamma (>30 Hz) (Canolty & Knight, 2010; Zhang, 2019). Table 3.1 is a summary of the oscillatory rhythms and their associated cognitive functions. Changes in oscillatory rhythms reflect the synchronous and periodic shifting of the neuronal activity between high and low excitability states. This shift is what is thought to coordinate communication between neurons and regions of the brain (Schroeder & Lakatos, 2009; Schroeder et al., 2008; Wang, 2010). The use of oscillatory rhythms for the study of time perception will be discussed later in this chapter.

Introduction to the study of time perception through EEG

Before going into the details of interesting EEG studies dedicated to the understanding of explicit and implicit timing and time perception, it is useful to return to the review of Macar and Vidal (2004) on event-related potentials (ERPs) about relevant components of the processing of temporal information. ERPs are electrical potentials in the brain that are related to internal (cognitive or perception) or external events (Luck, 2012). This review outlines important results of past studies using ERPs and different timing tasks with intervals or durations in the range of milliseconds to minutes. One of the most compelling components identified as playing a key role in time perception is the contingent negative variation (CNV). When a subsequent action or stimulus depends on a previous stimulus, a slow negative wave is elicited following the initial stimulus. This slow negative wave is called the CNV (Kononowicz & Penney, 2016). This slow negative wave is generally fronto-central and resorbs to baseline when the second stimulus occurs. There are some distinct interpretations of what the CNV reflects in timing and time perception, but generally, the CNV seems to be associated with the encoding of a duration in memory. At the time, Macar and Vidal (2004) proposed four interpretations of the CNV arising from a task involving temporal processing: (1) The CNV's amplitude generated over the supplementary motor area (SMA) might be an index revealing its contribution as a temporal accumulator. (2) The peak of the CNV's amplitude might reflect a time-related decision-making process. (3) For long time intervals (i.e., 5 s), the CNV's observed shifts toward positivity might reflect temporal encoding and, to some extent, the functioning of long-term memory processes. (4) The CNV's amplitude might be an index to timing performance as it is lower for better performance.

In their review, Macar and Vidal (2004) also reported the relevance of using the mismatch negativity (MMN) component, in several timing and time perception studies. The MMN is elicited in the auditory cortex and consists of negativity in the signal occurring 120–200 ms after the presence of a rare (deviant) auditory event (Näätänen et al., 2007; Näätänen et al., 2004). It is well established that this component is elicited by the isochrony of an expected event; in other words, the MMN reflects sensitivity to temporal disturbance. This component is elicited automatically; it does not require the subject's attention to be directed to a stimulus (Näätänen, 1990). This automatic response shows that changes in duration are detected at the early stages of information processing. Macar and Vidal (2004) also briefly described the error-related negativity component (Ne/ERN: negativity), which reaches a peak of about 100 ms after the error commitment. This Ne/ERN component is elicited when a subject is conscious that they committed an error in a given task. When the subject is conscious of the committed error, a large negativity component, the Ne/ERN, is prompted at fronto-central sites. Accuracy in behavioral tasks is thus subject to this Ne/ERN component, with its amplitude linked to the importance of the committed error (Luu et al., 2000). Time perception studies using behavioral responses may also provoke Ne/ERN, as discussed in the following sections.

The present chapter will review markers such as the *N1–P2 complex*, MMN, CNV, *P300*, and late positive component of timing (LPCt) components as indexes of temporal information processing. Their implications differ depending on the nature of the task. Regarding the frequency domain, fluctuation of power in the delta, theta, and beta frequencies have also been reported as a possible index of timing and time perception. In the following sections, we propose to brief the readers about how ERPs and frequency bands are related to cognitive engagement in tasks requiring timing or time perception.

Studying explicit time perception with EEG

For cognitive neurosciences, questioning timing and time perception remains an important topic considering that the fundamental specificity of all cognitive mechanisms is inherently time-dependent. In recent years, a growing body of work has tried to identify the neural mechanisms underlying the ability to process different types of timing processes such as estimating durations or expecting forthcoming events. In the present chapter, we distinguish explicit and implicit timing, opening with the former.

Explicit timing refers to the domain of time perception in which participants must provide time judgments with an overt and accurate estimate of elapsed time. The most often used tasks involving time judgments are temporal generalization, bisection, production, and reproduction (Grondin, 2010).

The early ERPs

Early ERP components have been identified in the study of explicit timing and time perception. Among them is P1 (see Figure 3.1), a fronto-central response-locked component typically elicited about 100 ms after movement or behavioral response. The functional implication of this component is

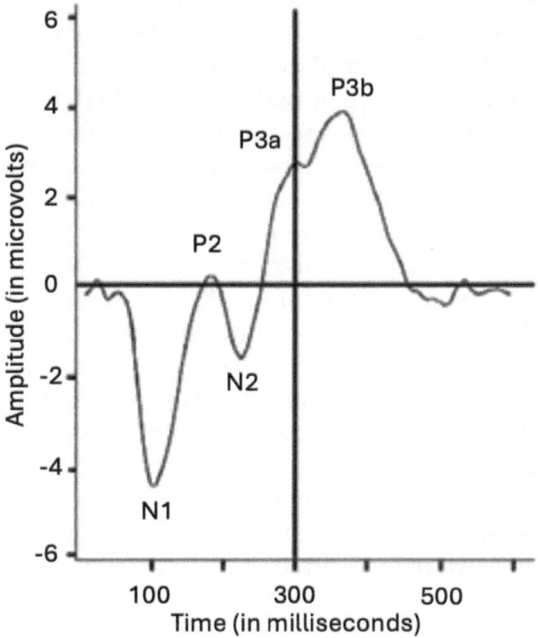

FIGURE 3.1 Illustration of the N1, P1, N2, P2, P3a, and P3b ERP components (Samima et al., 2017).

generally related to the action consequences and sensory feedback. In timing and time perception, the P1 component has also been considered as an index of time estimation when a participant produces a timed behavior (e.g., producing a duration by pressing a button). The higher the amplitude of the P1 component following the end of the interval produced by the participant, the shorter the estimated interval (Zhao et al., 2014). The P1 component is also generated for perceptual processes. The P1 component is a positive deflection peaking close to 50 ms for auditory and 100 ms for visual stimuli and reflects the earlier stage of sensory integration.

The N1–P2 complex (Figure 3.1) is a biphasic component with a negative phase (N1) and positive phase (P2) spreading from 50 to 300 ms following an event. Considering the hypothesis of SMA involvement in different temporal tasks (Coull, 2009; Schwartze et al., 2012), the N1–P2 complex is traditionally measured at FCz during a timing task. The N1–P2 complex amplitude corresponds to the peak-to-peak absolute amplitude between N1 and P2 components and has been described as an appropriate marker of perception in temporal generalization tasks.

Temporal generalization consists of asking participants to identify whether a duration is the same length as a previously repeatedly presented duration. In a temporal generalization task conducted by Kononowicz and van Rijn (2014), participants learned a standard duration to be compared with the duration of test intervals. The researchers reported that the amplitude of the N1–P2 complex evoked by the offset of the test intervals accurately reflected the subjective time experienced by participants. Indeed, the amplitude of the measured N1-P2 complex at the end of the test intervals was significantly correlated with the difference between the perceived duration of the test intervals and the duration of the standard interval previously learned.

The N1–P2 complex has also been studied in other explicit timing tasks. Notably, it has been used to analyze the discrimination of time intervals marked by brief signals delivered from different sensory modalities (intermodal intervals). Typically, brief intermodal intervals are much more difficult to discriminate than intramodal intervals (Azari et al., 2020, 2023). The amplitude of the N1 and P2 components is larger for the processing of intermodal intervals than for the processing of intramodal intervals (Gontier et al., 2013). Because the N1–P2 complex is associated with attentional processes, the weak performances observed in intermodal conditions were reported by Gontier et al. to depend on a bias induced by the necessity to switch attention between modalities.

The contingent negative variation

The CNV is a negative climbing wave generally elicited over the SMA when the participant anticipates an event or, in the specific case of time perception,

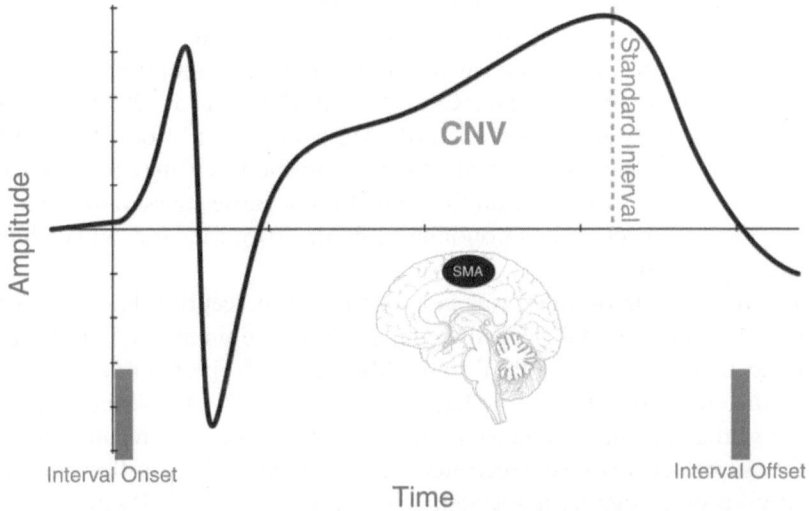

Current Opinion in Behavioral Sciences

FIGURE 3.2   Illustration of the CNV when participants are asked to compare a remembered interval to a target interval. The *y-axis* is inverted; the positivity is below the *x-axis* and the negativity over it (from Kononowicz & Penney, 2016).

is producing or estimating durations (Figure 3.2). In a typical explicit timing task, the duration produced or reproduced by the participants seems to be reflected in the amplitude of the CNV component. More precisely, the CNV can be involved in the subjectively perceived duration and/or produced duration. Thus, the longer the duration perceived, the greater the amplitude of the component (Schlichting et al., 2020; Vallet et al., 2019; Zhang et al., 2021). The CNV amplitude has been used as an index of pacemaker–accumulator activity, referring to pioneer computational models of internal clocks, like the *scalar expectancy theory* (SET; Gibbon, 1977). These models address the role of attentional control on time perception if a pacemaker, or an oscillator process, emitted neural pulses that entered an accumulator modulated by a switch process. The number of pulses provides the unit for estimating time with more pulses resulting in longer perceived duration (Gibbon et al., 1984; Grondin, 2010). The count of the accumulated pulses is then stored in memory for an eventual comparison. A subsequent model, the *attentional gate model* (AGM), later added a gate component to the SET to account for the role of attention in time perception tasks (Zakay & Block, 1997). The AGM model posits that the pulse count is under the control of a gate mechanism managing the allocation of attention, with more attention dedicated to the temporal feature of a stimulus resulting in longer perceived duration. It has been suggested that the modulation in the CNV's amplitude when following a stimulus may reflect the impact of the attentional gate on the pulse emission/accumulation processes. Indeed, greater amplitude of the CNV component is associated with situations where attention is focused toward arousing temporal events (Chiu et al., 2004; Gan et al., 2009; Hosseini Houripasand et al., 2023; Li et al., 2019; Ma et al., 2021; Mioni et al., 2020; Tamm et al., 2014). This effect of attention on the CNV may serve as evidence of the CNV's influence on the regulation of pulse emission/accumulation by the gate mechanism (Kononowicz et al., 2015; Macar & Vidal, 2009; Vallet et al., 2019).

A potential role of the CNV in explicit time perception has also been proposed in multimodal neuroimaging studies combining PET and EEG (Pouthas et al. (2000). In a PET study, Maquet et al. (1996) contrasted two visual discrimination tasks: one based on the duration of a visual stimulus and the other on the luminance of the same stimulus. The results showed that the same networks were recruited for both tasks, but the authors did not identify specific regions associated with time perception. Pouthas et al. (2000) used a similar method but combined PET with EEG. While the PET results failed to identify time perception-specific networks, differences in EEG waveforms were observed between time-related and non-timing-related tasks. A right frontal CNV peaking around 600 ms after the stimulus onset and a late positive component (LPC) peaking around 900–1,300 ms after the stimulus onset were observed for the duration task only. Interestingly, the PET results showed that the same networks (right prefrontal cortex, right inferior parietal lobule, anterior cingulate cortex, and left and right fusiform gyri) were activated in both tasks. However, for the duration task, the networks were still active for several hundred milliseconds following the end of the stimulus. These CNV findings revealed the importance of right frontal areas in time perception and time judgments.

In a related vein, Monfort et al. (2000) have further described the implication of the CNV in time perception, by investigating ERPs during encoding and recognition for sub-second (700 ms) and supra-second (2,500 ms) visual intervals. They observed a CNV over the right frontal electrodes after the visual stimulus for both sub- and supra-second intervals and this, during both the encoding and recognition phases of the task. This finding suggests that both left and right brain structures play an important role during both encoding and recognition. Mitsudo et al. (2009) also used the ERPs to investigate the mechanisms involved in a bisection task of short intervals. Their participants had to judge whether the first auditory empty time interval was the same as, or different from, the second time interval; the intervals were presented sequentially. ERP analysis revealed a CNV in the frontal area after the first tone. Similarly to Monfort et al. (2000), the authors postulated that an increase in the CNV amplitude in the frontal area reflects the memorization (encoding) of the first time interval.

Alternatively, numerous studies investigating CNV modulations during timing tasks relate this component to the pulse accumulator of the AGM model (Macar & Vidal, 2009; Vallet et al., 2019). However, the specific factors that modulate this component still require further clarification. The major criticism regarding the CNV component as the neural signature of the temporal accumulator is that the amplitude of the CNV does not necessarily reflect the behavioral production or reproduction of durations. For example, in the timing reproduction task, the amplitude of the CNV component can be positively correlated with the duration of reproduction, with shorter duration reproduction being linked to smaller CNV (Macar et al., 1999). One can thus question why a shorter reproduction, which represents an accelerated passage of time (i.e., more pulse accumulated in a shorter time period), is not related to higher CNV amplitude (van Rijn et al., 2011). Undeniably, several studies failed to replicate the association between the amplitude of the CNV (as a direct reflection of the pulse accumulation process) and the duration of reproduction (Gibbons & Rammsayer, 2005; Ng et al., 2011). Consequently, there could be alternative explanations to fully explain the generation of the CNV component. One hypothesis, from the theoretical framework of perceptual decision-making, posits that an increase in cortical excitability results in faster, but more error-prone responses (Forstmann et al., 2008). The amplitude or the peak of the CNV component could reflect this modulation of cortical excitability due to the adjustment of response alertness (Boehm et al., 2014; Zhang et al., 2021). The processes of motor preparation are also a putative source of CNV generation. Several studies have suggested that CNV amplitudes increase with the number of specified movement parameters and the calibration complexity required to execute a motor response (Leuthold et al., 2004). In summary, the functional implication of CNV seems to integrate dissociable and additive neural activities acting for the processing of upcoming events. The studies presented above did not identify the CNV as a pure neural process dedicated to temporal processing (i.e., the accumulator) but acknowledged that its involvement cannot be ruled out either. An important consideration is that it may be difficult to differentiate motor processes from timing processes with tasks such as interval production or reproduction that require motor actions. As such, the involvement of the CNV may become clearer with implicit timing studies, which will be further discussed later in this chapter.

Late positive and negative components

The P300, an LPC, has been used to investigate the mechanisms involved in a bisection task, another way of studying explicit timing and time perception. In such tasks, participants are generally asked to categorize a stimulus duration as shorter or longer than a previously presented standard duration.

Gontier et al. (2007) used a visual temporal bisection task in the context of a dual task. The dual task allowed for control of cognitive load (i.e., adding either a multiplication or a line orientation task to the temporal task). They found that increasing the cognitive load when a visual bisection task was performed, reduced the amplitudes of CNV at frontal sites and of the P300 wave in the parietal areas. In contrast, the dual task condition increased the amplitude of an LPC and a later negative component (LNC). The authors proposed that these two later components act as a warning signal to mitigate error during high cognitive load.

Gibbons and Rammsayer (2005) used a dissociation paradigm, contrasting a pitch discrimination task and a temporal discrimination task to isolate the neural correlates proper to temporal processes. When contrasted, the difference between the ERPs observed within each of these tasks is assumed to mirror the brain mechanisms associated with the processing of temporal information. Their findings, obtained with brief intervals (circa 200 ms), suggest that the processing of temporal information involved two steps. In their two-process model, the first process, called evaluation, starts at the onset of the tested stimulus, and is reflected by the P300 component. The second process starts at the expected offset of the standard duration and is reflected by the P500 component. In their bisection task, Mitsudo et al. (2009) presented two empty time intervals. They asked their participants whether the second time interval was the same or different from the first one. They observed a P300 component in parietal areas (a positivity around 300 ms) after the second tone of the first-time interval. This component is known to reflect attentional processes (Gray et al., 2004). Mitsudo et al. (2009) interpret the P300 component observed in the parietal area as reflecting the participant's attention and monitoring of the passage of time during the interval for future comparison.

Interestingly, in their temporal generalization task, Gibbons and Rammsayer (2005) observed that the P300 component appeared to be elicited solely for brief durations that were shorter than the standard. This sensibility of the P3 toward the explicit processing of brief durations has been interpreted as a marker of *time perception*, as defined by durations <300 ms (in opposition to *time estimation*, for durations >1,000 ms; see Fraisse, 1984; Gibbons, 2022).

There seems to exist a purely time-process-related LPC named the "Late Positive Component of Timing" (LPCt). This late component typically appears after the presentation of a duration and is generated in prefrontal areas. Optimal behavior performance in time discrimination tasks coincided with a high amplitude and lower latency of the LPCt (Gontier et al., 2008; Paul et al., 2003). The LPCt appeared to reflect decision-making processes related to timing tasks (Paul et al., 2011).

Considering the similarity between the bisection task and the generalization task, Bannier et al. (2019) investigated how decision-making modulates time perception performances. Their participants performed both a bisection task and a temporal generalization task. In their generalization task, participants had to indicate whether the duration (target) presented corresponded to a previously presented standard lasting 800 ms. In their bisection task, participants had to indicate if the presented stimulus (target) was closer to the short (200 ms) or to the long (1,400 ms) standard. They found that participants responded faster in the bisection task for long durations, accompanied by distinctions in the ERPs. Between 500 and 600 ms after the offset of the stimulus marking the interval, a centro-parietal LPCt had a lower amplitude in the bisection task than in the generalization task. Reaction times and the LPCt amplitudes suggest that categorization (bisection) of durations activates fewer cognitive regions than their identification (generalization paradigm). Another finding of the study by Bannier et al. (2019) is that the latency of the LPCt's peak had shorter latency for long durations than for short ones, indicating that the decision for long durations was quicker than for short ones. The LPCt located around centro-parietal regions has also been linked to attentional and working memory networks assigned to time perception, and it is generally elicited after 200–600 ms (Gladhill et al., 2022; Tarantino et al., 2010). Other researchers believe that this late positivity component if around the prefrontal and frontal electrodes, might be typical of successful decision-making or retrieval during time estimation (Paul et al., 2003).

Different methodological contexts, like the proximity between the standard duration and the target duration, will influence the magnitude of the LPCt. The LPCt has a higher amplitude for durations that are closer to the standard duration, likely because the task is harder (Paul et al., 2011; Thompson et al., 2015). Under situations of uncertainty such as discrimination between a standard and a close target duration, the participants are generally biased to classify a target duration as shorter when compared to the standard (Lieving et al., 2006; Wearden, 1991). This is known as the "choose-short effect." Gladhill et al. (2022) reported that this bias in decision-making can be reflected by a larger amplitude of the LPCt. Interestingly, an LPCt with a larger amplitude has also been associated with higher performance levels and more difficult timing tasks (Paul et al., 2003; Gontier et al., 2008). The LPCt is also sensitive to visual size magnitude (discriminating circles of different diameters) with larger amplitude for small magnitude size comparisons (Gontier et al., 2008; Gontier et al., 2009). Interestingly though, the authors also reported that the amplitude remains specifically and solely modulated by the performance for temporal tasks and not for other tasks such as the visual size comparison. Because the involvement of the frontal and pre-frontal cortices has previously been iden-tified to contribute to the encoding of durations (Constantinidis et al., 2002; Gruber et al., 2000; Macar et al., 2002), Gontier et al. (2009) have postulated that the link between decision-making and the temporal context is due to the LPCt generated in those regions.

To clarify the role of the LPCt and to distinguish it from the CNV, Baykan et al. (2023) varied contexts to influence participants in their responses. Their first context included more long durations than short durations (named long context) and the opposite for the short context. Behavioral data suggest that the same duration was perceived as longer in the short context than in the long context. ERP data of CNV and LPCt were sensitive to context modulation. CNV climbing rate increased in the short context, and the amplitude and latency of LPCt were lower when compared to the long context. The authors suggest that CNV reflects expectancy for upcom-ing decision-making, while LPCt reflects the decision-making itself in a bisection task (see Table 3.2 for a summary).

Oscillatory markers in explicit timing

As mentioned previously, with EEG, not only can we study ERPs, but we can also perform spectral analyses. Spectral analysis is applied to EEG record-ings to calculate the power of oscillatory rhythms. Spectral analysis can also be carried out using magnetoencephalography (MEG). Therefore, the func-tional role of oscillatory rhythms in time perception will also be discussed in the MEG chapter (see Chapter 4). Production and reproduction tasks have also been used paired with EEG recordings. Typically, in a production task, participants are required to produce a rhythm, or a duration determined by chronometric units. In a reproduction task, the participant is asked to repro-duce a rhythm, or a duration previously presented. Previous research with animals has shown that beta power originating from the putamen indexes temporal durations. In their experiment, Bartolo and Merchant (2015) used a synchronization-continuation task, in which they essentially presented a rhythmic sequence to macaques and rewarded them when they reproduced the rhythmic sequence (synchronization). During the continuation phase, macaques must reproduce the same sequence without stimuli. They observed that, compared to the synchronization phase, macaques had enhanced beta power oscillation during the continuation (reproduction) phase of a rhyth-mic sequence. Their analyses also revealed that larger beta power appeared to reflect longer produced durations (Bartolo & Merchant, 2015). Kononowicz and van Rijn (2015) have also found similar evidence for the predictive value of beta power in reproduced durations with human partici-pants.[1] Other studies have also reported evidence for the involvement of beta

TABLE 3.2 Explicit timing and time perception components and oscillatory markers of timing and time perception

ERP	Topography/generators	Timing (approximate)	Task	Role	References
P1	Fronto-central areas	100 ms after movement or behavioral response	Temporal generalization	Index of attentional allocation in time perception.	Zhao et al. (2014)
N1–P2	Fronto-central areas (SMA)	50–300 ms following the offset of the test interval	Temporal generalization	Reflects the subjective experience of time.	Kononowicz & van Rijn (2014)
N1–P2	Fronto-central areas	Peaking at around 100 ms (after the auditory signal) and 150 ms (after the visual stimulus)	Bisection	The increase of the amplitude of N1–P2 reflects the amount of attention necessary to process intermodal intervals.	Gontier et al. (2013)
P300	Parietal areas	400 ms after the second tone burst	Temporal generalization	Index of the participants' attention to the first interval and reflects the monitoring of the passage of time.	Mitsudo et al. (2009)
P300	Centro-parietal areas	300 ms after the stimulus onset	Bisection	Index of time estimation and crucial components of effective information processing.	Gontier et al. (2007)

(Continued)

TABLE 3.2 (Continued)

ERP	Topography/generators	Timing (approximate)	Task	Role	References
P300	Parietal areas	300 ms after the second tone of the interval	Temporal generalization	Reflects the participant's attention and monitoring of the passage of time comparison.	Mitsudo et al. (2009)
CNV	Fronto-central areas	After stimulus offset	Bisection/temporal discrimination	Index of pacemaker–accumulator activity.	Schlichting et al. (2020), Vallet et al. (2019), Zhang et al. (2021)
CNV	Fronto-central areas (right frontal areas)	Peaking at 600 ms after stimulus offset	Temporal generalization	Index of time judgment.	Pouthas et al. (2000)
CNV	Right frontal and left frontal areas	After stimulus offset	Temporal generalization	Variation in CNV reflects temporal information processing related to attentional demands.	Monfort et al. (2000)
CNV	Fronto-central areas	After stimulus onset	Bisection	Index of time estimation and crucial components of effective information processing.	Gontier et al. (2007)
CNV	Frontal areas	After offset of first tone of the interval	Bisection	Reflects the memorization (encoding) of the first-time interval.	Mitsudo et al. (2009)

Component	Area	Timing	Task	Function	Reference
CNV	Fronto-central areas	After stimulus offset	Bisection	Reflects expectancy for upcoming decision-making.	Baykan et al. (2023)
LNC (N500)	Frontal areas	500 ms after the stimulus onset	Bisection	Acts as a warning signal to mitigate error during high cognitive load.	Gontier et al. (2007)
LPC (P500)	Parietal areas	500 ms after the stimulus onset	Bisection	Acts as a warning signal to mitigate error during high cognitive load.	Gontier et al. (2007)
LPCt	Prefrontal areas	700–1,250 ms after the onset of stimulus	Temporal generalization & bisection	Index of time related decision-making processes.	Gontier et al. (2008), Paul et al. (2003, 2011)
LPCt	Centro-parietal areas	200–600 ms after stimulus offset	Temporal generalization & bisection	Attentional and working memory networks assigned to time perception.	Bannier et al. (2019), Gladhill et al. (2022)
LPCt	Fronto-central	300–500 ms after stimulus offset	Bisection	Reflects the decision-making.	Baykan et al. (2023)

frequency bands in explicit timing tasks. Indeed, in a bisection task, longer prior durations were associated with larger amplitudes of the CNV and increased beta oscillations in a fronto-central cluster (Wiener et al., 2018; Wiener & Thompson, 2015). These results suggest that beta oscillations are pertinent for the temporal bisection task, more specifically in the encoding and retention of standard temporal intervals which are to be compared with a target interval. The results also suggest that the CNV and beta power can be modulated by recently perceived durations. Makhin and Pavlenko (2003) reported a negative correlation between variations in time productions and the beta rhythm power. They postulate that in their experiment, beta power might reflect the speed of the internal clock of the participant. Animal studies lead to comparable results. Sun et al. (2019) trained two macaques to reach, grasp, and hold items for fixed time intervals (short: 500 ms; or long: 1,500 ms) and observed how beta power behaved in the primary motor cortex, primary somatosensory cortex, and the posterior parietal cortex. The peak latency of beta power was shorter in the short condition than in the long condition. Thus, beta power appears to be modulated by subjective duration. They also proposed that beta power modulations reflect the maintenance of the current state of working memory for temporal information and the attention directed toward the following task procedure. This is in line with the results of Makhin and Pavlenko (2003) on the association between beta power and subjective time. Further evidence supporting the contribution of beta oscillations lies in their role in brain regions related to timing and time perception. Indeed, beta oscillations play a key role in the communication between the SMA, the basal ganglia, and the thalamus, which are all part of timing and time perception networks, during temporal tasks (Bartolo & Merchant, 2015; Bartolo et al., 2014; Fujioka et al., 2012).

In this section, we approached explicit timing and time perception and their different ERPs. We have discussed different methods to study explicit timing and time perception with the EEG, such as bisection, generalization, and reproduction tasks. Different ERPs have been linked to the vast processes of timing and time perception. Notably, we discussed the implications of the CNV in the recognition and encoding of time intervals, and its potential role as the temporal accumulator. Late positive and negative components have also been considered. These components appear to reflect the cognitive processes required for time perception to either mitigate superfluous information or evaluate temporal context. Another LPC, specific to timing and time perception processes (LPCt), has been discussed. This component seems to reflect time-related decision-making processes. Lastly, the crucial role of beta power oscillations for precise time interval reproduction and production has been described. Beta oscillations may reflect the speed of the internal clock of individuals.

Studying implicit time perception with EEG

In implicit *timing*, participants must make covert assumptions, either automatic or not, toward forthcoming events related to the temporal structure of a given stimulus. The tasks that use this goal-directed behavior can be classified into various paradigms, such as *temporal cueing, foreperiod, oddball,* and *probability paradigm*. These paradigms generally ask participants to make expectations in time following a cue sharing probabilistic temporal information.

The early ERPs

Among the variety of early evoked components, the literature in the field of timing and time perception has focused on three major components: P50, N1, and P2 (Figure 3.1).

A very early positive deflection, named the P50 component, is described in the field of implicit timing and time perception. The short latency of this component (from 35 to 75 ms after the stimulus offset) is particularly indicative regarding predictive processing information in the early stages of sensory integration. In the context of temporal expectation, the amplitude of the P50 component decreases for predictable events (Baess et al., 2009; Pinheiro et al., 2019; Schwartze et al., 2013; White & Yee, 2006). Furthermore, the fact that the P50 component can be modulated by temporal information confirms that early sensory integration is sensitive to temporal information.

In the explicit timing section, we described the N1–P2 complex that can be decomposed into two components (the N1 and the P2). The N1 component is also of interest to implicit timing; it is an early negative deflection that occurs between 80 and 150 ms after the onset of a stimulus. The amplitude of the sensory N1 component can also be modulated by temporal expectation. For example, studies have reported that the amplitude of N1 is smaller when auditory events are expected (Rimmele et al., 2011). Moreover, the modulation of this component by exogenous temporal expectation alone is not found in the visual modality but reported when combined with spatial expectation (Correa & Nobre, 2008; Doherty et al., 2005; Nobre et al., 2007; Pollux & Guo, 2009; Praamstra et al., 2010; Rimmele et al., 2011). In an oddball study aimed at distinguishing the different neural mechanisms involved in passive short- and long-interval detection, Thibault et al. (2023) investigated the N1 and P2 components. They demonstrated that N1 over centro-frontal electrodes served as a marker of supra-second (>1.2 s) interval detection. N1 was evoked for deviants that arrived later than the anticipated standard in the supra-second condition and had parietal and motor generators. They postulated that in an oddball context, the N1 component may function as a marker of attention and cognitive control essential to process temporal information. On the other hand, the P2 was exclusively elicited during the processing of sub-second (<1.2 s) intervals over centro-frontal electrodes for deviants that arrived later than the anticipated standard. P2 was generated by the SMA, the primary auditory cortex, the inferior frontal gyrus (IFG), the cingulate cortex, and the parietal cortex. Thibault et al. (2023) identified the P2's crucial role in sensory gating, a process that filters irrelevant information from reaching higher cognitive processes.

The N1–P2 complex is often reported as a potential marker of internal and subjective time and temporal expectation. The N1–P2 complex is also proposed to reflect timing response when behavior needs to be synchronized or driven by external stimulation, such as rhythmic pace (Duzcu, 2019; Sanabria & Correa, 2013). When the brain processes temporal regularities, a larger amplitude of the N1–P2 complex following deviant temporal events correlates with behavioral synchronization ability (D'Andrea-Penna et al., 2020). More precisely, the participants with larger amplitude of the N1–P2 complex are better at processing temporal regularities and irregularities to adapt their timing behavior with external stimulation.

In implicit timing and time perception, some specific associations have been made between the amplitude of N1 and P2, and subjective time. These components are closely related to the early stages of perceptual processing. Still, independently of the type of timing tasks, that is, explicit (interval comparison or production) or implicit (temporal orienting), studies have linked the amplitude of the component to subjective time. When the task goal involves temporal orienting through temporal expectation (e.g., using temporal cues), the N1 component's amplitude is specifically modulated by the attention to time and other prediction mechanisms. The model proposed by Lange (2013) explained the modulation of the N1 component by opposing the effects of attention and the prediction mechanisms. Indeed, when a priori expectation led to the temporal prediction of a valid target, the amplitude of the N1 component decreased (Breska & Deouell, 2014; Lange, 2010; Pinheiro et al., 2019). Conversely, if a stimulus requires additional processing of the information relevant to the task goal, expectations may also lead to attention being directed to the expected stimuli (e.g., a tone predicted the temporal onset of the target; Herbst & Obleser, 2019). In this situation, the attentional orienting toward the target then increases the N1 amplitude (Hsu et al., 2013; Jones et al., 2017; Summerfield & Egner, 2009).

The temporal expectations can also modulate the N1 component even when the expectations are endogenously generated. Duzcu et al. (2019) proposed that the amplitude of the N1 component elicited during a trisection task can be a marker of endogenous temporal expectation. In their experiment, when an endogenous temporal expectation is congruent with the most plausible response (i.e., when duration increases the probability of identifying longer intervals as being longer than a previously learned standard interval), the amplitude of the N1 component decreases. Using the numerical magnitudes (Arabic digits) as stimuli in a bisection task, Xuan et al. (2009) reported a modulation of subjective time. During the early stage of perceptual-level processing, the N1 component amplitude elicited by small digits was enhanced compared to the amplitude elicited by larger digits. Participants discriminated time intervals more accurately when small time durations were indexed by digits of small magnitude and when longer durations were indexed by higher magnitude digits. The authors postulate that the magnitude of the digits perceived negatively modulates the N1, which may reflect selective attention and expectation toward temporal information.

The late ERPs

As discussed in the explicit timing section, late ERPs generally involve higher cognitive processes. Different LPCs such as P3 or the LPC have been associated with distinct cognitive processes involved in implicit timing and time perception (e.g., selective attention, discrimination, and decision-making). The P3 component is traditionally identified through the subcomponents P3a and P3b, depending on the solicited neural organizations and cognitive functions eliciting the P3. The P3a component was described for the first time in an oddball paradigm (Sutton et al., 1965). It is generated over central-parietal scalp sites, with a latency of around 250–350 ms, while detecting rare/infrequent stimuli. The P3b component, with a latency of around 300–600 ms, which is elicited over the temporo-parietal area during target stimulus processing (Comerchero & Polich, 1999; Polich, 2007), is associated with processes relevant to timing and time perception tasks such as attention and memory. In most cases, the P3b is the P3 of interest in studies exploring the mechanisms of timing and time perception. For example, when the task goal involves temporal orienting through temporal expectation, selective attention and memory are engaged to prioritize the processing of upcoming stimuli relevant to the task's goal (i.e., predicting when a target appears). The studies that explore temporal expectation have well documented the modulation of the P3b component in a variety of implicit timing tasks (Mento, 2017; Rimmele et al., 2011; Rufener et al., 2020; Shen & Alain, 2012); consequently, the P3b has been referred to as a locked component of "when" a target occurs (Schwartze et al., 2013). The latency of the P3b is the primary dimension of the component that is affected by temporal expectation, with a reduction of the latency when the cue drives temporal information with high probability. This effect is particularly powerful in temporal cueing or foreperiod paradigms (where an important stimulus is preceded by a warning stimulus) where the cue shares a high probability of information to expect in the upcoming probe (Breska & Deouell, 2014; Correa & Nobre, 2008; Schmidt-Kassow & Kotz, 2009). From a functional implication, the reduced latency of the P3b component has been proposed to reflect processing facilitation allowed by the predictive temporal cues.

As for its latency, the amplitude of the P3 component is also sensitive to temporal expectation. When a cue drives temporal expectation, the amplitude of the P3 component increases when the probe appears in a valid cued timing when compared to a non-valid one. A valid cued timing corresponds to the situation when the cue accurately predicts the incoming time interval length (if the cue for a short interval precedes a short interval). This pattern is interpreted as evidence that temporal orientating is reflected by the P3b amplitude and appears to be related to stimulus processing, memory, and decision-making (Lampar & Lange, 2011).

In addition to its latency, numerous studies have also linked the amplitude of the P3b component to timing and time perception (Benau et al., 2018; Breska & Deouell, 2014; Capizzi et al., 2013; Ernst et al., 2017; Mitsudo et al., 2012). The amplitude of the P3b component appears to be increased when the stimulus duration is increased. More precisely, when the task consists of comparing the duration or the physical magnitude of a standard and a target stimulus, the amplitude of the P3b component generated by the target varied with the duration but not with another physical magnitude comparison (Gontier et al., 2009). This modulation of the P3b component's amplitude has been linked with the processes underlying the estimation of duration. Indeed, previous studies using oddball paradigms have reported a larger amplitude of P3 following the overestimation of duration, when compared to correct estimations (Ernst et al., 2017).

As for the CNV, Praamstra et al. (2006) used an implicit motor timing task to investigate its effects on the CNV. Participants were required to press one of two buttons depending on whether an arrow was pointing to the left or the right. The authors showed that the standard duration does not need to be learned by explicit memorization but can consist of a duration acquired through simple repeated presentations via the expectancy/anticipation of a forthcoming stimulus. A major difference for the CNV between explicit and implicit timing lies in the localization of this component. Authors have suggested that, for explicit timing, the CNV emerges in the medial premotor cortex or the SMA (Macar et al., 1999; Pfeuty et al., 2005), but in implicit timing, the CNV is predominantly localized in the bilateral premotor cortex (Praamstra et al., 2006).

In their oddball study, Thibault et al. (2023) studied the differences between sub- (<1.2 s) and supra-second (>1.2 s) intervals based on the CNV. They observed that the CNV was only generated in conditions involving supra-second delayed intervals. The supra-second CNV originated from temporo-parietal, SMA, and motor regions. They identified the CNV's role as a marker of temporal accumulation, attention, and working memory for temporal information, specifically for long (supra-second) intervals, aligning with findings from Liu et al. (2013).

Oscillatory markers

For this section, it is imperative to first present the concept of cross-frequency coupling (CFC). CFC refers to the interaction of different frequency bands. It is known that high-frequency brain activity reflects local processes and that low frequencies reflect activity in distant distributed brain regions (Von Stein & Sarnthein, 2000). Accordingly, CFC serves as a mechanism to transfer information from large-scale brain networks to local cortical processing. This coupling of frequency integrates brain processes across multiple spatio-temporal scales because local processes are generally faster than their large-scale counterpart (Canolty & Knight, 2010). Wiener and Kanai (2016) suggest that distinct frequency bands are associated with the different networks engaged in various timing and time perception tasks. One oscillatory rhythm reported for time perception is the delta range (1–4 Hz). More precisely, subjective time seems to be reflected by the phase of delta oscillations, which could be used to keep track of time (Arnal & Kleinschmidt, 2017). Furthermore, it is understood that regular or rhythmic events promote low-frequency rhythms (delta: 1–4 Hz; theta: 4–8 Hz). By CFC (see Canolty and Knight 2010), low-frequency rhythms regulate higher-frequency rhythms (beta and gamma). This coupling boosts neuron excitability and further processes pertinent forthcoming pertinent events (Lakatos et al., 2008; Lakatos et al., 2005; Schroeder & Lakatos, 2009). In their experiment, Cravo et al. (2011) used a Go/No-go task and manipulated temporal expectations (by changing the probability of a target presentation) to investigate whether low-frequency oscillations could regulate temporal expectations. They found that theta (4–8 Hz) power allows for the regulation of temporal expectation through power coupling between theta and beta rhythms.

In this section, we approached implicit timing and time perception by studying its ERPs and frequency bands. We discussed different methods (temporal cueing, foreperiod, and oddball paradigm) to study implicit timing and time perception with the EEG. Different ERPs have been linked to the numerous processes involved in implicit timing and time perception.

Notably, we discussed the implications of the P50 component in the context of temporal expectation. The early P1 component has also been discussed as an index of subjective representation of time needed for reproducing time intervals. Furthermore, the N1 and P2 components have also been considered. The N1 component has been identified as the key to temporal expectation. As for the N1–P2 complex, it appears to be associated with the process of temporal regularities and irregularities. Later components, which are associated with higher cognitive processes, have also been identified. The P3 component seems to reflect temporal orientation and time estimation processes and is associated with processes such as attention and memory which are frequently used in temporal tasks (see Table 3.3). We also identified delta (1–4 Hz) oscillations as a potential key frequency band to keep track of time. CFC between beta and delta oscillatory rhythms could enhance temporal expectation and time estimation performance.

Conclusions and future research directions

To conclude, numerous markers for different components of explicit and implicit timing and time perception were presented and discussed. These markers are reported in Tables 3.2 and 3.3. In explicit timing, there is notably the CNV, which appears to play a role in the encoding, recognition, and expectancy of decision-making when comparing time intervals. The N1–P2 complex has also been identified as a marker for the regulation of timing behavior in response to external stimulation. There is also the LPCt, which has been linked to attentional processes, working memory, decision-making, and retrieval of time durations during timing and time perception tasks. In the implicit timing domain, distinct frequency bands are associated with the different networks engaged in various timing and time perception tasks. Notably, beta power originating from the putamen appears to index temporal durations. Larger beta power reflects longer durations when producing time intervals. Moreover, the delta oscillations could be used to keep track of the passage of time and the theta oscillations allow for the regulation of temporal expectation through power coupling between theta and beta rhythms. Other ERPs such as the N1–P2 complex are often reported as a potential marker of internal and subjective time and temporal expectation. In implicit timing, similarly to the LPCt, the P3 component may be related to certain cognitive processes required for timing and time perception such as memory and decision-making.

EEG could also be useful for investigating the effect of different factors on time perception. One factor is the expertise of musicians in timing and rhythm: do they differ from non-musicians in terms of brain signal, CNV amplitude, oscillation rhythm, or LPCt? There are very few studies using

TABLE 3.3 Implicit timing and time perception components and oscillatory markers of timing and time perception

ERP	Topography/generators	Timing (approximate)	Task	Role	References
P50	Fronto-central	35–75 ms after stimulus onset	Temporal expectation	Index of predictive temporal processing in the early stages of sensory integration. The amplitude of the P50 component decreases for predictable events.	Pinheiro et al. (2019), Schwartze et al. (2013)
N1–P2 complex	Central areas	128–255 ms after the offset of target stimulus	Temporal expectation	Index of internal and subjective time and temporal expectation.	Sanabria & Correa (2013), D'Andrea-Penna et al. (2020)
N1	Occipital areas/fronto-central (parietal and motor)	80–210 ms after the offset of the target stimulus	Temporal expectation and oddball	Index of the allocated attention to time and other prediction mechanisms.	Breska & Deouell (2014), Pinheiro et al. (2019), Thibault et al. (2023)
P2	Fronto-central areas (SMA, IFG, auditory, cingulate, parietal cortices)	164–254 ms after the onset of the deviant stimulus	Oddball	Index of sensory gating of temporal information.	Thibault et al. (2023)
P3a	Centro-parietal areas	250–350 ms after stimulus offset	Probability/oddball task	Index of detection of rare or more difficult stimulus.	Comerchero & Polich (1999)

(Continued)

TABLE 3.3 (Continued)

ERP	Topography/generators	Timing (approximate)	Task	Role	References
P3b	Temporo-parietal areas/ central areas	300–600 ms after stimulus offset	Probability/oddball task and temporal cueing	Index of attention and memory related to temporal processes. Index of "when" a target occurs (temporal orienting).	Comerchero & Polich (1999), Schwartze et al. (2013), Lampar & Lange (2011)
P300	Parieto-occipital	375 ms after stimulus onset	Oddball	Index of phasic norepinephrine response affecting the subjective experience of time.	Ernst et al. (2017)
CNV	Fronto-central areas (bilateral premotor cortex)	After stimulus offset	Temporal expectation	Index of the duration of a stored standard interval.	Praamstra et al. (2006)
CNV	Temporo-parietal, SMA, and motor regions	1,400–1,900 ms after stimulus onset	Oddball	Index of temporal accumulation, attention, and working memory for temporal information only in the range of supra-second (>1.2 s) intervals.	Thibault et al. (2023).

EEG that attempt to identify key differences in time perception between experts and non-experts. For example, the study of Habibi et al. (2014) indicates that musical training enhances the cortical activity (increased N1 and P2 amplitudes) used in the processing of temporal irregularities. Another frequently omitted factor is the interval length or duration of the perceived time. It is a known fact in psychophysics that time is processed by different mechanisms depending on its length (Gibbon et al., 1997; Grondin, 2012, 2014; Grondin et al., 2004, 2015). Notably, a vast portion of the literature has emphasized a sub-second (<1 s) versus supra-second (>1 s) distinction (Lewis & Miall, 2003; Rammsayer, 1997, 1999, 1993; Rammsayer & Lima, 1991), though the cutoff for two different timing and time perception mechanisms would most likely appear around 1.2–1.5 s. EEG would certainly be a useful tool to identify a transition point, or window, from one timing mechanism to another.

Finally, to the best of our knowledge, there are currently no studies using intracranial EEG (iEEG) to study time or rhythmic perception. iEEG consists of electrocorticography when using grid electrodes, or stereotactic EEG when using depth electrodes. Intracranial recordings are usually obtained from neurosurgical patients with drug-resistant focal epilepsy. iEEG has comparable time resolution with EEG or MEG with high (but limited in coverage) spatial resolution. The current literature on timing and time perception could benefit from this instrument, by combining the spatial resolution of fMRI and the time resolution of EEG for different time perception and timing tasks.

Note

1 Further data on oscillatory rhythms will be discussed in the implicit section of this chapter since spectral analysis is most frequently paired with implicit timing tasks.

References

Albouy, P., Benjamin, L., Morillon, B., & Zatorre, R. J. (2020). Distinct sensitivity to spectrotemporal modulation supports brain asymmetry for speech and melody. *Science, 367*(6481), 1043–1047. https://doi.org/10.1126/science.aaz3468

Albouy, P., Weiss, A., Baillet, S., & Zatorre, R. J. (2017). Selective entrainment of theta oscillations in the dorsal stream causally enhances auditory working memory performance. *Neuron, 94*(1), 193–206.e195. https://doi.org/10.1016/j.neuron.2017.03.015

Amzica, F., & Steriade, M. (1998). Electrophysiological correlates of sleep delta waves. *Electroencephalography and Clinical Neurophysiology, 107*(2), 69–83. https://doi.org/10.1016/S0013-4694(98)00051-0

Arnal, L. H., & Kleinschmidt, A. K. (2017). Entrained delta oscillations reflect the subjective tracking of time. *Communicative & Integrative Biology, 10*(5-6), 3077–3085. https://doi.org/10.1080/19420889.2017.1349583

Awan, F. G., Saleem, O., & Kiran, A. (2019). Recent trends and advances in solving the inverse problem for EEG source localization. *Inverse Problems in Science and Engineering*, *27*(11), 1521–1536. https://doi.org/10.1080/17415977.2018.1490279

Azari, L., Drouin, J., Plante, G. & Grondin, S. (2023). Discrimination of brief empty time intervals when the first marker is tactile. *Timing & Time Perception*, *11*, 343–361.

Azari, L., Mioni, G., Rousseau, R., & Grondin, S. (2020). An analysis of the processing of intramodal and intermodal time intervals. *Attention, Perception, & Psychophysics*, *82*(3), 1473–1487. https://doi.org/10.3758/s13414-019-01900-7

Baess, P., Widmann, A., Roye, A., Schröger, E., & Jacobsen, T. (2009). Attenuated human auditory middle latency response and evoked 40-Hz response to self-initiated sounds. *European Journal of Neuroscience*, *29*(7), 1514–1521. https://doi.org/10.1111/j.1460-9568.2009.06683.x

Bannier, D., Wearden, J., Le Dantec, C. C., & Rebaï, M. (2019). Differences in the temporal processing between identification and categorization of durations: A behavioral and ERP study. *Behavioural Brain Research*, *356*, 197–203. https://doi.org/10.1016/j.bbr.2018.08.027

Bartolo, R., & Merchant, H. (2015). β oscillations are linked to the initiation of sensory-cued movement sequences and the internal guidance of regular tapping in the monkey. *Journal of Neuroscience*, *35*(11), 4635–4640. https://doi.org/10.1523/JNEUROSCI.4570-14.2015

Bartolo, R., Prado, L., & Merchant, H. (2014). Information processing in the primate basal ganglia during sensory-guided and internally driven rhythmic tapping. *Journal of Neuroscience*, *34*(11), 3910–3923. https://doi.org/10.1523/JNEUROSCI.2679-13.2014

Başar, E., Schürmann, M., Başar-Eroglu, C., & Karakaş, S. (1997). Alpha oscillations in brain functioning: An integrative theory. *International Journal of Psychophysiology*, *26*(1–3), 5–29. https://doi.org/10.1016/S0167-8760(97)00753-8

Baykan, C., Zhu, X., Zinchenko, A., Mueller, H. J., & Shi, Z. (2023). Electrophysiological signatures of temporal context in the bisection task. https://doi.org/10.1101/2023.03.15.532795

Benau, E. M., DeLoretta, L. C., & Moelter, S. T. (2018). The time is "right:" Electrophysiology reveals right parietal electrode dominance in time perception. *Brain and Cognition*, *123*, 92–102. https://doi.org/10.1016/j.bandc.2018.03.008

Boehm, U., van Maanen, L., Forstmann, B., & van Rijn, H. (2014). Trial-by-trial fluctuations in CNV amplitude reflect anticipatory adjustment of response caution. *Neuroimage*, *96*, 95–105. https://doi.org/10.1016/j.neuroimage.2014.03.063

Breska, A., & Deouell, L. Y. (2014). Automatic bias of temporal expectations following temporally regular input independently of high-level temporal expectation. *Journal of Cognitive Neuroscience*, *26*(7), 1555–1571. https://doi.org/10.1162/jocn_a_00564

Canolty, R. T., & Knight, R. T. (2010). The functional role of cross-frequency coupling. *Trends in Cognitive Sciences*, *14*(11), 506–515. https://doi.org/10.1016/j.tics.2010.09.001

Capizzi, M., Correa, A., & Sanabria, D. (2013). Temporal orienting of attention is interfered by concurrent working memory updating. *Neuropsychologia*, *51*(2), 326–339. https://doi.org/10.1016/j.neuropsychologia.2012.10.005

Chiu, P., Ambady, N., & Deldin, P. (2004). Contingent negative variation to emotional in-and out-group stimuli differentiates high-and low-prejudiced individuals. *Journal of Cognitive Neuroscience*, *16*(10), 1830–1839. https://doi.org/10.1162/0898929042947946

Comerchero, M. D., & Polich, J. (1999). P3a and P3b from typical auditory and visual stimuli. *Clinical Neurophysiology*, *110*(1), 24–30. https://doi.org/10.1016/S0168-5597(98)00033-1

Constantinidis, C., Williams, G. V., & Goldman-Rakic, P. S. (2002). A role for inhibition in shaping the temporal flow of information in prefrontal cortex. *Nature Neuroscience, 5*(2), 175–180. https://doi.org/10.1038/nn799

Correa, A., & Nobre, A. C. (2008). Neural modulation by regularity and passage of time. *Journal of Neurophysiology, 100*(3), 1649–1655. https://doi.org/10.1152/jn.90656.2008

Coull, J. T. (2009). Neural substrates of mounting temporal expectation. *PLoS Biology, 7*(8), e1000166. https://doi.org/10.1371/journal.pbio.1000166

Cravo, A. M., Rohenkohl, G., Wyart, V., & Nobre, A. C. (2011). Endogenous modulation of low frequency oscillations by temporal expectations. *Journal of Neurophysiology, 106*(6), 2964–2972. https://doi.org/10.1152/jn.00157.2011

D'Andrea-Penna, G. M., Iversen, J. R., Chiba, A. A., Khalil, A. K., & Minces, V. H. (2020). One tap at a time: Correlating sensorimotor synchronization with brain signatures of temporal processing. *Cerebral Cortex Communications, 1*(1), 1-8 https://doi.org/10.1093/texcom/tgaa036

de Peralta Menendez, R. G., Murray, M. M., Michel, C. M., Martuzzi, R., & Andino, S. L. G. (2004). Electrical neuroimaging based on biophysical constraints. *Neuroimage, 21*(2), 527–539. https://doi.org/10.1016/j.neuroimage.2003.09.051

Doherty, J. R., Rao, A., Mesulam, M. M., & Nobre, A. C. (2005). Synergistic effect of combined temporal and spatial expectations on visual attention. *Journal of Neuroscience, 25*(36), 8259–8266. https://doi.org/10.1523/JNEUROSCI.1821-05.2005

Duzcu, H. (2019). A neural marker of the start-gun in interval timing: Onset N1P2. *Behavioral Neuroscience, 133*(4), 414. https://doi.org/10.1037/bne0000325

Duzcu, H., Özkurt, T., Mapelli, I., & Hohenberger, A. (2019). N1-P2: Neural markers of temporal expectation and response discrimination in interval timing. *Acta Neurobiologiae Experimentalis, 79*(2), 193–204. https://doi.org/10.21307/ane-2019-017

Ernst, B., Reichard, S. M., Riepl, R. F., Steinhauser, R., Zimmermann, S. F., & Steinhauser, M. (2017). The P3 and the subjective experience of time. *Neuropsychologia, 103*, 12–19. https://doi.org/10.1016/j.neuropsychologia.2017.06.033

Fatourechi, M., Bashashati, A., Ward, R. K., & Birch, G. E. (2007). EMG and EOG artifacts in brain computer interface systems: A survey. *Clinical Neurophysiology, 118*(3), 480–494. https://doi.org/10.1016/j.clinph.2006.10.019

Forstmann, B. U., Dutilh, G., Brown, S., Neumann, J., Von Cramon, D. Y., Ridderinkhof, K. R., & Wagenmakers, E.-J. (2008). Striatum and pre-SMA facilitate decision-making under time pressure. *Proceedings of the National Academy of Sciences, 105*(45), 17538–17542. https://doi.org/10.1073/pnas.0805903105

Fraisse, P. (1984). Perception and estimation of time. *Annual Review of Psychology, 35*(1), 1–37.

Fujioka, T., Trainor, L. J., Large, E. W., & Ross, B. (2012). Internalized timing of isochronous sounds is represented in neuromagnetic beta oscillations. *Journal of Neuroscience, 32*(5), 1791–1802. https://doi.org/10.1016/j.clinph.2006.10.019

Gan, T., Wang, N., Zhang, Z., Li, H., & Luo, Y.-J. (2009). Emotional influences on time perception: Evidence from event-related potentials. *NeuroReport, 20*(9), 839–843. https://doi.org/10.1097/WNR.0b013e32832be7dc

Ge, S., Yang, Q., Wang, R., Lin, P., Gao, J., Leng, Y., Yang, Y., & Wang, H. (2017). A brain-computer interface based on a few-channel EEG-fNIRS bimodal system. *IEEE Access, 5*, 208–218. https://doi.org/10.1109/ACCESS.2016.2637409

Gibbon, J. (1977). Scalar expectancy theory and Weber's law in animal timing. *Psychological Review, 84*(3), 279–325. https://doi.org/10.1037/0033-295X.84.3.279

Gibbon, J., Church, R. M., & Meck, W. H. (1984). Scalar timing in memory. *Annals of the New York Academy of Sciences, 423*(1), 52–77. https://doi.org/10.1111/j.1749-6632.1984.tb23417.x

Gibbon, J., Malapani, C., Dale, C. L., & Gallistel, C. R. (1997). Toward a neurobiology of temporal cognition: Advances and challenges. *Current Opinion in Neurobiology, 7*(2), 17–184. https://doi.org/10.1016/S0959-4388(97)80005-0

Gibbons, H. (2022). Event-related brain potentials of temporal generalization: The P300 span marks the transition between time perception and time estimation. *Behavioral Neuroscience.* https://doi.org/10.1037/bne0000530

Gibbons, H., & Rammsayer, T. H. (2005). Electrophysiological correlates of temporal generalization: Evidence for a two-process model of time perception. *Cognitive Brain Research, 25*(1), 195–209. https://doi.org/10.1016/j.cogbrainres.2005.05.009

Gladhill, K. A., Mioni, G., & Wiener, M. (2022). Dissociable effects of emotional stimuli on electrophysiological indices of time and decision-making. *PloS One, 17*(11), e0276200. https://doi.org/10.1371/journal.pone.0276200

Gontier, E., Hasuo, E., Mitsudo, T., & Grondin, S. (2013). EEG investigations of duration discrimination: The intermodal effect is induced by an attentional bias. *PloS One, 8*(8). https://doi.org/10.1371/journal.pone.0074073

Gontier, E., Le Dantec, C., Leleu, A., Paul, I., Charvin, H., Bernard, C., Lalonde, R., & Rebaï, M. (2007). Frontal and parietal ERPs associated with duration discriminations with or without task interference. *Brain Research, 1170*, 79–89. https://doi.org/10.1016/j.brainres.2007.07.022

Gontier, E., Le Dantec, C., Paul, I., Bernard, C., Lalonde, R., & Rebai, M. (2008). A prefrontal ERP involved in decision making during visual duration and size discrimination tasks. *International Journal of Neuroscience, 118*(1), 149–162. https://doi.org/10.1080/00207450601046798

Gontier, E., Paul, I., Le Dantec, C., Pouthas, V., Jean-Marie, G., Bernard, C., Lalonde, R., & Rebaï, M. (2009). ERPs in anterior and posterior regions associated with duration and size discriminations. *Neuropsychology, 23*(5), 668. https://doi.org/10.1037/a0015757

Gorjan, D., Gramann, K., De Pauw, K., & Marusic, U. (2022). Removal of movement-induced EEG artifacts: Current state of the art and guidelines. *Journal of Neural Engineering.* https://doi.org/10.1088/1741-2552/ac542c

Gosseries, O., Demertzi, A., Noirhomme, Q., Tshibanda, J., Boly, M., Op de Beeck, M., Hustinx, R., Maquet, P., Salmon, E., & Moonen, G. (2008). Functional neuroimaging (fMRI, PET and MEG): What do we measure? *Revue Médicale de Liège, 63*(5–6), 231–237.

Gray, H. M., Ambady, N., Lowenthal, W. T., & Deldin, P. (2004). P300 as an index of attention to self-relevant stimuli. *Journal of Experimental Social Psychology, 40*(2), 216–224. https://doi.org/10.1016/S0022-1031(03)00092-1

Grondin, S. (2010). Timing and time perception: A review of recent behavioral and neuroscience finding and theoretical directions. Attention. *Perception, & Psychophysics, 72*, 561–582. https://doi.org/10.3758/APP.72.3.561

Grondin, S. (2012). Violation of the scalar property for time perception between 1 and 2 seconds: Evidence from interval discrimination, reproduction, and categorization. *Journal of Experimental Psychology: Human Perception and Performance, 38*(4), 880. https://doi.org/10.1037/a0027188

Grondin, S. (2014). About the (non) scalar property for time perception. *Neurobiology of Interval Timing, 17*–32. https://doi.org/10.1007/978-1-4939-1782-2_2

Grondin, S., Laflamme, V., & Mioni, G. (2015). Do not count too slowly: Evidence for a temporal limitation in short-term memory. *Psychonomic Bulletin & Review, 22*(3), 863–868. https://doi.org/10.3758/s13423-014-0740-0

Grondin, S., Ouellet, B., & Roussel, M.-E. (2004). Benefits and limits of explicit counting for discriminating temporal intervals. *Canadian Journal of Experimental Psychology/Revue canadienne de psychologie expérimentale, 58*(1), 1. https://doi. org/10.1037/h0087436

Gruber, O., Kleinschmidt, A., Binkofski, F., Steinmetz, H., & von Cramon, D. Y. (2000). Cerebral correlates of working memory for temporal information. *NeuroReport, 11*(8), 1689–1693.

Habibi, A., Wirantana, V., & Starr, A. (2014). Cortical activity during perception of musical rhythm: Comparing musicians and nonmusicians. *Psychomusicology: Music, Mind, and Brain, 24*(2), 125. https://doi.org/10.1037/pmu0000046

Hallez, H., Vanrumste, B., Grech, R., Muscat, J., De Clercq, W., Vergult, A., D'Asseler, Y., Camilleri, K. P., Fabri, S. G., & Van Huffel, S. (2007). Review on solving the forward problem in EEG source analysis. *Journal of Neuroengineering and Rehabilitation, 4*(1), 1–29. https://doi.org/10.1186/1743-0003-4-46

Hanslmayr, S., Gross, J., Klimesch, W., & Shapiro, K. L. (2011). The role of alpha oscillations in temporal attention. *Brain Research Reviews, 67*(1–2), 331–343. https://doi.org/10.1016/j.brainresrev.2011.04.002

Harmony, T., Fernández, T., Silva, J., Bernal, J., Díaz-Comas, L., Reyes, A., Marosi, E., Rodríguez, M., & Rodríguez, M. (1996). EEG delta activity: An indicator of attention to internal processing during performance of mental tasks. *International Journal of Psychophysiology, 24*(1–2), 161–171. https://doi.org/10.1016/S0167-8760(96)00053-0

He, B., Musha, T., Okamoto, Y., Homma, S., Nakajima, Y., & Sato, T. (1987). Electric dipole tracing in the brain by means of the boundary element method and its accuracy. *IEEE Transactions on Biomedical Engineering*, (6), 406–414. https://doi.org/10.1109/TBME.1987.326056

He, B., Wang, Y., & Wu, D. (1999). Estimating cortical potentials from scalp EEGs in a realistically shaped inhomogeneous head model by means of the boundary element method. *IEEE Transactions on Biomedical Engineering, 46*(10), 1264–1268. https://doi.org/10.1109/10.790505

Herbst, S. K., & Obleser, J. (2019). Implicit temporal predictability enhances pitch discrimination sensitivity and biases the phase of delta oscillations in auditory cortex. *Neuroimage, 203*, 116198. https://doi.org/10.1016/j.neuroimage.2019.116198

Hosseini Houripasand, M., Sabaghypour, S., Farkhondeh Tale Navi, F., & Nazari, M. A. (2023). Time distortions induced by high-arousing emotional compared to low-arousing neutral faces: An event-related potential study. *Psychological Research*, 1–12. https://doi.org/10.1007/s00426-022-01789-2

Hsu, Y.-F., Hämäläinen, J. A., & Waszak, F. (2013). Temporal expectation and spectral expectation operate in distinct fashion on neuronal populations. *Neuropsychologia, 51*(13), 2548–2555. https://doi.org/10.1016/j.neuropsychologia. 2013.09.018

İnce, R., Adanır, S. S., & Sevmez, F. (2021). The inventor of electroencephalography (EEG): Hans Berger (1873–1941). *Child's Nervous System, 37*, 2723–2724. https:// doi.org/10.1007/s00381-020-04564-z

Jain, S. N., & Rai, C. (2012). Blind source separation and ICA techniques: A review. *International Journal of Engineering Science and Technology, 4*(4), 1490–1503.

Jenkinson, N., & Brown, P. (2011). New insights into the relationship between dopamine, beta oscillations and motor function. *Trends in Neurosciences, 34*(12), 611–618. https://doi.org/10.1016/j.tins.2011.09.003

Jones, A., Hsu, Y.-F., Granjon, L., & Waszak, F. (2017). Temporal expectancies driven by self-and externally generated rhythms. *Neuroimage, 156*, 352–362. https://doi.org/10.1016/j.neuroimage.2017.05.042

Klimesch, W., Vogt, F., & Doppelmayr, M. (1999). Interindividual differences in alpha and theta power reflect memory performance. *Intelligence, 27*(4), 347–362. https://doi.org/10.1016/S0160-2896(99)00027-6

Kononowicz, T. W., & Penney, T. B. (2016). The contingent negative variation (CNV): Timing isn't everything. *Current Opinion in Behavioral Sciences, 8*, 231–237.

Kononowicz, T. W., Sander, T., & van Rijn, H. (2015). Neuroelectromagnetic signatures of the reproduction of supra-second durations. *Neuropsychologia, 75*, 201–213. https://doi.org/10.1016/j.neuropsychologia.2015.06.001

Kononowicz, T. W., & van Rijn, H. (2014). Decoupling interval timing and climbing neural activity: A dissociation between CNV and N1P2 amplitudes. *Journal of Neuroscience, 34*(8), 2931–2939. https://doi.org/10.1523/JNEUROSCI.2523-13.2014

Kononowicz, T. W., & van Rijn, H. (2015). Single trial beta oscillations index time estimation. *Neuropsychologia, 75*, 381–389. https://doi.org/10.1016/j.neuropsychologia.2015.06.014

Lakatos, P., Karmos, G., Mehta, A. D., Ulbert, I., & Schroeder, C. E. (2008). Entrainment of neuronal oscillations as a mechanism of attentional selection. *Science, 320*(5872), 110–113. https://doi.org/10.1126/science.1154735

Lakatos, P., Shah, A. S., Knuth, K. H., Ulbert, I., Karmos, G., & Schroeder, C. E. (2005). An oscillatory hierarchy controlling neuronal excitability and stimulus processing in the auditory cortex. *Journal of Neurophysiology, 94*(3), 1904–1911. https://doi.org/10.1152/jn.00263.2005

Lampar, A., & Lange, K. (2011). Effects of temporal trial-by-trial cuing on early and late stages of auditory processing: Evidence from event-related potentials. *Attention, Perception, & Psychophysics, 73*, 1916–1933. https://doi.org/10.3758/s13414-011-0149-z

Lange, K. (2010). Can a regular context induce temporal orienting to a target sound? *International Journal of Psychophysiology, 78*(3), 231–238. https://doi.org/10.1016/j.ijpsycho.2010.08.003

Lange, K. (2013). The ups and downs of temporal orienting: a review of auditory temporal orienting studies and a model associating the heterogeneous findings on the auditory N1 with opposite effects of attention and prediction. *Frontiers in Human Neuroscience, 7*, 263. https://doi.org/10.3389/fnhum.2013.00263

Leuthold, H., Sommer, W., & Ulrich, R. (2004). Preparing for action: Inferences from CNV and LRP. *Journal of Psychophysiology, 18*(2/3), 77–88. https://doi.org/10.1027/0269-8803.18.23.77

Lewis, P., & Miall, C. (2003). Brain activation patterns during measurement of sub- and supra-second intervals. *Neuropsychologia, 41*(12), 1583–1592. https://doi.org/10.1016/s0028-3932(03)00118-0

Li, X., Zhang, G., Zhou, C., & Wang, X. (2019). Negative emotional state slows down movement speed: Behavioral and neural evidence. *PeerJ, 7*, e7591. https://doi.org/10.7717/peerj.7591

Lieving, L. M., Lane, S. D., Cherek, D. R., & Tcheremissine, O. V. (2006). Effects of delays on human performance on a temporal discrimination procedure: Evidence of a choose-short effect. *Behavioural Processes, 71*(2–3), 135–143. https://doi.org/10.1016/j.beproc.2005.10.002

Liu, Y., Zhang, D., Ma, J., Li, D., Yin, H., & Luo, Y. (2013). The attention modulation on timing: An event-related potential study. *PloS One, 8*(6), e66190.

Luck, S. J. (2012). Event-related potentials. In *APA handbook of research methods in psychology, Vol 1: Foundations, Planning, measures, and psychometrics* (pp. 523–546). https://doi.org/10.1037/13619-028

Luu, P., Flaisch, T., & Tucker, D. M. (2000). Medial frontal cortex in action monitoring. *Journal of Neuroscience*, *20*(1), 464–469. https://doi.org/10.1523/JNEUROSCI. 20-01-00464.2000

Ma, J., Lu, J., & Li, X. (2021). The influence of emotional awareness on time perception: Evidence from event-related potentials. *Frontiers in Psychology*, 6008. https://doi.org/10.3389/fpsyg.2021.704510

Macar, F., Lejeune, H., Bonnet, M., Ferrara, A., Pouthas, V., Vidal, F., & Maquet, P. (2002). Activation of the supplementary motor area and of attentional networks during temporal processing. *Experimental Brain Research*, *142*, 475–485. https:// doi.org/10.1007/s00221-001-0953-0

Macar, F., & Vidal, F. (2004). Event-related potentials as indices of time processing: A review. *Journal of Psychophysiology*, *18*(2/3), 89–104. https://doi.org/10. 1027/0269-8803.18.23.89

Macar, F., & Vidal, F. (2009). Timing processes: An outline of behavioural and neural indices not systematically considered in timing models. *Canadian Journal of Experimental Psychology/Revue canadienne de psychologie expérimentale*, *63*(3), 227. https://doi.org/10.1037/a0014457

Macar, F., Vidal, F., & Casini, L. (1999). The supplementary motor area in motor and sensory timing: Evidence from slow brain potential changes. *Experimental Brain Research*, *125*, 271–280. https://doi.org/10.1007/s002210050683

Makhin, S., & Pavlenko, V. (2003). EEG Activity in the process of measuring-off of time Intervals by humans. *Neurophysiology*, *35*(2), 143–148. https://doi.org/10. 1023/A:1026072910466

Maquet, P., Lejeune, H., Pouthas, V., Bonnet, M., Casini, L., Macar, F., Timsit-Berthier, M., Vidal, F., Ferrara, A., & Degueldre, C. (1996). Brain activation induced by estimation of duration: A PET study. *Neuroimage*, *3*(2), 119–126. https://doi.org/10.1006/nimg.1996.0014

Mento, G. (2017). The role of the P3 and CNV components in voluntary and automatic temporal orienting: A high spatial-resolution ERP study. *Neuropsychologia*, *107*, 31–40. https://doi.org/10.1016/j.neuropsychologia.2017.10.037

Michel, C. M., & He, B. (2019). EEG source localization. In *Handbook of clinical neurology* (Vol. 160, pp. 85–101). https://doi.org/10.1016/B978-0-444-64032-1.00006-0

Michel, C. M., & Murray, M. M. (2012). Towards the utilization of EEG as a brain imaging tool. *Neuroimage*, *61*(2), 371–385. https://doi.org/10.1016/j.neuroimage. 2011.12.039

Mioni, G., Shelp, A., Stanfield-Wiswell, C. T., Gladhill, K. A., Bader, F., & Wiener, M. (2020). Modulation of individual alpha frequency with tacs shifts time perception. *Cerebral Cortex Communications*, *1*(1), tgaa064. https://doi.org/10.1093/ texcom/tgaa064

Mitsudo, T., Gagnon, C., Takeichi, H., & Grondin, S. (2012). An electroencephalographic investigation of the filled-duration illusion. *Frontiers in Integrative Neuroscience*, *5*, 84. https://doi.org/10.3389/fnint.2011.00084

Mitsudo, T., Nakajima, Y., Remijn, G. B., Takeichi, H., Goto, Y., & Tobimatsu, S. (2009). Electrophysiological evidence of auditory temporal perception related to the assimilation between two neighboring time intervals. *NeuroQuantology*, *7*(1), 114–127.

Monfort, V., Pouthas, V., & Ragot, R. (2000). Role of frontal cortex in memory for duration: An event-related potential study in humans. *Neuroscience Letters*, *286*(2), 91–94. https://doi.org/10.1016/S0304-3940(00)01097-1

Morillon, B., & Baillet, S. (2017). Motor origin of temporal predictions in auditory attention. *Proceedings of the National Academy of Sciences, 114*(42), E8913–E8921. https://doi.org/10.1073/pnas.1705373114

Näätänen, R. (1990). The role of attention in auditory information processing as revealed by event-related potentials and other brain measures of cognitive function. *Behavioral and Brain Sciences, 13*(2), 201–233. https://doi.org/10.1017/S0140525X00078407

Näätänen, R., Paavilainen, P., Rinne, T., & Alho, K. (2007). The mismatch negativity (MMN) in basic research of central auditory processing: A review. *Clinical Neurophysiology, 118*(12), 2544–2590. https://doi.org/10.1016/0304-3940(89)90513-2

Näätänen, R., Pakarinen, S., Rinne, T., & Takegata, R. (2004). The mismatch negativity (MMN): Towards the optimal paradigm. *Clinical Neurophysiology, 115*(1), 140–144. https://doi.org/10.1016/j.clinph.2003.04.001

Nani, A., Manuello, J., Liloia, D., Duca, S., Costa, T., & Cauda, F. (2019). The neural correlates of time: A meta-analysis of neuroimaging studies. *Journal of Cognitive Neuroscience, 31*(12), 1796–1826. https://doi.org/10.1162/jocn_a_01459

Ng, K. K., Tobin, S., & Penney, T. B. (2011). Temporal accumulation and decision processes in the duration bisection task revealed by contingent negative variation. *Frontiers in Integrative Neuroscience, 5*, 77. https://doi.org/10.3389/fnint.2011.00077

Nobre, A. C., Correa, A., & Coull, J. T. (2007). The hazards of time. *Current Opinion in Neurobiology, 17*(4), 465–470. https://doi.org/10.1016/j.conb.2007.07.006

Nunez, P. L., & Srinivasan, R. (2006). *Electric fields of the brain: The neurophysics of EEG*. Oxford University Press.

Paul, I., Le Dantec, C., Bernard, C., Lalonde, R., & Rebaï, M. (2003). Event-related potentials in the frontal lobe during performance of a visual duration discrimination task. *Journal of Clinical Neurophysiology, 20*(5), 351–360.

Paul, I., Wearden, J., Bannier, D., Gontier, E., Le Dantec, C., & Rebaï, M. (2011). Making decisions about time: Event-related potentials and judgements about the equality of durations. *Biological Psychology, 88*(1), 94–103. https://doi.org/10.1016/j.biopsycho.2011.06.013

Pfeuty, M., Ragot, R., & Pouthas, V. (2005). Relationship between CNV and timing of an upcoming event. *Neuroscience Letters, 382*(1–2), 106–111. https://doi.org/10.1016/j.neulet.2005.02.067

Pinheiro, A. P., Schwartze, M., Gutierrez, F., & Kotz, S. A. (2019). When temporal prediction errs: ERP responses to delayed action-feedback onset. *Neuropsychologia, 134*, 107200. https://doi.org/10.1016/j.neuropsychologia.2019.107200

Polich, J. (2007). Updating P300: An integrative theory of P3a and P3b. *Clinical Neurophysiology, 118*(10), 2128–2148. https://doi.org/10.1016/j.clinph.2007.04.019

Pollux, P. M., & Guo, K. (2009). Event-related potential correlates of spatiotemporal regularities in vision. *NeuroReport, 20*(5), 525–530. https://doi.org/10.1097/WNR.0b013e32832770a5

Pouthas, V., Garnero, L., Ferrandez, A. M., & Renault, B. (2000). ERPs and PET analysis of time perception: Spatial and temporal brain mapping during visual discrimination tasks. *Human Brain Mapping, 10*(2), 49–60. https://doi.org/10.1002/(SICI)1097-0193(200006)10:2<49::AID-HBM10>3.0.CO;2-8

Praamstra, P. (2010). Electrophysiological markers of foreperiod effects. In K. Nobre and J. T. Coull (Eds.) *Attention and Time* (New York, NY: Oxford University Press), 331–345.

Praamstra, P., Kourtis, D., Kwok, H. F., & Oostenveld, R. (2006). Neurophysiology of implicit timing in serial choice reaction-time performance. *Journal of Neuroscience, 26*(20), 5448–5455. https://doi.org/10.1523/JNEUROSCI.0440-06.2006

Radüntz, T., Scouten, J., Hochmuth, O., & Meffert, B. (2017). Automated EEG artifact elimination by applying machine learning algorithms to ICA-based features. *Journal of Neural Engineering, 14*(4), 046004. https://doi.org/10.1088/1741-2552/aa69d1

Rammsayer, T. H. (1993). On dopaminergic modulation of temporal information processing. *Biological Psychology, 36*(3), 209–222. https://doi.org/10.1016/0301-0511(93)90018-4

Rammsayer, T. H. (1997). Are there dissociable roles of the mesostriatal and mesolimbocortical dopamine systems on temporal information processing in humans? *Neuropsychobiology, 35*(1), 36–45. https://doi.org/10.1159/000119328

Rammsayer, T. H. (1999). Neuropharmacological evidence for different timing mechanisms in humans. *The Quarterly Journal of Experimental Psychology: Section B, 52*(3), 273–286. https://doi.org/10.1080/713932708

Rammsayer, T. H., & Lima, S. D. (1991). Duration discrimination of filled and empty auditory intervals: Cognitive and perceptual factors. *Perception & Psychophysics, 50*(6), 565–574. https://doi.org/10.3758/BF03207541

Rimmele, J., Jolsvai, H., & Sussman, E. (2011). Auditory target detection is affected by implicit temporal and spatial expectations. *Journal of Cognitive Neuroscience, 23*(5), 1136–1147. https://doi.org/10.1162/jocn.2010.21437

Rufener, K. S., Husemann, A.-M., & Zaehle, T. (2020). The internal time keeper: Causal evidence for the role of the cerebellum in anticipating regular acoustic events. *Cortex, 133*, 177–187. https://doi.org/10.1016/j.cortex.2020.09.021

Samima, S., Sarma, M., & Samanta, D. (2017). Detecting vigilance in people performing continual monitoring task. In P. Horain, C. Achard, & M. Mallem (Eds.), *Intelligent human computer interaction.* Cham.

Sanabria, D., & Correa, Á. (2013). Electrophysiological evidence of temporal preparation driven by rhythms in audition. *Biological Psychology, 92*(2), 98–105. https://doi.org/10.1016/j.biopsycho.2012.11.012

Schacter, D. L. (1977). EEG theta waves and psychological phenomena: A review and analysis. *Biological Psychology, 5*(1), 47–82. https://doi.org/10.1016/0301-0511(77)90028-X

Schlichting, N., de Jong, R., & van Rijn, H. (2020). Performance-informed EEG analysis reveals mixed evidence for EEG signatures unique to the processing of time. *Psychological Research, 84*, 352–369. https://doi.org/10.1007/s00426-018-1039-y

Schmidt-Kassow, M., & Kotz, S. A. (2009). Attention and perceptual regularity in speech. *NeuroReport, 20*(18), 1643–1647. https://doi.org/10.1097/WNR.0b013e328333b0c6

Schroeder, C. E., & Lakatos, P. (2009). Low-frequency neuronal oscillations as instruments of sensory selection. *Trends in Neurosciences, 32*(1), 9–18. https://doi.org/10.1016/j.tins.2008.09.012

Schroeder, C. E., Lakatos, P., Kajikawa, Y., Partan, S., & Puce, A. (2008). Neuronal oscillations and visual amplification of speech. *Trends in Cognitive Sciences, 12*(3), 106–113. https://doi.org/10.1016/j.tics.2008.01.002

Schwartze, M., Farrugia, N., & Kotz, S. A. (2013). Dissociation of formal and temporal predictability in early auditory evoked potentials. *Neuropsychologia, 51*(2), 320–325. https://doi.org/10.1016/j.neuropsychologia.2012.09.037

Schwartze, M., Rothermich, K., & Kotz, S. A. (2012). Functional dissociation of pre-SMA and SMA-proper in temporal processing. *Neuroimage, 60*(1), 290–298. https://doi.org/10.1016/j.neuroimage.2011.11.089

Shen, D., & Alain, C. (2012). Implicit temporal expectation attenuates auditory attentional blink. *PloS One, 7*(4), e36031. https://doi.org/10.1371/journal.pone.0036031

Summerfield, C., & Egner, T. (2009). Expectation (and attention) in visual cognition. *Trends in Cognitive Sciences*, *13*(9), 403–409. https://doi.org/10.1016/j.tics.2009.06.003

Sun, H., Ma, X., Tang, L., Han, J., Zhao, Y., Xu, X., Wang, L., Zhang, P., Chen, L., & Zhou, J. (2019). Modulation of beta oscillations for implicit motor timing in primate sensorimotor cortex during movement preparation. *Neuroscience Bulletin*, *35*, 826–840. https://doi.org/10.1007/s12264-019-00387-4

Sutton, S., Braren, M., Zubin, J., & John, E. (1965). Evoked-potential correlates of stimulus uncertainty. *Science*, *150*(3700), 1187–1188. https://doi.org/10.1126/science.150.3700.1187

Tamm, M., Uusberg, A., Allik, J., & Kreegipuu, K. (2014). Emotional modulation of attention affects time perception: Evidence from event-related potentials. *Acta Psychologica*, *149*, 148–156. https://doi.org/10.1016/j.actpsy.2014.02.008

Tarantino, V., Ehlis, A.-C., Baehne, C., Boreatti-Huemmer, A., Jacob, C., Bisiacchi, P., & Fallgatter, A. J. (2010). The time course of temporal discrimination: An ERP study. *Clinical Neurophysiology*, *121*(1), 43–52. https://doi.org/10.1016/j.clinph.2009.09.014

Thibault, N., Albouy, P., & Grondin, S. (2023). Distinct brain dynamics and networks for processing short and long auditory time intervals. *Scientific Reports*, *13*(1), 22018.

Thompson, J., Wiener, M., & Michaelis, K. (2015). Distinct spatial and temporal discounting during decision making in humans. *Journal of Vision*, *15*(12), 411–411. https://doi.org/10.1167/15.12.411

Tudor, M., Tudor, L., & Tudor, K. I. (2005). Hans Berger (1873-1941)--The history of electroencephalography. *Acta medica Croatica: casopis Hravatske akademije medicinskih znanosti*, *59*(4), 307–313.

Vallet, W., Laflamme, V., & Grondin, S. (2019). An EEG investigation of the mechanisms involved in the perception of time when expecting emotional stimuli. *Biological Psychology*, *148*, 107777. https://doi.org/10.1016/j.biopsycho.2019.107777

Van Rijn, H., Kononowicz, T. W., Meck, W. H., Ng, K. K., & Penney, T. B. (2011). Contingent negative variation and its relation to time estimation: a theoretical evaluation. *Frontiers in Integrative Neuroscience*, *5*, 91. https://doi.org/10.3389/fnint.2011.00091

Von Stein, A., & Sarnthein, J. (2000). Different frequencies for different scales of cortical integration: From local gamma to long range alpha/theta synchronization. *International Journal of Psychophysiology*, *38*(3), 301–313. https://doi.org/10.1016/S0167-8760(00)00172-0

Wang, X.-J. (2010). Neurophysiological and computational principles of cortical rhythms in cognition. *Physiological Reviews*, *90*(3), 1195–1268. https://doi.org/10.1152/physrev.00035.2008

Wearden, J. (1991). Human performance on an analogue of an interval bisection task. *The Quarterly Journal of Experimental Psychology Section B*, *43*(1b), 59–81. https://doi.org/10.1080/14640749108401259

White, P. M., & Yee, C. M. (2006). P50 sensitivity to physical and psychological state influences. *Psychophysiology*, *43*(3), 320–328. https://doi.org/10.1111/j.1469-8986.2006.00408.x

Wiener, M., & Kanai, R. (2016). Frequency tuning for temporal perception and prediction. *Current Opinion in Behavioral Sciences*, *8*, 1–6. https://doi.org/10.1016/j.cobeha.2016.01.001

Wiener, M., Parikh, A., Krakow, A., & Coslett, H. B. (2018). An intrinsic role of beta oscillations in memory for time estimation. *Scientific Reports*, *8*(1), 7992. https://doi.org/10.1038/s41598-018-26385-6

Wiener, M., & Thompson, J. C. (2015). Repetition enhancement and memory effects for duration. *Neuroimage*, *113*, 268–278. https://doi.org/10.1016/j.neuroimage. 2015.03.054

Xuan, B., Chen, X.-C., He, S., & Zhang, D.-R. (2009). Numerical magnitude modulates temporal comparison: An ERP study. *Brain Research*, *1269*, 135–142. https://doi.org/10.1016/j.brainres.2009.03.016

Zakay, D., & Block, R. A. (1997). Temporal cognition. *Current Directions in Psychological Science*, *6*(1), 12–16. https://doi.org/10.1111/1467-8721.ep11512604

Zhang, M., Zhang, K., Zhou, X., Zhan, B., He, W., & Luo, W. (2021). Similar CNV neurodynamic patterns between sub-and supra-second time perception. *Brain Sciences*, *11*(10), 1362. https://doi.org/10.3390/brainsci11101362

Zhang, Z. (2019). Spectral and Time-Frequency Analysis. In L. Hu & Z. Zhang (Eds.), *EEG Signal Processing and Feature Extraction*. Singapore: Springer. 89–116. https://doi.org/10.1007/978-981-13-9113-2_6

Zhao, K., Gu, R., Wang, L., Xiao, P., Chen, Y.-H., Liang, J., Hu, L., & Fu, X. (2014). Voluntary pressing and releasing actions induce different senses of time: evidence from event-related brain responses. *Scientific Reports*, *4*(1), 6047. https://doi.org/10.1038/srep06047

4

STUDYING TIME PERCEPTION WITH MAGNETOENCEPHALOGRAPHY

Sophie K. Herbst

NeuroSpin, CEA Saclay, DRF/Joliot Inserm Cognitive Neuroimaging Unit, Université Paris-Saclay, France

Introduction

There are several tools in neuroscience for studying the cerebral mechanisms involved in the processing of temporal information. Magnetoencephalography (MEG) research has contributed important insights, especially regarding the identification the dynamical brain signals and their anatomical sources, to the vast literature on the neuroimaging of timing and time perception. This chapter describes the methodological foundations related to MEG and its complementary usefulness when studying time perception from a neuroscientific perspective.

Magnetoencephalography – overview of the technique

Magnetoencephalography (MEG) measures the magnetic fields generated by electrical currents in the brain (Baillet, 2017; Cohen, 1968, 1972). The physiological origin of MEG signals is thus similar to the related technique, electroencephalography (EEG). Both techniques provide wide-field measures of local field potentials, which mainly result from postsynaptic potentials traveling along the apical dendrites of pyramidal neurons in the neocortex (Buzsáki et al., 2012). If a sufficiently large and spatially aligned population of neurons is activated synchronously, it forms a hypothetical current dipole, whose electrical activity and the resulting magnetic fields are measurable on or outside the skull.

In a pioneering MEG recording of a human participant with a single sensor and concurrent recording of the scalp EEG, the magnetic field of alpha

DOI: 10.4324/9781003449546-4

oscillations was reported at the Massachusetts Institute for Technology in 1968 (Cohen, 1968; see also Hari & Salmelin, 2012 for an extended historical perspective). The development of MEG was fueled by the motivation to overcome distortions of electric recordings by the conductivity of different tissues (skull, muscle, and skin) surrounding the brain, as well as technical advancements from telecommunication. Magnetic induction travels through the air, and the magnetic field strength of the brain is in the range of 10–100 femtotesla (10–15 T), which is ten orders of magnitudes smaller than the earth's magnetic field. Critically, the development of the superconducting quantum interference device (SQUID) allowed to measure the faint magnetic signals from the brain (Cohen, 1972). SQUIDs are still commonly used in today's MEG systems, which have around 300 sensors to cover the whole head. However, to achieve their superconducting properties, SQUIDs need to be cooled at 4 K, achieved by liquid helium, which requires for the sensors to be embedded in a rigid helmet. To shield the recordings from environmental noise, the MEG is installed in a magnetically shielded room. Recent developments of so-called optically pumped magnetometers (OPMs) provide more flexible and mobile recordings that will likely become standard in a few years from now (see below).

The temporal resolution of MEG, like EEG, is in the range of milliseconds, with common sampling rates of 1,000 Hz, but even 3,000 Hz are routinely possible on state-of-the-art machines. Conventional MEG systems allow to record distributed signals from the human brain with a spatial resolution of a few millimeters. The higher spatial resolution of MEG compared to EEG allows to disentangle brain activity from several, possibly synchronized sources (see below), supposedly reflecting the synchronized activation of distinct neural populations. Today, MEG is already used for retinotopy in vision (Nasiotis et al., 2017) and tonotopy in audition (Falet et al., 2021), requiring very high and reliable spatial resolution. Methodological advances in recording and analysis techniques are likely to push the spatial resolution even further (Bonaiuto et al., 2021).

The apical dendrites of cortical pyramidal neurons are mostly aligned perpendicular to the cortical surface, and the resulting magnetic fields are oriented perpendicular to the current flow, that is, the hypothetical dipole formed by neuronal dendrites. Therefore, MEG activity mainly reflects tangential currents mainly found in sulcal walls, while perpendicular currents (mainly found along gyral crowns) cancel out (Ahlfors et al., 2009). This is one of the main reasons why MEG and EEG are complementary, EEG also having access to radial sources. The best spatial resolution is achieved when combining MEG and EEG in the same recording, which entails fitting an MEG-compatible EEG cap to the participant before they enter the MEG and recording both signals concurrently. The signals can then be combined on the source level, which increases the spatial resolution and reliability (Aydin et al., 2015; Hauk et al., 2019).

Recording MEG puts more constraints on the experimenter, even though the lengthy electrode preparation necessary for EEG is not applicable. Crucially, the participant's placement and immobility during the recording determines the quality of the data and the resulting spatial resolution. Given that the helmet is rigid, bad placement or movement of the participant will lead to reduced signal quality, more so than in the EEG where the cap moves with the participant's head. Therefore, head movements are usually tracked with dedicated coils and recorded at a low sampling rate, to be taken into account during preprocessing. Concerning the preprocessing stages, MEG data are very similar to EEG, with the exception that some recording systems apply active shielding during the recording, which require additional spatial filtering of the data afterwards. Frequency-filtering, removal of biological artifacts (ocular, cardiac, and muscular), resampling, and epoching are routinely applied, and several software packages exist for the analysis of MEG data (e.g., Fieldtrip (Oostenveld et al., 2011), Brainstorm (Baillet et al., 2000; Tadel et al., 2011), and MNE Python (Gramfort et al., 2013, 2014)).

In addition to the signal processing in *sensor space*, a host of techniques are available to infer the **anatomical sources** of the signals measured at the sensors. Determining the sources of the observed brain signals is motivated not only by the scientific question at hand but also by the necessity to align all participants to a common space: as mentioned above, the use of a standard size helmet does not account for interindividual differences in brain anatomy, which potentially increases the noise when averaging over participants at the sensor level. To achieve alignment to a common anatomical space, the data are projected onto the individual brain anatomy, and then aligned on a common brain template. To obtain the best source reconstruction, individual anatomical MRIs are usually obtained along with the MEG data, and used to construct a *forward model*, which is a detailed 3D representation of the individual anatomy along with the biophysical properties of the underlying sources. Uncovering the sources of the signals recorded at the sensor level represents an ill-conditioned inverse problem, because the number of sources usually exceeds the number of sensors by at least one order of magnitude. Several techniques are available, such as dipole fitting, minimum norm estimates, and beamforming, which are described in detail elsewhere (Baillet, 2017; Gross, 2019; Hari & Puce, 2017). There is no unique source solution; the obtained mappings rely on model assumptions and parameter choices and are thus better described as an estimation or *source reconstruction*, rather than source imaging, which assumes a veridical mapping as in fMRI.

In sum, MEG gives access to brain signals with the highest combination of spatial and temporal resolution that is currently available for non-invasive techniques, allowing to measure transient responses to sensory stimuli, or neural oscillatory signals across different brain regions. These properties make MEG an excellent modality for the study of temporal cognition in humans, involving distributed brain regions and intricate temporally structured signaling (Hari & Parkkonen, 2015). While the signal-to-noise-ratio of MEG and spatial resolution is superior, EEG is more widely available, and thus the number of timing studies using EEG is much higher. The decision for one or the other technique often depends on the availability and prior training of the researcher, and for many scientific questions, both techniques can be used to provide important and often converging insights.

MEG signatures of implicit and explicit timing

To interact with a dynamic environment, it is crucial to apprehend its temporal structure and build internal representations of temporal features like duration, order, or synchronization. The cognitive neuroscience literature distinguishes between **implicit** and **explicit timing** (Coull & Nobre, 2008; Michon, 1990), which coexist in ecological contexts and are naturally embedded in the way we interact with dynamic environments.

Implicit Timing: Extraction of temporal contingencies between external events to form temporal predictions, resulting in facilitation of behavior (indirect measurement).

Explicit Timing: Deliberate engagement in timekeeping, resulting in overt temporal estimates (direct measurement).

Parsimoniously, one could posit that the brain treats temporal information as a single representation, rather than maintaining separate representations for different purposes, that is, implicit and explicit timing situations. Yet, the existing literature has approached implicit and explicit timing mainly as separate processes, even though their neural dynamics of implicit and explicit timing overlap to a considerable extent. In the following, I will discuss neural dynamics revealed by MEG research with respect to implicit and explicit timing processes. Due to the wider availability of EEG and the considerable overlap between both techniques, no clear boundary can be drawn between results obtained with EEG versus MEG. Thus, EEG work is cited along with MEG work where applicable in the following sections, but with an emphasis on the specific insights gained from MEG. A summary of MEG studies addressing human timing is provided in Table 4.1.

TABLE 4.1 Summary of the main MEG findings in the field of timing and time perception

Author/year	Implicit/explicit	Task	Stimulus modality	Neural signature	Brain structures	Main finding
Andersen and Dalal (2021)	implicit, rhythmic	no task	tactile	beta power	cerebellum	increased omission response after regular stimulation
Arnal et al. (2014)	explicit, rhythmic	temporal expectation judgment	auditory	delta–beta phase–amplitude coupling	auditory/motor cortices	higher pre-stimulus coupling predicts correct responses
Auksztulewicz et al. (2019)	implicit, rhythmic	auditory categorization	auditory	evoked response	auditory cortex	increased sensitivity to tone frequency in rhythmic stimuli
Azizi and van Wassenhove (2023)	explicit	retrospective verbal duration estimate	none	alpha power and bursts	parietal, visual	positive correlation between power/burst counts and estimated duration
Barne et al. (2022)	implicit, rhythmic	frequency & orientation discrimination	auditory and visual	delta phase	auditory and visual cortices	rhythmic pre-activation of the relevant sensory cortices
Carver et al. (2012)	explicit, interval	duration and pitch discrimination	auditory	beta band desynchronization in explicit timing	sensorimotor, inferior frontal gyrus, auditory	

Chen et al. (2023)	implicit, interval	working-memory task, orientation discrimination	visual	alpha power	left parietal, visual	reduced power in temporally variable conditions during retention
Daume et al. (2021)	explicit, interval	temporal expectation judgment	visual, tactile	delta phase consistency	sensory, parietal, frontal, cerebellum	increased pre-stimulus phase consistency in predictive conditions
ElShafei et al. (2018)	implicit, interval	frequency discrimination	visual, auditory	alpha power	sensory	decrease in auditory (9 Hz), increase in visual (13 Hz)
Fujioka et al. (2012)	explicit, rhythmic	discrimination of target position with respect to the beat	auditory	beta power	auditory, motor regions	anticipatory beta power increases prior, increased coherence pre-stimulus
Fujioka et al. (2015)	explicit, rhythmic	discrimination of target position with respect to the beat	auditory	beta power	auditory, sensorimotor	anticipatory beta power increase pre-stimulus
Gauthier et al. (2020)	explicit, interval	mental time travel	visual	evoked activity	hippocampus	temporal ordinality engaged mostly the left hippocampus, spatial ordinality engaged mostly the right hippocampus

(Continued)

TABLE 4.1 (Continued)

Author/year	Implicit/ explicit	Task	Stimulus modality	Neural signature	Brain structures	Main finding
Grabot, Kononowicz et al. (2019)	explicit, interval	production	none	alpha–beta phase–amplitude coupling	left sensorimotor	positive correlation between coupling and production precision
Gunasekaran et al. (2023)	explicit	rest, tapping, counting	none	delta band oscillations	motor	spontaneous oscillations present only during rest
Herbst et al. (2022b)	implicit and explicit	pitch and duration discrimination	auditory	alpha and beta power	visual/parietal sensorimotor	enhanced alpha power with implicit predictability, reduced beta power in explicit task
Herrmann et al. (2016)	implicit, rhythmic	intensity deviant detection	auditory	delta power	auditory	Temporal expectations modulated target-detection performance, when delta power was high
Kononowicz et al. (2018)	explicit	duration production	none	beta power	sensorimotor	beta power after production initiation predicts produced duration

Kononowicz et al. (2018)	explicit, interval	production	none	alpha–beta phase–amplitude coupling	sensorimotor	positive correlation between coupling and production precision during *encoding* and reproduction
Kösem et al. (2014)	implicit, rhythmic	temporal order judgment	visual, auditory	delta phase	auditory, visual	auditory delta phase shift observed when temporal order perception shifted
Kösem et al. (2018)	implicit, rhythmic	syllable discrimination	auditory	delta power	auditory	frequency of entrained oscillations predicts perception
Kulashekhar et al. (2016)	explicit, interval	duration or color discrimination	visual	beta power	sensorimotor	higher beta power in duration discrimination
Lerousseau et al. (2021)	implicit, rhythmic	passive	auditory	delta, gamma amplitude	auditory	higher gamma but not delta phase coherence after rhythmic stimulation
Martin et al. (2006)	implicit, rhythmic	speeded response	auditory	evoked activity	cerebellum, sensorimotor	pre-stimulus amplitude in both regions predicts reaction time

(Continued)

TABLE 4.1 (Continued)

Author/year	Implicit/ explicit	Task	Stimulus modality	Neural signature	Brain structures	Main finding
Martin et al. (2008)	implicit, rhythmic	speeded response	visual	evoked activity, delta phase	cerebellum, sensorimotor, parietal	pre-stimulus amplitude predicts reaction time, cerebellar and sensorimotor 2 Hz phase correlates with RT
Meindertsma et al. (2018)	implicit, quasi-rhythmic	orientation discrimination	visual	beta power	motor, prefrontal, parietal	negative correlation between beta power and RT post target
Morillon and Baillet (2017)	implicit, quasi-rhythmic	frequency discrimination (average across tones)	auditory	delta phase-phase and beta phase-amplitude coupling with stimulation	auditory, left sensorimotor	delta-beta coupling predicts behavior, directed functional connectivity from sensorimotor (beta) to auditory (delta) regions
Sohoglu and Chait (2016)	implicit, rhythmic	passive and interference detection	auditory, parietal	evoked activity		increased sustained activity in regular contexts

Spitzer et al. (2014)	explicit, interval	WM, intensity or duration discrimination	somatosensory prefrontal	beta power		higher somatosensory beta power for duration task, correlation between prefrontal beta power and stimulus magnitude
Todorovic et al. (2015)	implicit, interval	deviant detection	auditory	beta power	auditory	decrease in auditory beta power before expected tones (only if unattended)
Todorovic A, Auksztulewicz et al. (2021) (same data as above)	implicit, interval	deviant detection	auditory (A1, sup. temp. gyri), inferior parietal	evoked, dynamic causal modeling (DCM)	auditory	amplitude sensitive to passage of time (early effect, early auditory regions and superior temporal gyri) and contextual probability (later effect, additional effects in inf. parietal), DCM: passage of time effects explained by modulations at low levels, context at higher levels

(Continued)

TABLE 4.1 (Continued)

Author/year	Implicit/explicit	Task	Stimulus modality	Neural signature	Brain structures	Main finding
van Ede et al. (2020)	implicit, interval	speeded RT, orientation discrimination	visual	alpha, beta power	sensorimotor, visual	visual task: anticipatory modulation of alpha lateralization, RT task: anticipatory modulation of beta lateralization
van Wassenhove and Lecoutre (2015)	explicit, interval	duration discrimination	auditory	evoked	auditory	positive correlation between offset response and perceived duration, peak latencies of the onset and ramping predicted subjective time perception
Wilsch et al. (2018)	implicit, interval	same/different judgment of auditory sequence	auditory, visual	alpha power	auditory	increase of alpha power in auditory and decrease in visual areas during delay phase, lower alpha power in predictive conditions

Study	Type	Task	Modality	Measure	Brain areas	Findings
Wilsch et al. (2020)	implicit, rhythmic	direction discrimination	auditory, visual	delta phase coherence	auditory, visual, frontal	increased coherence in rhythmic condition: broadband in vision, frequency-specific in audition
Kononowicz et al., 2015	explicit	duration reproduction	auditory	evoked activity	supplementary motor area	Pre-movement evoked field, but no clear ramping activity (CMV) as observed in the concurrently recorded EEG (CNV) for supra-second duration reproduction
N'Diaye et al., 2004	explicit, interval	bisection	visual, auditory	evoked activity	sensory cortices, frontal, parietal areas	sustained fields in modality-dependent sensory areas, and modality independent frontal and parietal areas

Note: This list of studies was compiled by searching for the keywords ('interval timing' or 'temporal prediction' or 'temporal expectation') and ('magneto-encephalography' or 'MEG') in the title or abstract of papers indexed on PubMed. The list was completed manually, but it is likely still not exhaustive.

Time is both an essential dimension of brain function and a product thereof (Hari & Parkkonen, 2015; van Wassenhove et al., 2019), which is difficult to disentangle empirically. Several neural signatures have been reported and interpreted in relation to theoretical accounts of timing. Here it is important to distinguish between *time-locked* or *evoked* neural dynamics, which reliably occur with the same temporal profile in response to a triggering event, for instance, the sensory response or the phase of a slow neural oscillation, versus *non time-locked* dynamics, also referred to as *induced* dynamics, such as amplitude modulations of relatively fast oscillations which relate to specific aspects of the task such as memory processes or the motor response (Tallon-Baudry & Bertrand, 1999).

While no clear one-to-one mapping between different signatures and mechanistic models of timing could be established, some principles can be laid out. One of the most prominent groups of explicit timing models, pacemaker–accumulator models, assume that during a to-be-timed interval, and internal pacemaker emits temporal pulses, which are then stored and compared between intervals (Gibbon et al., 1984). E/MEG studies have searched for the **signals of the pacemaker**, supposed to reside in time-locked neural dynamics that evolve at temporal scales that match the most common durations for interval timing (several hundreds of milliseconds to seconds). **Ramping evoked potentials** (Macar et al., 1999; Walter et al., 1964) and **neural oscillations** (Matell & Meck, 2004; Treisman, 1984) have been examined for pacemaker-like properties, which per definition should occur from the onset of the to-be-timed interval and covary with subjective time estimates. These signatures have provided valuable insights on timing processes (detailed in the following paragraphs), but no unique neural correlate of the hypothesized pacemaker has been found.

More recently, distributed nonlinear clocking signals have become the dominant model, namely population clocks, also known as state-space models (Buonomano & Laje, 2010; Paton & Buonomano, 2018). While single-cell recordings have provided important evidence for these models, E/MEG research is likely not the best method to observe the neural trajectories directly. However, higher-order signatures such as the initiation of said trajectories or their readout can be addressed (Kononowicz et al., 2018; Ofir & Landau, 2022).

Slow ramping potentials

As described in the previous chapter, EEG research has identified components of sensory and motor evoked responses that relate specifically to timing, the most prominent being the contingent negative variation (CNV;

Walter et al., 1964), observed during a relevant time interval *in implicit and explicit timing tasks* (Macar et al., 1999; Praamstra et al., 2006). The ramping activity reflected by the CNV has been associated with the accumulation of temporal pulses in pacemaker–accumulator models (Gibbon et al., 1984), and is localized in the supplementary motor area (Brunia et al., 2012; Herbst et al., 2018; Macar et al., 1999; Praamstra et al., 2006; Walter et al., 1964).

The CNV has its corresponding magnetic response, the contingent *magnetic* variation (CMV). Compared to the relatively equivocal sources reported for the CNV, there is considerable variability in the observed sources of the CMV, both between participants and between tasks. Early studies mainly used Go/No-go tasks in the supra-second range (Dammers & Ioannides, 2000; Elbert et al., 1994; Hultin et al., 1996), in which timing is implicit, and careful single-subject analyses found sources in premotor areas consistent with the generators of the CNV (Basile et al., 1994; Hultin et al., 1996), but also temporal, prefrontal, and parietal areas (Dammers & Ioannides, 2000; Elbert et al., 1994). Studies targeting explicit duration discrimination reveal a similar picture of distributed CMV sources: for example, a duration oddball task with concurrent recording of EEG and MEG showed an enhanced sustained potential in auditory areas when stimuli were attended, and the MEG data revealed another source sensitive to attention to time in the posterior cingulate cortex, which correlated with behavioral duration sensitivity (Sieroka et al., 2003). This distinction is not visible in EEG where source reconstructions are performed less commonly, and where it is difficult to resolve distributed sources. Further MEG studies that investigated slow ramping potentials in timing tasks also show distributed sources over sensory areas, notably auditory ones, corresponding to the task at hand (Hari et al., 1980; Herbst et al., 2022a; N'Diaye et al., 2004; van Wassenhove & Lecoutre, 2015). While there are reports of a robust CMV with motor origins, its amplitude showed no relationship with supra-second duration judgments (Kononowicz et al., 2015; N'Diaye et al., 2004).

In sum, MEG research on slow ramping potentials supports the notion of 'many CN/MVs' (Basile et al., 1994), or distributed ramping signals for the timing of task-relevant intervals. The variability of their localization and time course of the slow ramping response across studies likely results from the higher sensitivity of the MEG, allowing to address more veridically the sources of the response, yet also leading to more difficulty in its interpretation. These ramping potentials cannot be interpreted as the neural signature of a centralized pacemaker, and understanding their functional roles will require further research with systematic variations of timing tasks and sensory modalities.

Interval offset responses

As time itself is not a physical stimulus, sensory events are used in timing studies to delimit time intervals. Evoked responses typically reflect the sensory processing of a sensory stimulus and its modulation by cognitive processing such as prediction, attention, and task goals. Indeed, in timing responses to the stimulus that signals the end of a relevant time interval are indicative of temporal surprise and temporal decision-making (Bueno & Cravo, 2021; Kononowicz & van Rijn, 2014; Ofir & Landau, 2022; van Wassenhove & Lecoutre, 2015 (MEG, all others EEG); Visalli et al., 2021).

In *implicit timing*, early sensory evoked responses reflect the observer's expectation of the currently unfolding interval, which is built and updated over time (Nobre & van Ede, 2018), surfacing as amplitude modulations of early components of the evoked response (N1, P2) (Auksztulewicz et al., 2019; Herbst et al., 2022a; Herbst & Obleser, 2019 (EEG); Lange, 2009, 2013 (EEG); Sohoglu & Chait, 2016; Todorovic & Auksztulewicz, 2021). The direction of the modulation can inform about the cognitive process: while predictive coding accounts postulate *reduced* amplitudes for predicted signals (Friston, 2005), attentional orienting would rather be reflected by *enhanced* amplitudes (Summerfield & Egner, 2016). Furthermore, later components of the evoked response, and in particular the P300, have been related to temporal surprise and prediction update (EEG; Visalli et al., 2021, 2023), in line with their known interpretation for higher-level cognitive processing (Sergent et al., 2005).

In *explicit timing*, several EEG studies have emphasized that the evoked response to the end of the interval indexes the subjectively elapsed duration and the decision about it made by the observer (Bueno & Cravo, 2021; Kononowicz & van Rijn, 2014; Ofir & Landau, 2022). However, to the best of my knowledge, these findings have not yet been assessed with MEG.

Oscillations

An influential and long-standing theoretical proposal holds that neural oscillations serve as the temporal scaffold that allows the brain to internally keep time (Herbst & Landau, 2016; Matell & Meck, 2004; Pöppel, 1972; Treisman, 1984; van Wassenhove, 2016; van Wassenhove et al., 2019; Wiener & Kanai, 2016). As discussed in an increasingly complex literature (for a review: Wiener & Kanai, 2016), a role of neural oscillations in temporal cognition is appealing, not least because of the inherent temporal structure conveyed by oscillatory dynamics. The role of a central pacemaker was initially attributed to alpha oscillations, but this hypothesis could not be confirmed empirically (Kononowicz & van Wassenhove, 2016; Treisman, 1984; Treisman et al., 1994; see van Wassenhove et al., 2019 for an in-depth discussion). Yet, depending on the timing task under consideration, many roles for oscillations have been conceived, and a unique functional account of neural oscillations for timing in different situations seems unlikely. The neural oscillatory signatures of implicit and explicit timing have mainly been addressed separately, but the reported neural dynamics show some interesting overlaps.

Two main accounts will be recounted in more detail in the following: first, a neural phase code implements temporal predictions derived from temporal regularities in the sensory environment via phase shifts and phase alignment in the delta and the theta (0.5–3/4–7 Hz) bands, mainly in early sensory and motor areas. For visual perception, this proposal can be extended to include the phase of alpha oscillations. Second, transient modulations of oscillatory power, notably in the beta band (15–35 Hz), but also in the alpha band (8–13 Hz), relate to various aspects of temporal cognition, spanning implicit and explicit timing.

Slow oscillatory phase codes for implicit timing

Low frequency activity (delta/theta range) evolves at timescales that are directly relevant to behavior, such as active sensing across species (Barczak et al., 2019), and human perception and production of speech or music (Ding et al., 2016; Poeppel, 2003), modulated on the timescales of few hundreds of milliseconds to seconds. Slow neural oscillations, notably in the delta (0.5–3 Hz) and theta (4–7 Hz) bands, have been suggested to form an endogenous temporal structure that could implement temporal predictions derived from those regularities in the sensory environment (Kayser, 2009; Lakatos et al., 2008; Schroeder & Lakatos, 2009), a hypothesis that can be linked back to the influential proposal of Dynamic Attending in Time (Jones, 1976; Morillon & Schroeder, 2015). Slow neural oscillations, prominently found across species (Buzsáki & Draguhn, 2004), are also particularly important in coordinating neural activity within and across brain areas via a nested frequency regime (Canolty et al., 2006; Fries, 2005). Slow neural oscillations thus have the capacity to synchronize large brain networks to the natural temporal structure of the dynamic sensory environment.

Seminal work in non-human primates has shown that slow neural oscillations emulate the temporal structure of sensory input by *entraining* to an external temporal structure (Lakatos et al., 2005). Aligning the phase of slow oscillations to an external rhythm tunes the excitability of the relevant sensory areas to the predicted onset of the to-be attended input, thereby implementing its attentional selection. Building on the initial work in non-human primates, an important body of research demonstrated that slow oscillations measured with M/EEG in humans can lock to the phase of periodic inputs, and modulate human behavior, surfacing as phase coherence between the measured neural oscillations in auditory, motor, and frontal areas and the external rhythms (Barne et al., 2022 (MEG); Besle et al., 2011 (intracranial EEG); Cravo et al., 2013 (EEG); Henry & Obleser, 2012 (EEG); Herrmann et al., 2016; Kösem et al., 2014; Morillon & Baillet, 2017; Park et al., 2015; Wilsch et al., 2020 (all using MEG)).

Crucially, entrainment in a narrow sense assumes an *endogenous*, physiological predisposition of the system to oscillate at a particular frequency range, which is then modulated to match the external input through phase shifts and frequency entrainment (Lakatos et al., 2019; Obleser & Kayser, 2019). Evidence for an endogenous role of delta oscillations in implementing temporal predictions comes from studies reporting relationships between delta phase and human behavior in absence of rhythmic input (Daume et al., 2021 (MEG); Henry et al., 2016 (EEG); Herbst & Obleser, 2019 (EEG); Kayser et al., 2015 (intracranial recordings in rats)). An active role of delta phase entrainment is also suggested by its susceptibility to top-down influences such as the attended sensory modality (Keil et al., 2016; Lakatos et al., 2008), task demands (Lakatos et al., 2013), perceptual grouping (Barczak et al., 2018), and hierarchical rhythmic structure of inputs (Nozaradan et al., 2011). Furthermore, entrainment can occur selectively at one frequency present in the input (Morillon & Baillet, 2017) and be sustained after the offset of the periodic stimulus yet still profoundly affecting perception (Bouwer et al., 2023 (EEG); Kösem et al., 2018; but see Lerousseau et al., 2021 (both MEG)).[1] Finally, previous studies have shown that entrainment scales with the strength of temporal predictions (Cravo et al., 2011, 2013; Stefanics et al., 2010; Herbst et al. 2022b).

Critically, very few studies have tested for the presence of endogenously periodic neural dynamics in the delta band (Henry et al., 2016; Herbst & Obleser, 2019), and none have unequivocally shown the existence of an **endogenous delta oscillation** underlying the observed effects in human M/EEG recordings. The superposition of various signals in the human M/EEG makes it difficult to separate ongoing oscillatory from stimulus-evoked activity and bears the risk to conflate pre-stimulus activity (i.e., delta phase) with post-stimulus evoked activity (Doelling et al., 2019; Zoefel & Heil, 2013). Indeed, observing endogenous slow oscillations in human MEG is not trivial, and requires specific signal processing pipelines (Gunasekaran et al., 2023; Keitel & Gross, 2016). In a recent study, we showed that endogenous delta oscillations can be observed in human resting state MEG recordings

(Gunasekaran et al., 2023), with heterogeneous peak frequencies across individuals. This result confirms that relatively local oscillators can be observed non-invasively with MEG (Hari et al., 1997), and the sparseness and distributed nature of the underlying neural sources can likely explain the divergences across studies.

While the above-cited studies mainly focus on implicit temporal predictions and not on explicit time judgments, it has also been shown that subjective event timing is encoded in the phase of neural oscillations, selectively in auditory cortex (Kösem et al., 2014; Piper, 2019; van Wassenhove, 2016). The study by Kösem et al. (2014) demonstrates benefits from recording MEG over EEG in spatial resolution: targeted region-of-interest analyses in source space allowed to separate the auditory and visual delta oscillations.

Phase effects have also been observed in the **alpha band** (8–13 Hz), mainly in the visual modality (Busch et al., 2009; Mathewson et al., 2009; but see Ruzzoli et al., 2019).[2] Here, visual stimulus perception fluctuates with alpha phase, which can be driven by temporal constraints or predictions, derived from the sensory stimulation (Solís-Vivanco et al., 2018; Spaak et al., 2014). However, the authors of these later studies interpret their findings as automatically driven by the temporal properties of the stimuli (visual flicker), compared to more active entrainment mechanisms observed in the delta frequency range.

In sum, a number of studies provide evidence that the phase of slow neural oscillations is under endogenous control and can serve a as a mechanism to entrain or attune brain dynamics to the temporalities of the sensory environment.

Transient modulations of oscillatory power

The many roles of beta oscillations for temporal cognition

Switching now from phase coding to **amplitude fluctuations**, beta-band activity (15–35 Hz) has been related to temporal processing in a variety of timing tasks, including rhythmic, implicit, and explicit timing. Convergence over studies exists toward the sources of beta power modulations, reported to originate from sensorimotor, but also frontal and parietal brain areas. Divergence can be found with respect to the temporal occurrence of the functionally relevant modulations: following the start of the time interval (self-initiated or not), during the to-be-timed interval (implicit and explicit timing), as well as around the target stimulus, to update temporal predictions (implicit timing). While observed at distinct moments during the timing process, beta power modulations are not phase-locked, but rather occur as transient episodes of bursting activity (Shin et al., 2017). To date, there exists no single unifying account to integrate all the findings regarding beta, and it is likely that different functional oscillators might implement distinct timing processes.

Concerning *implicit timing*, an influential line of research reports that when monkeys or humans tap along and continue a rhythmic beat (Bartolo et al., 2014 (intracranial recordings in non-human primates); Fujioka et al., 2012, 2015 (both MEG)), the beta-power time course recorded in auditory and motor-related areas (sensorimotor cortex, inferior-frontal gyrus, supplementary motor area, and cerebellum) between two beats is indicative of the anticipated onset of the next beat, both for internally and externally generated rhythms. In the same line, but in the absence of strict periodicity in the input signals, implicit timing studies have found beta band power modulations prior to temporally predicted target onsets (Cravo et al., 2011 (EEG); Todorovic et al., 2015 (MEG)). A recent study found that temporal surprise in a quasi-rhythmic temporal prediction task was correlated with power suppression in the beta band in parietal and prefrontal areas, after the target occurred (Meindertsma et al., 2018).

In line with these findings, in a recent study (Herbst et al., 2022a), we observed increased beta power over sensorimotor areas when participants used implicit timing in a pitch discrimination task, compared to when they explicitly judged the foreperiod intervals based on the same auditory stimuli. These results point to the importance of beta oscillations in extracting temporal predictability from sensory input, in line with the general role in predictive processing assigned to the beta band (Bastos et al., 2015), suggested to be achieved through temporal synchronization of inputs within and between sensory modalities (Hipp et al., 2011; La Rocca et al., 2020). Indeed, studies which empirically manipulated implicit temporal predictability reported enhanced functional connectivity in the beta band in temporally predictive conditions, between frontal, parietal, motor, and sensory areas (Keil et al., 2016 (EEG); Morillon & Baillet, 2017 (MEG)).

In *explicit* timing tasks, beta power has been found to increase in particular when comparing an explicit temporal discrimination task to a control task that did not require duration discrimination, but had the same physical stimuli (Kulashekhar et al., 2016; Spitzer et al., 2014). Using auditory stimulation, Carver et al. (2012) observed the opposite relationship, namely reduced beta power in the explicit timing task, in line with our own findings reported above. Furthermore, in explicit non-rhythmic motor timing, a transient increase in beta power after the onset-button press has been found to reflect the duration the participant will produce on a single trial (Kononowicz, 2015; Kononowicz et al., 2018). Kononowicz et al. (2018) further found that beta power was indicative of participant's metacognitive evaluation of their produced intervals, suggesting that transient modulations of beta power in motor areas initiate an internal timing process along a ballistic trajectory.

In sum, beta power is functionally implicated in various aspects of temporal cognition, spanning rhythmic and interval prediction, as well as explicit timing with and without motor components.

And how about alpha power?

It shall not be neglected that many studies also report alpha power modulations during timing. Alpha oscillations are often too fast to emulate the timescales of temporal prediction or explicit timing, but they have characteristic functions for attentional orienting and motor control. In implicit timing, the modulation of alpha power prior to temporally predicted stimuli is a classical finding, commonly shown by a decrease when the stimuli are visual (van Ede et al., 2016; Rohenkohl & Nobre, 2011; van Diepen et al., 2015), and an over visual/ parietal areas increase when the stimuli are auditory (Herbst et al., 2022a; Herbst & Obleser, 2017; Schneider et al., 2022; Strauß et al., 2014). The parietal sources of the alpha power effects are in line with a previously reported involvement of these areas in temporal prediction (Coull & Nobre, 2008 (fMRI); ElShafei et al., 2018 (MEG); Meindertsma et al., 2018 (MEG); Visalli et al., 2019 (fMRI)), and more generally attention (Corbetta & Shulman, 2002) and working memory (Chen et al., 2023; Wilsch et al., 2018).

In line with its roles in top-down attention, alpha is likely not the timing signal per se, but reflects the allocation of attentional resources in time. In line with this interpretation, a recent MEG study by van Ede et al. (2020) demonstrated the differential dynamics of alpha power during a temporally predictive interval with respect to the task that was carried out by participant (speeded motor response vs. visual discrimination). It is currently still unclear which brain dynamics time and trigger the onset of these preparatory responses, whose generators notably include subcortical areas such as the basal ganglia, thalamus, and the cerebellum (Schwartze & Kotz, 2013). In *explicit timing*, a recent MEG study found that the power of alpha during a resting state period, and relatedly the number of alpha burst episodes, correlated with participants' retrospective time estimates[3] (Azizi et al., 2023).

Multiplexing of slow and fast oscillatory dynamics

Bringing together the phase effects in the slower delta/theta bands, and the power fluctuations in the alpha and beta bands in both implicit and explicit timing, leads to the hypothesis that slow oscillations might drive transient power fluctuations of faster oscillations in a **phase–amplitude coupling regime** (Arnal & Giraud, 2012; Buzsáki & Draguhn, 2004; Canolty et al., 2006; Giraud & Poeppel, 2012; Palva & Palva, 2018). The biophysical properties of slow oscillations allow for the synchronization of larger networks of brain areas to a common temporal regime (Chartove et al., 2020; Fries, 2015; Kopell et al., 2000; Nácher et al., 2013). Previous studies indeed suggested that slow oscillations, which allow to bridge longer time windows like foreperiods in the second range, could control more short-lived brain states, critical for the processing of temporal and sensory information, using intracranial recordings (Saleh et al., 2010) and E/MEG (Arnal et al., 2014 (MEG); Cravo et al., 2011 (EEG); Mento et al., 2017 (EEG); Morillon & Baillet, 2017 (MEG)). For instance, Arnal et al. (2014) found that pre-stimulus coupling of delta phase and beta power indicated correlated with the performance of the participant when judging whether the target was presented at the predicted time point (inferred from a preceding rhythmic sequence), or too late. Using MEG, the authors were able to distinguish an inverse relationship of the modulation of beta power between auditory and motor regions (i.e., high beta power in auditory vs. low beta in motor sensors in the pre-stimulus time window) predicting performance. While strictly speaking, this task involves a form of explicit timing, the task structure aligns with implicit timing studies. Another study by Morillon and Baillet (2017) found frequency-specific directional interactions between right auditory and left sensorimotor cortices, where the auditory regions track the stimulus frequency (3 Hz), and the sensorimotor regions control the temporal predictions through beta bursts, whose amplitude is coupled to the frequency relevant for temporal predictions (1.5 Hz). Again, this study emphasizes the power of MEG to resolve spatially distinct sources with temporally correlated activity.

In an *explicit temporal reproduction task*, phase–amplitude coupling between alpha and beta in sensorimotor and parietal regions accounted for precision of temporal productions, with strongest coupling observed within the self-initiated reproduction interval (Grabot et al., 2019). Kononowicz et al. (2020) replicated the association between alpha–beta phase–amplitude coupling and timing precision using a reproduction task, and provided novel evidence that it occurs both during encoding and reproduction of temporal intervals, thus arguing for an effect on timing precision rather than motor control.

In sum, coupling regimes between low-frequency phase and high-frequency amplitudes appear to be a general scheme for the implementation of timing within and across brain regions. Thereby the involvement of particular frequency bands likely depends on the sensory contingencies, such as the relevant intervals and modalities.

Comparison of implicit and explicit timing

Few studies have directly compared the neural signatures of implicit and explicit timing using EEG or MEG. The anatomical sources identified by studies on implicit timing (Herbst et al., 2018; Martin et al., 2006 (MEG); Mento, 2013; Praamstra et al., 2006 (all EEG)) partially converge with those of explicit timing principally localized in pre-motor and motor areas (Kononowicz & van Rijn, 2015 (EEG); Kulashekhar et al., 2016 (MEG); see review by Wiener et al., 2010), but also include brain networks more generally associated with attention and predictive processing in frontal and parietal cortices (Meindertsma et al., 2018; Visalli et al., 2019). These findings are confirmed by the few but influential studies that directly compared the anatomical correlates of an implicit (temporal prediction) and explicit timing (temporal production) task using fMRI (Coull et al., 2013; Coull & Nobre, 2008), reporting a common substrate of implicit and explicit timing in pre-motor areas, but distinct contributions to implicit timing from inferior parietal, versus, for explicit timing, from pre-frontal areas, cerebellum, and basal ganglia. Furthermore, visual and auditory cortices have been reported to process information related to temporal hazard (Bueti et al., 2010; Visalli et al., 2019 (both fMRI)), not necessarily restricted to the modality in which the relevant stimuli are presented. From this mostly fMRI-based literature, it is difficult to infer the succession of timing-related processes throughout the unfolding of an interval and distinguish timing proper (toward an anticipated event) from prediction updating (once the predicted time has elapsed, or the stimulus occurred), motor responses, and post-stimulus processing like surprise and decision making.

Several studies compared explicit timing to judgments of other sensory features based on the very same stimuli and temporal statistics, such as auditory tone frequency, tactile vibration rate, or color (Carver et al., 2012 (MEG); Kulashekhar et al., 2016 (MEG); Spitzer et al., 2014 (EEG)). These studies all reported effects in the beta band, albeit in different directions: in comparison with visual and tactile judgments, explicit duration judgments were accompanied by increased beta power, while the study by Carver et al. used auditory stimuli and found beta band desynchronization.

In a recent MEG study (Herbst et al., 2022a), we directly contrasted explicit and implicit timing, in an auditory foreperiod design. In the implicit timing tasks, participants judged the relative pitch of the two tones delimiting the foreperiod, with implicit timing indexed by enhanced pitch discrimination after predictive foreperiods (fixed interval throughout a block). The explicit timing task consisted in performing a temporal bisection task on the foreperiods, with the same auditory stimuli. During the foreperiod, we observed stronger induced beta power over sensorimotor and parietal areas, including the SMA during pitch discrimination compared to duration discrimination. Surprisingly, we did not observe a positive task difference, that is, distinct brain dynamics emerging when attention is directed to time during explicit timing. This divergence from the previous fMRI work (Coull et al., 2013) is likely due to the different neural dynamics targeted by fMRI versus MEG, and the focus on a fine-grained temporal resolution in our study. Taken together, the results of these studies on implicit and explicit timing using M/EEG suggest that the brain flexibly uses overlapping time coding mechanisms in implicit and explicit timing situations, depending on the temporal statistics of the inputs and the task requirements (see also van Ede et al., 2020), rather than deploying specialized modules for either timing task.

Deep sources

An exciting outlook for the future of MEG in timing research is the ability to access non-cortical sources well known to be involved in timing (Buhusi & Meck, 2005; Coull et al., 2011, 2013; Matell & Meck, 2004; Merchant et al., 2013; Schwartze et al., 2012; Schwartze & Kotz, 2013) at a high temporal resolution. Advances in source reconstruction methods, combined with realistic neurophysiological characterization derived from individual MRIs, have allowed to create distributed source models specifying the locations and orientations of hypothesized current dipoles in cortical and non-cortical areas (Attal et al., 2012; Vrba & Robinson, 2001). Simulations allow to verify the ability of the source models to distinguish deep sources from cortical ones (Attal & Schwartz, 2013; Meyer et al., 2017). Although the signal strength is, on average, about one tenth weaker than signals from cortical sources, these techniques have permitted to observe temporal dynamics from the auditory thalamus (Coffey et al., 2016), hippocampus (Meyer et al., 2017; Riggs et al., 2009), and the basal ganglia, notably the striatum (Hinault et al., 2023). Concurrent recordings of MEG and intracranial activity in treatment resistant epilepsy patients further confirm the reliable and specific source reconstruction of activity from the hippocampus and the amygdala (López-Madrona et al., 2022; Pizzo et al., 2019). These techniques have already been used to study temporal cognition, revealing activity in hippocampus, entorhinal cortex, and parahippocampal gyrus during mental time travel (Gauthier et al., 2020).

Less challenging than the deep sources is to reconstruct activity from the cerebellum (Andersen et al., 2020; Samuelsson et al., 2020), which is well known to be crucial for timing (Breska & Ivry, 2020; Kotz et al., 2014; Martin et al., 2008). Andersen and Dalal (2021) recently reported cerebellar beta band power to increase during temporal prediction, indexing a time point at which a stimulus is most likely to occur within an isochronous stream of tactile stimuli. They further found temporal omission responses in the thalamus and putamen. Interestingly, activity in cerebellum and inferior parietal cortex to expected stimuli was observed earlier than the response of the basal ganglia. While these findings are yet to be confirmed in other sensory domains, they provide novel insights in the temporal interplay of non-cortical and cortical brain regions in temporal processing and underline the excellent spatial resolution of MEG.

Outlook: Technical enhancements to improve SNR and flexibility

MEG hardware and analyses techniques are subject to active developments that will result in important advancements in the future. Currently, novel sensor types, so-called OPMs, measuring magnetic fields through laser lights which do not require cooling (Boto et al., 2016), are being developed. This relieves the constraint of a rigid helmet and allows to place the sensors closer to the head, which increases the signal-to-noise ratio. Flexible sensor layouts and portable devices will allow to record brain activity in more diverse populations, such as young children or patients. Furthermore, the sensors can be positioned closely to the areas of interest, for instance, over the cerebellum. It has even been shown that placing a sensor in the mouth allows to capture hippocampal activity (Tierney et al., 2021). These technologies are already used in research and will likely improve the spatial resolution of MEG even further, which together with its temporal resolution makes it a very interesting tool to study human brain dynamics related to timing.

Conclusions

Overall, MEG research has contributed important insights to the understanding of timing and time perception (see a summary in Table 4.1). This research partially overlaps with and replicates findings from EEG work, but is also complementary, notably when it comes to understanding the anatomical sources. For example, slow ramping potentials (CNV) are likely better observed in EEG in the SMA proper, while MEG work has shown a distributed nature of these potentials in motor and sensory areas. Neural oscillations relate to many aspects of temporal cognition, such as implicit temporal prediction, explicit encoding, and (re-) production. In particular, multiplexing schemes of oscillations at different frequency bands have the potential to synchronize brain networks to internal or external temporal motives. Exciting developments in the last decade have shown that MEG has the power to access cortical and subcortical brain regions, which paves the path to studying timing networks known through fMRI (Wiener, 2023), and reveal the temporal dynamics of the signals emerging from individual regions. MEG and EEG should not be considered as opposing methods; rather, they should be used flexibly and jointly to obtain the best signals for the scientific question at hand.

Acknowledgments

I would like to thank the Cognition and Brain Dynamics Team at Neurospin, and in particular Virginie van Wassenhove, for ongoing discussions of all aspects of timing and MEG. Furthermore, I thank Maximilien Chaumon, Simon Grandin, and Philippe Albouy for comments on a previous version of this chapter.

Notes

1 The study by Lerousseau et al. reports that no sustained entrainment (oscillatory activity) was found in the delta band *after* the rhythmic stimulation. The conditions under which sustained entrainment can be observed require further investigation.
2 In a recent registered report, Ruzzoli et al. were not able to replicate the relationship between alpha phase and visual perception reported in earlier studies.
3 In retrospective timing, participants are informed about the timing task only after the to-be-judged interval has elapsed, and hence cannot deliberately engage in timing, unlike in prospective timing where participants are informed up-front that a temporal judgment will have to be made.

References

Ahlfors, S. P., Han, J., Lin, F., Witzel, T., Belliveau, J. W., Hämäläinen, M. S., & Halgren, E. (2009). Cancellation of EEG and MEG signals generated by extended and distributed sources. *Human Brain Mapping*, *31*(1), 140–149. https://doi.org/10.1002/hbm.2085

Andersen, L. M., & Dalal, S. S. (2021). The cerebellar clock: Predicting and timing somatosensory touch. *NeuroImage*, *238*, 118202. https://doi.org/10.1016/j.neuroimage.2021.118202

Andersen, L. M., Jerbi, K., & Dalal, S. S. (2020). Can EEG and MEG detect signals from the human cerebellum? *NeuroImage*, *215*, 116817. https://doi.org/10.1016/j.neuroimage.2020.116817

Arnal, L. H., Doelling, K. B., & Poeppel, D. (2014). Delta–beta coupled oscillations underlie temporal prediction accuracy. *Cerebral Cortex*, bhu103. https://doi.org/10.1093/cercor/bhu103

Arnal, L. H., & Giraud, A.-L. (2012). Cortical oscillations and sensory predictions. *Trends in Cognitive Sciences*, *16*(7), 390–398. https://doi.org/10.1016/j.tics.2012.05.003

Attal, Y., Maess, B., Friederici, A., & David, O. (2012). Head models and dynamic causal modeling of subcortical activity using magnetoencephalographic/electroencephalographic data. *Reviews in the Neurosciences*, *23*(1), 85–95. https://doi.org/10.1515/rns.2011.056

Attal, Y., & Schwartz, D. (2013). Assessment of subcortical source localization using deep BRAIN activity imaging model with minimum norm operators: A MEG study. *PloS one*, *8*(3), e59856. https://doi.org/10.1371/journal.pone.0059856

Auksztulewicz, R., Myers, N. E., Schnupp, J. W., & Nobre, A. C. (2019). Rhythmic temporal expectation boosts neural activity by increasing neural gain. *Journal of Neuroscience*. https://doi.org/10.1523/JNEUROSCI.0925-19.2019

Aydin, Ü., Vorwerk, J., Dümpelmann, M., Küpper, P., Kugel, H., Heers, M., Wellmer, J., Kellinghaus, C., Haueisen, J., Rampp, S., Stefan, H., & Wolters, C. H. (2015). Combined EEG/MEG can outperform single modality EEG or MEG source reconstruction in presurgical epilepsy diagnosis. *PloS one*, *10*(3), e0118753. https://doi.org/10.1371/journal.pone.0118753

Azizi, L., Polti, I., & van Wassenhove, V. (2023). Spontaneous α Brain Dynamics Track the Episodic "When." *Journal of Neuroscience*, *43*(43), 7186–7197. https://doi.org/10.1523/JNEUROSCI.0816-23.2023

Baillet, S. (2017). Magnetoencephalography for brain electrophysiology and imaging. *Nature Neuroscience*, *20*(3), 3. https://doi.org/10.1038/nn.4504

Baillet, S., Mosher, J., & Leahy, R. (2000). BrainStorm beta release: A Matlab software package for MEG signal processing and source localization and visualization. *NeuroImage*, *11*. https://doi.org/10.1016/S1053-8119(00)91843-3

Barczak, A., Haegens, S., Ross, D. A., McGinnis, T., Lakatos, P., & Schroeder, C. E. (2019). Dynamic modulation of cortical excitability during visual active sensing. *Cell Reports*, *27*(12), 3447–3459.

Barczak, A., O'Connell, M. N., McGinnis, T., Ross, D., Mowery, T., Falchier, A., & Lakatos, P. (2018). Top-down, contextual entrainment of neuronal oscillations in the auditory thalamocortical circuit. *Proceedings of the National Academy of Sciences*, *115*(32), E7605–E7614. https://doi.org/10.1073/pnas.1714684115

Barne, L. C., Cravo, A. M., de Lange, F. P., & Spaak, E. (2022). Temporal prediction elicits rhythmic preactivation of relevant sensory cortices. *The European Journal of Neuroscience*, *55*(11–12), 3324–3339. https://doi.org/10.1111/ejn.15405

Bartolo, R., Prado, L., & Merchant, H. (2014). Information processing in the primate basal Ganglia during sensory-guided and internally driven rhythmic tapping. *Journal of Neuroscience*, *34*(11), 3910–3923. https://doi.org/10.1523/JNEUROSCI.2679-13.2014

Basile, L. F., Rogers, R. L., Bourbon, W. T., & Papanicolaou, A. C. (1994). Slow magnetic flux from human frontal cortex. *Electroencephalography and Clinical Neurophysiology*, *90*(2), 157–165. https://doi.org/10.1016/0013-4694(94)90007-8

Bastos, A. M., Vezoli, J., Bosman, C. A., Schoffelen, J.-M., Oostenveld, R., Dowdall, J. R., De Weerd, P., Kennedy, H., & Fries, P. (2015). Visual areas exert feedforward and feedback influences through distinct frequency channels. *Neuron*, *85*(2), 390–401. https://doi.org/10.1016/j.neuron.2014.12.018

Besle, J., Schevon, C. A., Mehta, A. D., Lakatos, P., Goodman, R. R., McKhann, G. M., Emerson, R. G., & Schroeder, C. E. (2011). Tuning of the human neocortex to the temporal dynamics of attended events. *The Journal of Neuroscience*, *31*(9), 3176–3185. https://doi.org/10.1523/JNEUROSCI.4518-10.2011

Bonaiuto, J. J., Little, S., Neymotin, S. A., Jones, S. R., Barnes, G. R., & Bestmann, S. (2021). Laminar dynamics of high amplitude beta bursts in human motor cortex. *NeuroImage*, *242*, 118479. https://doi.org/10.1016/j.neuroimage.2021.118479

Boto, E., Bowtell, R., Krüger, P., Fromhold, T. M., Morris, P. G., Meyer, S. S., Barnes, G. R., & Brookes, M. J. (2016). On the potential of a new generation of magnetometers for MEG: A beamformer simulation study. *PloS one*, *11*(8), e0157655. https://doi.org/10.1371/journal.pone.0157655

Bouwer, F. L., Fahrenfort, J. J., Millard, S. K., Kloosterman, N. A., & Slagter, H. A. (2023). A silent disco: Differential effects of beat-based and pattern-based temporal expectations on persistent entrainment of low-frequency neural oscillations. *Journal of Cognitive Neuroscience*, 1–31. https://doi.org/10.1162/jocn_a_01985

Breska, A., & Ivry, R. B. (2020). Context-specific control over the neural dynamics of temporal attention by the human cerebellum. *Science Advances*, *6*(49). https://doi.org/10.1126/sciadv.abb1141

Brunia, C. H. M., van Boxtel, G. J. M., & Böcker, K. B. E. (2012). Negative slow waves as indices of anticipation: The Bereitschaftspotential, the contingent negative variation, and the stimulus-preceding negativity. In *The Oxford handbook of event-related potential components* (pp. 189–207). Oxford University Press.

Bueno, F. D., & Cravo, A. M. (2021). Post-interval EEG activity is related to task-goals in temporal discrimination. *PloS one, 16*(9), e0257378. https://doi.org/10.1371/journal.pone.0257378

Bueti, D., Bahrami, B., Walsh, V., & Rees, G. (2010). Encoding of temporal probabilities in the human brain. *The Journal of Neuroscience, 30*(12), 4343–4352. https://doi.org/10.1523/JNEUROSCI.2254-09.2010

Buhusi, C. V., & Meck, W. H. (2005). What makes us tick? Functional and neural mechanisms of interval timing. *Nature Reviews Neuroscience, 6*(10), 755–765.

Buonomano, D. V., & Laje, R. (2010). Population clocks: Motor timing with neural dynamics. *Trends in Cognitive Sciences, 14*(12), 520–527. https://doi.org/10.1016/j.tics.2010.09.002

Busch, N. A., Dubois, J., & VanRullen, R. (2009). The phase of ongoing EEG oscillations predicts visual perception. *The Journal of Neuroscience, 29*(24), 7869–7876. https://doi.org/10.1523/JNEUROSCI.0113-09.2009

Buzsáki, G., Anastassiou, C. A., & Koch, C. (2012). The origin of extracellular fields and currents – EEG, ECoG, LFP and spikes. *Nature Reviews Neuroscience, 13*(6), 407–420. https://doi.org/10.1038/nrn3241

Buzsáki, G., & Draguhn, A. (2004). Neuronal oscillations in cortical networks. *Science, 304*(5679), 1926–1929.

Canolty, R. T., Edwards, E., Dalal, S. S., Soltani, M., Nagarajan, S. S., Kirsch, H. E., Berger, M. S., Barbaro, N. M., & Knight, R. T. (2006). High gamma power is phase-locked to theta oscillations in human neocortex. *Science, 313*(5793), 1626–1628. https://doi.org/10.1126/science.1128115

Carver, F. W., Elvevåg, B., Altamura, M., Weinberger, D. R., & Coppola, R. (2012). The neuromagnetic dynamics of time perception. *PloS one, 7*(8), e42618. https://doi.org/10.1371/journal.pone.0042618

Chartove, J. A. K., McCarthy, M. M., Pittman-Polletta, B. R., & Kopell, N. J. (2020). A biophysical model of striatal microcircuits suggests gamma and beta oscillations interleaved at delta/theta frequencies mediate periodicity in motor control. *PLoS Computational Biology, 16*(2), e1007300. https://doi.org/10.1371/journal.pcbi.1007300

Chen, F.-W., Li, C.-H., & Kuo, B.-C. (2023). Temporal expectation based on the duration variability modulates alpha oscillations during working memory retention. *NeuroImage, 265*, 119789. https://doi.org/10.1016/j.neuroimage.2022.119789

Coffey, E. B. J., Herholz, S. C., Chepesiuk, A. M. P., Baillet, S., & Zatorre, R. J. (2016). Cortical contributions to the auditory frequency-following response revealed by MEG. *Nature Communications, 7*(1), 1. https://doi.org/10.1038/ncomms11070

Cohen, D. (1968). Magnetoencephalography: Evidence of magnetic fields produced by alpha-rhythm currents. *Science, 161*(3843), 784–786. https://doi.org/10.1126/science.161.3843.784

Cohen, D. (1972). Magnetoencephalography: Detection of the brain's electrical activity with a superconducting magnetometer. *Science, 175*(4022), 664–666. https://doi.org/10.1126/science.175.4022.664

Corbetta, M., & Shulman, G. L. (2002). Control of goal-directed and stimulus-driven attention in the brain. *Nature Reviews Neuroscience, 3*(3), 3. https://doi.org/10.1038/nrn755

Coull, J. T., Cheng, R.-K., & Meck, W. H. (2011). Neuroanatomical and neurochemical substrates of timing. *Neuropsychopharmacology*, *36*(1), 3–25. https://doi.org/10.1038/npp.2010.113

Coull, J. T., Davranche, K., Nazarian, B., & Vidal, F. (2013). Functional anatomy of timing differs for production versus prediction of time intervals. *Neuropsychologia*, *51*(2), 309–319. https://doi.org/10.1016/j.neuropsychologia.2012.08.017

Coull, J. T., & Nobre, A. C. (2008). Dissociating explicit timing from temporal expectation with fMRI. *Current Opinion in Neurobiology*, *18*(2), 137–144. https://doi.org/10.1016/j.conb.2008.07.011

Cravo, A. M., Rohenkohl, G., Wyart, V., & Nobre, A. C. (2011). Endogenous modulation of low frequency oscillations by temporal expectations. *Journal of Neurophysiology*, *106*(6), 2964–2972. https://doi.org/10.1152/jn.00157.2011

Cravo, A. M., Rohenkohl, G., Wyart, V., & Nobre, A. C. (2013). Temporal expectation enhances contrast sensitivity by phase entrainment of low-frequency oscillations in visual cortex. *The Journal of Neuroscience*, *33*(9), 4002–4010. https://doi.org/10.1523/JNEUROSCI.4675-12.2013

Dammers, J., & Ioannides, A. A. (2000). Neuromagnetic Localization of CMV generators using incomplete and full-head biomagnetometer. *NeuroImage*, *11*(3), 167–178. https://doi.org/10.1006/nimg.1999.0524

Daume, J., Wang, P., Maye, A., Zhang, D., & Engel, A. K. (2021). Non-rhythmic temporal prediction involves phase resets of low-frequency delta oscillations. *NeuroImage*, *224*, 117376. https://doi.org/10.1016/j.neuroimage.2020.117376

Ding, N., Melloni, L., Zhang, H., Tian, X., & Poeppel, D. (2016). Cortical tracking of hierarchical linguistic structures in connected speech. *Nature Neuroscience*, *19*(1), 158–164. https://doi.org/10.1038/nn.4186

Doelling, K. B., Assaneo, M. F., Bevilacqua, D., Pesaran, B., & Poeppel, D. (2019). An oscillator model better predicts cortical entrainment to music. *Proceedings of the National Academy of Sciences*, *116*(20), 10113–10121. https://doi.org/10.1073/pnas.1816414116

Elbert, T., Rockstroh, B., Hampson, S., Pantev, C., & Hoke, M. (1994). The magnetic counterpart of the contingent negative variation. *Electroencephalography and Clinical Neurophysiology/Evoked Potentials Section*, *92*(3), 262–272. https://doi.org/10.1016/0168-5597(94)90069-8

ElShafei, H. A., Bouet, R., Bertrand, O., & Bidet-Caulet, A. (2018). Two sides of the same coin: Distinct sub-bands in the α rhythm reflect facilitation and suppression mechanisms during auditory anticipatory attention. *eNeuro*, *5*(4). https://doi.org/10.1523/ENEURO.0141-18.2018

Falet, J.-P. R., Côté, J., Tarka, V., Martínez-Moreno, Z. E., Voss, P., & de Villers-Sidani, E. (2021). Mapping the human auditory cortex using spectrotemporal receptive fields generated with magnetoencephalography. *NeuroImage*, *238*, 118222. https://doi.org/10.1016/j.neuroimage.2021.118222

Fries, P. (2005). A mechanism for cognitive dynamics: Neuronal communication through neuronal coherence. *Trends in Cognitive Sciences*, *9*(10), 474–480. https://doi.org/10.1016/j.tics.2005.08.011

Fries, P. (2015). Rhythms for cognition: Communication through coherence. *Neuron*, *88*(1), 220–235.

Friston, K. J. (2005). A theory of cortical responses. *Philosophical Transactions of the Royal Society B: Biological Sciences*, *360*(1456), 815–836. https://doi.org/10.1098/rstb.2005.1622

Fujioka, T., Ross, B., & Trainor, L. J. (2015). Beta-band oscillations represent auditory beat and its metrical hierarchy in perception and imagery. *Journal of Neuroscience*, *35*(45), 15187–15198. https://doi.org/10.1523/JNEUROSCI.2397-15.2015

Fujioka, T., Trainor, L. J., Large, E. W., & Ross, B. (2012). Internalized timing of isochronous sounds is represented in neuromagnetic beta oscillations. *The Journal of Neuroscience*, *32*(5), 1791–1802. https://doi.org/10.1523/JNEUROSCI.4107-11.2012

Gauthier, B., Prabhu, P., Kotegar, K. A., & van Wassenhove, V. (2020). Hippocampal contribution to ordinal psychological time in the human brain. *Journal of Cognitive Neuroscience*, 1–15. https://doi.org/10.1162/jocn_a_01586

Gibbon, J., Church, R. M., & Meck, W. H. (1984). Scalar timing in memory. *Annals of the New York Academy of Sciences*, *423*, 52–77.

Giraud, A.-L., & Poeppel, D. (2012). Cortical oscillations and speech processing: Emerging computational principles and operations. *Nature Neuroscience*, *15*(4), 511–517. https://doi.org/10.1038/nn.3063

Grabot, L., Kononowicz, T. W., la Tour, T. D., Gramfort, A., Doyère, V., & van Wassenhove, V. (2019). The strength of alpha-beta oscillatory coupling predicts motor timing precision. *Journal of Neuroscience*, *39*, 3277–3291. https://doi.org/10.1523/JNEUROSCI.2473-18.2018

Gramfort, A., Luessi, M., Larson, E., Engemann, D. A., Strohmeier, D., Brodbeck, C., Goj, R., Jas, M., Brooks, T., Parkkonen, L., & Hämäläinen, M. (2013). MEG and EEG data analysis with MNE-Python. *Frontiers in Neuroscience*, *7*. https://doi.org/10.3389/fnins.2013.00267

Gramfort, A., Luessi, M., Larson, E., Engemann, D. A., Strohmeier, D., Brodbeck, C., Parkkonen, L., & Hämäläinen, M. S. (2014). MNE software for processing MEG and EEG data. *NeuroImage*, *86*, 446–460. https://doi.org/10.1016/j.neuroimage.2013.10.027

Gross, J. (2019). Magnetoencephalography in cognitive neuroscience: A primer. *Neuron*, *104*(2), 189–204. https://doi.org/10.1016/j.neuron.2019.07.001

Gunasekaran, H., Azizi, L., van Wassenhove, V., & Herbst, S. K. (2023). Characterizing endogenous delta oscillations in human MEG. *Scientific Reports*, *13*(1), 1. https://doi.org/10.1038/s41598-023-37514-1

Hari, R., Aittoniemi, K., Järvinen, M.-L., Katila, T., & Varpula, T. (1980). Auditory evoked transient and sustained magnetic fields of the human brain localization of neural generators. *Experimental Brain Research*, *40*(2), 237–240. https://doi.org/10.1007/BF00237543

Hari, R., & Parkkonen, L. (2015). The brain timewise: How timing shapes and supports brain function. *Philosophical Transactions of the Royal Society B: Biological Sciences*, *370*(1668), 20140170. https://doi.org/10.1098/rstb.2014.0170

Hari, R., & Puce, A. (2017). *MEG-EEG Primer* (1st ed.). Oxford University Press. https://doi.org/10.1093/med/9780190497774.001.0001

Hari, R., & Salmelin, R. (2012). Magnetoencephalography: From SQUIDs to neuroscience: Neuroimage 20th anniversary special edition. *NeuroImage*, *61*(2), 386–396. https://doi.org/10.1016/j.neuroimage.2011.11.074

Hari, R., Salmelin, R., Mäkelä, J. P., Salenius, S., & Helle, M. (1997). Magnetoencephalographic cortical rhythms. *International Journal of Psychophysiology*, *26*(1), 51–62. https://doi.org/10.1016/S0167-8760(97)00755-1

Hauk, O., Stenroos, M., & Treder, M. (2019). *Towards an objective evaluation of EEG/MEG source estimation methods: The linear tool kit* (672956). https://doi.org/10.1101/672956

Henry, M. J., Herrmann, B., & Obleser, J. (2016). Neural microstates govern perception of auditory input without rhythmic structure. *Journal of Neuroscience*, *36*(3), 860–871. https://doi.org/10.1523/JNEUROSCI.2191-15.2016

Henry, M. J., & Obleser, J. (2012). Frequency modulation entrains slow neural oscillations and optimizes human listening behavior. *Proceedings of the National Academy of Sciences*, *109*(49), 20095–20100. https://doi.org/10.1073/pnas.1213390109

Herbst, S. K., Fiedler, L., & Obleser, J. (2018). Tracking temporal hazard in the human electroencephalogram using a forward encoding model. *Eneuro, 5*(2), ENEURO.0017-18.2018. https://doi.org/10.1523/ENEURO.0017-18.2018

Herbst, S. K., & Landau, A. N. (2016). Rhythms for cognition: The case of temporal processing. *Current Opinion in Behavioral Sciences, 8*, 85–93. https://doi.org/10.1016/j.cobeha.2016.01.014

Herbst, S. K., & Obleser, J. (2017). Implicit variations of temporal predictability: Shaping the neural oscillatory and behavioural response. *Neuropsychologia, 101*, 141–152. https://doi.org/10.1016/j.neuropsychologia.2017.05.019

Herbst, S. K., & Obleser, J. (2019). Implicit temporal predictability enhances pitch discrimination sensitivity and biases the phase of delta oscillations in auditory cortex. *NeuroImage, 203*, 116198. https://doi.org/10.1016/j.neuroimage.2019.116198

Herbst, S. K., Obleser, J., & van Wassenhove, V. (2022a). Implicit versus explicit timing – Separate or shared mechanisms? *Journal of Cognitive Neuroscience, 34*(8), 1447–1466. https://doi.org/10.1162/jocn_a_01866

Herbst, S. K., Stefanics, G., & Obleser, J. (2022b). Endogenous modulation of delta phase by expectation–A replication of Stefanics et al., 2010. *Cortex, 149*, 226–245. https://doi.org/10.1016/j.cortex.2022.02.001

Herrmann, B., Henry, M. J., Haegens, S., & Obleser, J. (2016). Temporal expectations and neural amplitude fluctuations in auditory cortex interactively influence perception. *NeuroImage, 124*, Part A, 487–497. https://doi.org/10.1016/j.neuroimage.2015.09.019

Hinault, T., Baillet, S., & Courtney, S. M. (2023). Age-related changes of deep-brain neurophysiological activity. *Cerebral Cortex, 33*(7), 3960–3968. https://doi.org/10.1093/cercor/bhac319

Hipp, J. F., Engel, A. K., & Siegel, M. (2011). Oscillatory synchronization in large-scale cortical networks predicts perception. *Neuron, 69*(2), 387–396. https://doi.org/10.1016/j.neuron.2010.12.027

Hultin, L., Rossini, P., Romani, G. L., Högstedt, P., Tecchio, F., & Pizzella, V. (1996). Neuromagnetic localization of the late component of the contingent negative variation. *Electroencephalography and Clinical Neurophysiology, 98*(6), 435–448. https://doi.org/10.1016/0013-4694(96)95507-8

Jones, M. R. (1976). Time, our lost dimension: Toward a new theory of perception, attention, and memory. *Psychological Review, 83*(5), 323–355.

Kayser, C. (2009). Phase resetting as a mechanism for supramodal attentional control. *Neuron, 64*(3). https://pub.uni-bielefeld.de/publication/2914176

Kayser, C., Wilson, C., Safaai, H., Sakata, S., & Panzeri, S. (2015). Rhythmic auditory cortex activity at multiple timescales shapes stimulus–response gain and background firing. *Journal of Neuroscience, 35*(20), 7750–7762. https://doi.org/10.1523/JNEUROSCI.0268-15.2015

Keil, J., Pomper, U., & Senkowski, D. (2016). Distinct patterns of local oscillatory activity and functional connectivity underlie intersensory attention and temporal prediction. *Cortex; A Journal Devoted to the Study of the Nervous System and Behavior, 74*, 277–288. https://doi.org/10.1016/j.cortex.2015.10.023

Keitel, A., & Gross, J. (2016). Individual human brain areas can be identified from their characteristic spectral activation fingerprints. *PLOS Biology, 14*(6), e1002498. https://doi.org/10.1371/journal.pbio.1002498

Kononowicz, T. W. (2015). Dopamine-dependent oscillations in frontal cortex index 'start-gun' signal in interval timing. *Frontiers in Human Neuroscience, 9*, 331. https://doi.org/10.3389/fnhum.2015.00331

Kononowicz, T. W., Roger, C., & van Wassenhove, V. (2018). Temporal metacognition as the decoding of self-generated brain dynamics. *Cerebral Cortex (New York, N.Y.: 1991)*. https://doi.org/10.1093/cercor/bhy318

Kononowicz, T. W., Sander, T., & van Rijn, H. (2015). Neuroelectromagnetic signatures of the reproduction of supra-second durations. Neuropsychologia, 75, 201–213. https://doi.org/10.1016/j.neuropsychologia.2015.06.001

Kononowicz, T. W., Sander, T., van Rijn, H., & van Wassenhove, V. (2020). Precision timing with α–β oscillatory coupling: Stopwatch or motor control? *Journal of Cognitive Neuroscience, 32*(9), 1624–1636. https://doi.org/10.1162/jocn_a_01570

Kononowicz, T. W., & van Rijn, H. (2014). Decoupling interval timing and climbing neural activity: A dissociation between CNV and N1P2 amplitudes. *Journal of Neuroscience, 34*(8), 2931–2939. https://doi.org/10.1523/JNEUROSCI.2523-13.2014

Kononowicz, T. W., & van Rijn, H. (2015). Single trial beta oscillations index time estimation. *Neuropsychologia, 75,* 381–389. https://doi.org/10.1016/j.neuropsychologia.2015.06.014

Kononowicz, T. W., & van Wassenhove, V. (2016). In search of oscillatory traces of the internal clock. *Frontiers in Psychology, 7.* https://doi.org/10.3389/fpsyg.2016.00224

Kopell, N., Ermentrout, G. B., Whittington, M. A., & Traub, R. D. (2000). Gamma rhythms and beta rhythms have different synchronization properties. *Proceedings of the National Academy of Sciences, 97*(4), 1867–1872. https://doi.org/10.1073/pnas.97.4.1867

Kösem, A., Bosker, H. R., Takashima, A., Meyer, A., Jensen, O., & Hagoort, P. (2018). Neural entrainment determines the words we hear. *Current Biology.* https://doi.org/10.1016/j.cub.2018.07.023

Kösem, A., Gramfort, A., & van Wassenhove, V. (2014). Encoding of event timing in the phase of neural oscillations. *NeuroImage, 92,* 274–284. https://doi.org/10.1016/j.neuroimage.2014.02.010

Kotz, S. A., Stockert, A., & Schwartze, M. (2014). Cerebellum, temporal predictability and the updating of a mental model. *Philosophical Transactions of the Royal Society B: Biological Sciences, 369*(1658), 20130403. https://doi.org/10.1098/rstb.2013.0403

Kulashekhar, S., Pekkola, J., Palva, J. M., & Palva, S. (2016). The role of cortical beta oscillations in time estimation. *Human Brain Mapping, 37*(9), 3262–3281. https://doi.org/10.1002/hbm.23239

La Rocca, D., Ciuciu, P., Engemann, D.-A., & van Wassenhove, V. (2020). Emergence of β and γ networks following multisensory training. NeuroImage, 206, 116313. https://doi.org/10.1016/j.neuroimage.2019.116313

Lakatos, P., Gross, J., & Thut, G. (2019). A new unifying account of the roles of neuronal entrainment. *Current Biology, 29*(18), R890–R905. https://doi.org/10.1016/j.cub.2019.07.075

Lakatos, P., Karmos, G., Mehta, A. D., Ulbert, I., & Schroeder, C. E. (2008). Entrainment of neuronal oscillations as a mechanism of attentional selection. *Science, 320*(5872), 110–113. https://doi.org/10.1126/science.1154735

Lakatos, P., Musacchia, G., O'Connel, M. N., Falchier, A. Y., Javitt, D. C., & Schroeder, C. E. (2013). The spectrotemporal filter mechanism of auditory selective attention. *Neuron, 77*(4), 750–761. https://doi.org/10.1016/j.neuron.2012.11.034

Lakatos, P., Shah, A. S., Knuth, K. H., Ulbert, I., Karmos, G., & Schroeder, C. E. (2005). An oscillatory hierarchy controlling neuronal excitability and stimulus processing in the auditory cortex. *Journal of Neurophysiology, 94*(3), 1904–1911. https://doi.org/10.1152/jn.00263.2005

Lange, K. (2009). Brain correlates of early auditory processing are attenuated by expectations for time and pitch. *Brain and Cognition, 69*(1), 127–137. https://doi.org/10.1016/j.bandc.2008.06.004

Lange, K. (2013). The ups and downs of temporal orienting: A review of auditory temporal orienting studies and a model associating the heterogeneous findings on the auditory N1 with opposite effects of attention and prediction. *Frontiers in Human Neuroscience, 7*, 263. https://doi.org/10.3389/fnhum.2013.00263

Lerousseau, J. P., Trébuchon, A., Morillon, B., & Schön, D. (2021). Frequency selectivity of persistent cortical oscillatory responses to auditory rhythmic stimulation. *Journal of Neuroscience, 41*(38), 7991–8006. https://doi.org/10.1523/JNEUROSCI.0213-21.2021

López-Madrona, V. J., Medina Villalon, S., Badier, J.-M., Trébuchon, A., Jayabal, V., Bartolomei, F., Carron, R., Barborica, A., Vulliémoz, S., Alario, F.-X., & Bénar, C. G. (2022). Magnetoencephalography can reveal deep brain network activities linked to memory processes. *Human Brain Mapping, 43*(15), 4733–4749. https://doi.org/10.1002/hbm.25987

Macar, F., Vidal, F., & Casini, L. (1999). The supplementary motor area in motor and sensory timing: Evidence from slow brain potential changes. *Experimental Brain Research, 125*(3), 271–280. https://doi.org/10.1007/s002210050683

Martin, T., Houck, J. M., Bish, J. P., Kičić, D., Woodruff, C. C., Moses, S. N., Lee, D. C., & Tesche, C. D. (2006). MEG reveals different contributions of somatomotor cortex and cerebellum to simple reaction time after temporally structured cues. *Human Brain Mapping, 27*(7), 552–561. https://doi.org/10.1002/hbm.20200

Martin, T., Houck, J. M., Kicić, D., & Tesche, C. D. (2008). Interval timers and coupled oscillators both mediate the effect of temporally structured cueing. *NeuroImage, 40*(4), 1798–1806. https://doi.org/10.1016/j.neuroimage.2008.01.024

Matell, M. S., & Meck, W. H. (2004). Cortico-striatal circuits and interval timing: Coincidence detection of oscillatory processes. *Cognitive Brain Research, 21*(2), 139–170. https://doi.org/10.1016/j.cogbrainres.2004.06.012

Mathewson, K. E., Gratton, G., Fabiani, M., Beck, D. M., & Ro, T. (2009). To see or not to see: Prestimulus α phase predicts visual awareness. *The Journal of Neuroscience, 29*(9), 2725–2732. https://doi.org/10.1523/JNEUROSCI.3963-08.2009

Meindertsma, T., Kloosterman, N. A., Engel, A. K., Wagenmakers, E. J., & Donner, T. H. (2018). Surprise about sensory event timing drives cortical transients in the beta frequency band. *Journal of Neuroscience*, 0307–0318. https://doi.org/10.1523/JNEUROSCI.0307-18.2018

Mento, G. (2013). The passive CNV: Carving out the contribution of task-related processes to expectancy. *Frontiers in Human Neuroscience, 7*. https://doi.org/10.3389/fnhum.2013.00827

Mento, G., Astle, D. E., & Scerif, G. (2017). Cross-frequency phase–amplitude coupling as a mechanism for temporal orienting of attention in childhood. *Journal of Cognitive Neuroscience, Early Access, 66*, 1–9.

Merchant, H., Harrington, D. L., & Meck, W. H. (2013). Neural basis of the perception and estimation of time. *Annual Review of Neuroscience, 36*, 313–336. https://doi.org/10.1146/annurev-neuro-062012-170349

Meyer, S. S., Rossiter, H., Brookes, M. J., Woolrich, M. W., Bestmann, S., & Barnes, G. R. (2017). Using generative models to make probabilistic statements about hippocampal engagement in MEG. *NeuroImage, 149*, 468–482. https://doi.org/10.1016/j.neuroimage.2017.01.029

Michon, J. A. (1990). Implicit and explicit representations of time. *Cognitive Models of Psychological Time, 37*–58. https://doi.org/10.4324/9781315807898

Morillon, B., & Baillet, S. (2017). Motor origin of temporal predictions in auditory attention. *Proceedings of the National Academy of Sciences, 114*(42), E8913–E8921. https://doi.org/10.1073/pnas.1705373114

Morillon, B., & Schroeder, C. E. (2015). Neuronal oscillations as a mechanistic substrate of auditory temporal prediction. *Annals of the New York Academy of Sciences, 1337*(1), 26–31. https://doi.org/10.1111/nyas.12629

N'Diaye, K., Ragot, R., Garnero, L., & Pouthas, V. (2004). What is common to brain activity evoked by the perception of visual and auditory filled durations? A study with MEG and EEG co-recordings. *Cognitive Brain Research, 21*(2), 250–268. https://doi.org/10.1016/j.cogbrainres.2004.04.006

Nácher, V., Ledberg, A., Deco, G., & Romo, R. (2013). Coherent delta-band oscillations between cortical areas correlate with decision making. *Proceedings of the National Academy of Sciences, 110*(37), 15085–15090. https://doi.org/10.1073/pnas.1314681110

Nasiotis, K., Clavagnier, S., Baillet, S., & Pack, C. C. (2017). High-resolution retinotopic maps estimated with magnetoencephalography. *NeuroImage, 145*(Pt A), 107–117. https://doi.org/10.1016/j.neuroimage.2016.10.017

Nobre, A. C., & van Ede, F. (2018). Anticipated moments: Temporal structure in attention. *Nature Reviews. Neuroscience, 19*(1), 34–48. https://doi.org/10.1038/nrn.2017.141

Nozaradan, S., Peretz, I., Missal, M., & Mouraux, A. (2011). Tagging the neuronal entrainment to beat and meter. *Journal of Neuroscience, 31*(28), 10234–10240. https://doi.org/10.1523/JNEUROSCI.0411-11.2011

Obleser, J., & Kayser, C. (2019). Neural entrainment and attentional selection in the listening brain. *Trends in Cognitive Sciences.* https://doi.org/10.1016/j.tics.2019.08.004

Ofir, N., & Landau, A. N. (2022). Neural signatures of evidence accumulation in temporal decisions. *Current Biology, 32*(18), 4093–4100.e6. https://doi.org/10.1016/j.cub.2022.08.006

Oostenveld, R., Fries, P., Maris, E., & Schoffelen, J.-M. (2011). FieldTrip: Open source software for advanced analysis of MEG, EEG, and invasive electrophysiological data. *Computational Intelligence and Neuroscience, 2011*(3), 156869. https://doi.org/10.1155/2011/156869

Palva, S., & Palva, J. M. (2018). Roles of brain criticality and multiscale oscillations in temporal predictions for sensorimotor processing. *Trends in Neurosciences, 41*(10), 729–743. https://doi.org/10.1016/j.tins.2018.08.008

Park, H., Ince, R. A. A., Schyns, P. G., Thut, G., & Gross, J. (2015). Frontal top-down signals increase coupling of auditory low-frequency oscillations to continuous speech in human listeners. *Current Biology, 25*(12), 1649–1653. https://doi.org/10.1016/j.cub.2015.04.049

Paton, J. J., & Buonomano, D. V. (2018). The neural basis of timing: Distributed mechanisms for diverse functions. *Neuron, 98*(4), 687–705. https://doi.org/10.1016/j.neuron.2018.03.045

Piper, M. S. (2019). Neurodynamics of time consciousness: An extensionalist explanation of apparent motion and the specious present via reentrant oscillatory multiplexing. *Consciousness and Cognition, 73*, 102751. https://doi.org/10.1016/j.concog.2019.04.006

Pizzo, F., Roehri, N., Medina Villalon, S., Trébuchon, A., Chen, S., Lagarde, S., Carron, R., Gavaret, M., Giusiano, B., McGonigal, A., Bartolomei, F., Badier, J. M., & Bénar, C. G. (2019). Deep brain activities can be detected with magneto-encephalography. *Nature Communications, 10*(1), 1. https://doi.org/10.1038/s41467-019-08665-5

Poeppel, D. (2003). The analysis of speech in different temporal integration windows: Cerebral lateralization as 'asymmetric sampling in time'. *Speech Communication, 41*(1), 245–255. https://doi.org/10.1016/S0167-6393(02)00107-3

Pöppel, E. (1972). Oscillations as possible basis for time perception. In *The study of time* (pp. 219–241). Springer. https://doi.org/10.1007/978-3-642-65387-2_16

Praamstra, P., Kourtis, D., Kwok, H. F., & Oostenveld, R. (2006). Neurophysiology of implicit timing in serial choice reaction-time performance. *The Journal of Neuroscience*, *26*(20), 5448–5455. https://doi.org/10.1523/JNEUROSCI.0440-06.2006

Riggs, L., Moses, S. N., Bardouille, T., Herdman, A. T., Ross, B., & Ryan, J. D. (2009). A complementary analytic approach to examining medial temporal lobe sources using magnetoencephalography. *NeuroImage*, *45*(2), 627–642. https://doi.org/10.1016/j.neuroimage.2008.11.018

Rohenkohl, G., & Nobre, A. C. (2011). Alpha oscillations related to anticipatory attention follow temporal expectations. *The Journal of Neuroscience*, *31*(40), 14076–14084. https://doi.org/10.1523/JNEUROSCI.3387-11.2011

Ruzzoli, M., Torralba, M., Morís Fernández, L., & Soto-Faraco, S. (2019). The relevance of alpha phase in human perception. *Cortex; A Journal Devoted to the Study of the Nervous System and Behavior*, *120*, 249–268. https://doi.org/10.1016/j.cortex.2019.05.012

Saleh, M., Reimer, J., Penn, R., Ojakangas, C. L., & Hatsopoulos, N. G. (2010). Fast and slow oscillations in human primary motor cortex predict oncoming behaviorally relevant cues. *Neuron*, *65*(4), 461–471. https://doi.org/10.1016/j.neuron.2010.02.001

Samuelsson, J. G., Sundaram, P., Khan, S., Sereno, M. I., & Hämäläinen, M. S. (2020). Detectability of cerebellar activity with magnetoencephalography and electroencephalography. *Human Brain Mapping*, *41*(9), 2357–2372.

Schneider, D., Herbst, S. K., Klatt, L.-I., & Wöstmann, M. (2022). Target enhancement or distractor suppression? Functionally distinct alpha oscillations form the basis of attention. *European Journal of Neuroscience*, *55*(11–12), 3256–3265. https://doi.org/10.1111/ejn.15309

Schroeder, C. E., & Lakatos, P. (2009). Low-frequency neuronal oscillations as instruments of sensory selection. *Trends in Neurosciences*, *32*(1), 9–18. https://doi.org/10.1016/j.tins.2008.09.012

Schwartze, M., & Kotz, S. A. (2013). A dual-pathway neural architecture for specific temporal prediction. *Neuroscience & Biobehavioral Reviews*, *37*(10, Part 2), 2587–2596. https://doi.org/10.1016/j.neubiorev.2013.08.005

Schwartze, M., Tavano, A., Schröger, E., & Kotz, S. A. (2012). Temporal aspects of prediction in audition: Cortical and subcortical neural mechanisms. *International Journal of Psychophysiology*, *83*(2), 200–207. https://doi.org/10.1016/j.ijpsycho.2011.11.003

Sergent, C., Baillet, S., & Dehaene, S. (2005). Timing of the brain events underlying access to consciousness during the attentional blink. *Nature Neuroscience*, *8*(10), 1391–1400. https://doi.org/10.1038/nn1549

Shin, H., Law, R., Tsutsui, S., Moore, C. I., & Jones, S. R. (2017). The rate of transient beta frequency events predicts behavior across tasks and species. *eLife*, *6*, e29086. https://doi.org/10.7554/eLife.29086

Sieroka, N., Dosch, H. G., Specht, H. J., & Rupp, A. (2003). Additional neuromagnetic source activity outside the auditory cortex in duration discrimination correlates with behavioural ability. *NeuroImage*, *20*(3), 1697–1703. https://doi.org/10.1016/s1053-8119(03)00445-2

Sohoglu, E., & Chait, M. (2016). Detecting and representing predictable structure during auditory scene analysis. *eLife*, *5*, e19113. https://doi.org/10.7554/eLife.19113

Solís-Vivanco, R., Jensen, O., & Bonnefond, M. (2018). Top–down control of alpha phase adjustment in anticipation of temporally predictable visual stimuli. *Journal of Cognitive Neuroscience*, *30*(4), 594–602. https://doi.org/10.1162/jocn_a_01280

Spaak, E., de Lange, F. P., & Jensen, O. (2014). Local entrainment of alpha oscillations by visual stimuli causes cyclic modulation of perception. *The Journal of Neuroscience*, *34*(10), 3536–3544. https://doi.org/10.1523/JNEUROSCI.4385-13.2014

Spitzer, B., Gloel, M., Schmidt, T. T., & Blankenburg, F. (2014). Working memory coding of analog stimulus properties in the human prefrontal cortex. *Cerebral Cortex (New York, N.Y.: 1991)*, *24*(8), 2229–2236. https://doi.org/10.1093/cercor/bht084

Stefanics, G., Hangya, B., Hernádi, I., Winkler, I., Lakatos, P., & Ulbert, I. (2010). Phase entrainment of human delta oscillations can mediate the effects of expectation on reaction speed. *The Journal of Neuroscience*, *30*(41), 13578–13585.

Strauß, A., Wöstmann, M., & Obleser, J. (2014). Cortical alpha oscillations as a tool for auditory selective inhibition. *Frontiers in Human Neuroscience*, *8*, 350. https://doi.org/10.3389/fnhum.2014.00350

Summerfield, C., & Egner, T. (2016). Feature-based attention and feature-based expectation. *Trends in Cognitive Sciences*, *20*(6), 401–404. https://doi.org/10.1016/j.tics.2016.03.008

Tadel, F., Baillet, S., Mosher, J. C., Pantazis, D., & Leahy, R. M. (2011). Brainstorm: A user-friendly application for MEG/EEG analysis. *Computational Intelligence and Neuroscience*, *2011*, 879716. https://doi.org/10.1155/2011/879716

Tallon-Baudry, C., & Bertrand, O. (1999). Oscillatory gamma activity in humans and its role in object representation. *Trends in Cognitive Sciences*, *3*(4), 151–162. https://doi.org/10.1016/S1364-6613(99)01299-1

Tierney, T. M., Levy, A., Barry, D. N., Meyer, S. S., Shigihara, Y., Everatt, M., Mellor, S., Lopez, J. D., Bestmann, S., Holmes, N., Roberts, G., Hill, R. M., Boto, E., Leggett, J., Shah, V., Brookes, M. J., Bowtell, R., Maguire, E. A., & Barnes, G. R. (2021). Mouth magnetoencephalography: A unique perspective on the human hippocampus. *NeuroImage*, *225*, 117443. https://doi.org/10.1016/j.neuroimage.2020.117443

Todorovic, A., & Auksztulewicz, R. (2021). Dissociable neural effects of temporal expectations due to passage of time and contextual probability. *Hearing Research*, *399*, 107871. https://doi.org/10.1016/j.heares.2019.107871

Todorovic, A., Schoffelen, J.-M., van Ede, F., Maris, E., & de Lange, F. P. (2015). Temporal expectation and attention jointly modulate auditory oscillatory activity in the beta band. *PloS one*, *10*(3), e0120288. https://doi.org/10.1371/journal.pone.0120288

Treisman, M. (1984). Temporal rhythms and cerebral rhythms. *Annals of the New York Academy of Sciences*, *423*(1), 542–565.

Treisman, M., Cook, N., Naish, P. L., & MacCrone, J. K. (1994). The internal clock: Electroencephalographic evidence for oscillatory processes underlying time perception. *The Quarterly Journal of Experimental Psychology*, *47*(2), 241–289.

van Diepen, R. M., Cohen, M. X., Denys, D., & Mazaheri, A. (2015). Attention and temporal expectations modulate power, not phase, of ongoing alpha oscillations. *Journal of Cognitive Neuroscience*, *27*(8), 1573–1586. https://doi.org/10.1162/jocn_a_00803

van Ede, F., Niklaus, M., & Nobre, A. C. (2016). Temporal expectations guide dynamic prioritization in visual working memory through attenuated alpha oscillations. *Journal of Neuroscience*, 2272–2276. https://doi.org/10.1523/JNEUROSCI.2272-16.2016

van Ede, F., Rohenkohl, G., Gould, I., & Nobre, A. C. (2020). Purpose-dependent consequences of temporal expectations serving perception and action. *Journal of
Neuroscience*, *40*(41), 7877–7886. https://doi.org/10.1523/JNEUROSCI.1134-20.2020

van Wassenhove, V. (2016). Temporal cognition and neural oscillations. *Current
Opinion in Behavioral Sciences*, *8*, 124–130. https://doi.org/10.1016/j.cobeha.
2016.02.012

van Wassenhove, V., Herbst, S. K., & Kononowicz, T. W. (2019). Timing the Brain to
Time the Mind: Critical Contributions of Time-Resolved Neuroimaging for
Temporal Cognition. In S. Supek & C. J. Aine (Eds.), *Magnetoencephalography*,
855–905. Springer. https://doi.org/10.1007/978-3-030-00087-5_67

van Wassenhove, V., & Lecoutre, L. (2015). Duration estimation entails predicting
when. *NeuroImage*, *106*, 272–283. https://doi.org/10.1016/j.neuroimage.2014.
11.005

Visalli, A., Capizzi, M., Ambrosini, E., Kopp, B., & Vallesi, A. (2021). Electroencephalographic correlates of temporal Bayesian belief updating and surprise.
NeuroImage, *231*, 117867. https://doi.org/10.1016/j.neuroimage.2021.117867

Visalli, A., Capizzi, M., Ambrosini, E., Kopp, B., & Vallesi, A. (2023). P3-like signatures of temporal predictions: A computational EEG study. *Experimental Brain
Research*. https://doi.org/10.1007/s00221-023-06656-z

Visalli, A., Capizzi, M., Ambrosini, E., Mazzonetto, I., & Vallesi, A. (2019). Bayesian
modeling of temporal expectations in the human brain. *NeuroImage*, *202*, 116097.
https://doi.org/10.1016/j.neuroimage.2019.116097

Vrba, J., & Robinson, S. E. (2001). Signal processing in magnetoencephalography.
Methods (San Diego, CA.), *25*(2), 249–271. https://doi.org/10.1006/meth.2001.1238

Walter, W. G., Cooper, R., Aldridge, V. J., McCallum, W. C., & Winter, A. L. (1964).
Contingent negative variation: An electric sign of sensori-motor association and
expectancy in the human brain. *Nature*, *203*, 380–384. https://doi.org/10.1038/
203380a0

Wiener, M. (2023). Coordinate-Based Meta-Analyses of the Time Perception
Network. In H. Merchant & V. de Lafuente (Eds.), *Neurobiology of Interval
Timing* (pp. 215–226). Springer International Publishing. https://doi.org/10.
1007/978-3-031-60183-5_12

Wiener, M., & Kanai, R. (2016). Frequency tuning for temporal perception and prediction. *Current Opinion in Behavioral Sciences*, *8*, 1–6. https://doi.org/10.1016/j.
cobeha.2016.01.001

Wiener, M., Turkeltaub, P., & Coslett, H. B. (2010). The image of time: A voxel-wise
meta-analysis. *NeuroImage*, *49*(2), 1728–1740. https://doi.org/10.1016/j.
neuroimage.2009.09.064

Wilsch, A., Henry, M. J., Herrmann, B., Herrmann, C. S., & Obleser, J. (2018).
Temporal expectation modulates the cortical dynamics of short-term memory.
Journal of Neuroscience, 7428–7439. https://doi.org/10.1523/JNEUROSCI.2928-
17.2018

Wilsch, A., Mercier, M., Obleser, J., Schroeder, C. E., & Haegens, S. (2020). Spatial
attention and temporal expectation exert differential effects on visual and auditory discrimination. *Journal of Cognitive Neuroscience*, 1–15. https://doi.org/
10.1162/jocn_a_01567

Zoefel, B., & Heil, P. (2013). Detection of near-threshold sounds is independent of
EEG phase in common frequency bands. *Frontiers in Psychology*, *4*. https://doi.
org/10.3389/fpsyg.2013.00262

5

TEMPORAL PROCESSING AND NON-INVASIVE BRAIN STIMULATION TECHNIQUES

Sonia Betti
Università di Padova, Padova, Italy

Mariagrazia Capizzi
University of Granada, Spain

Giovanna Mioni
Università di Padova, Padova, Italy

Introduction

Accumulating evidence from functional magnetic resonance imaging (fMRI), and meta-analyses of fMRI studies, points to the involvement of various brain regions in temporal processing (Mondok & Wiener, 2023; Nani et al., 2019; Naghibi et al., 2023; Teghil et al., 2019; see also Chapter 2). However, the correlational nature of fMRI findings impedes drawing definitive conclusions about the causal role of a specific brain area in temporal processing. One way of probing the causal involvement of brain areas in temporal processing is to use non-invasive brain stimulation techniques, such as transcranial magnetic stimulation (TMS) and transcranial electric stimulation (tES). Studies with brain stimulation are critical to address the main open questions on the neural correlates of time processing.

In the present chapter, we offer a comprehensive summary of non-invasive brain stimulation studies of temporal processing. We start by briefly describing the most important features of TMS and tES techniques, and for each technique, we then present the timing studies using that specific one. Studies are organized according to the type of stimulation employed. For readers more interested in a review of non-invasive brain stimulation studies sorted as a function of the specific brain area involved in temporal processing, we suggest reading Mioni et al. (2020b) and Wiener (2014). The conclusion

DOI: 10.4324/9781003449546-5

section will highlight some critical aspects that should be considered in future research.

Two methods are the best known and most used to non-invasively intervene with brain activity: TMS and tES. Both techniques produce electrical stimulation of neurons in the brain; however, they rely on different mechanisms of action. While TMS depolarizes the axonal membrane of cortical neurons and triggers action potentials, low-intensity tES modulates membrane potentials and spontaneous firing rates, without directly inducing action potentials (Siebner et al., 2022; Klomjai et al. 2015; Liu et al., 2018). In general, the setting of parameters such as stimulation intensity, frequency, and pattern of stimulation is a crucial aspect to consider. Indeed, these factors affect the magnitude of the delivered electric field and the types of effects that can be obtained in neuronal populations. Understanding the main differences between TMS and tES techniques, as well as the most common stimulation protocols, would help researchers select the best tool to be used for their purposes.

TMS and tES are used in a variety of temporal studies. Among the tasks most often employed, there are interval discrimination and categorization (like the bisection and generalization methods), interval reproduction, and synchronization tasks, which involve the contribution of a motor component, and tasks involving an implicit processing of time. In the latter case, although no explicit judgment about durations is required, the implicit use of the temporal information afforded by the task context (i.e., either manipulating the foreperiod or the informative value of a temporal pre-cue) will benefit performance (Niemi & Näätänen 1981). These tasks are described in detail in Chapter 1.

Studies using transcranial magnetic stimulation

TMS uses a rapidly changing magnetic field to induce neural activity in the underlying neural tissue via the principle of electromagnetic induction. The magnetic field can penetrate the scalp without attenuation inducing a flow of electric current in neuronal axons non-invasively (Hallett, 2000). Behavioral modifications following TMS would indicate that the targeted brain area, or those connected to it, is involved in a specific task or function (Bergmann & Hartwigsen, 2021; Jahanshahi & Rothwell, 2000). The spatial resolution of the stimulated region for TMS, which depends on several physical and physiological factors including coil shape, size and orientation, coil-to-cortex distance, and tissue properties (Farzan, 2014), is in the order of ~1–3 cm^2 (Deng et al., 2013; but see Romero et al., 2019, for TMS effects at a single-cell level). As the magnetic field decreases rapidly with increasing distance, only regions on the cortical surface can be directly stimulated with TMS, whereas deep brain structures cannot be easily reached (Walsh & Cowey, 2000).

Coils of various sizes and shapes can be adopted (e.g., common circular and figure-of-eight coils; see also double-cone and H-coils used for deeper stimulation; e.g., Rossi et al., 2021), but it is important to consider that there is a trade-off between stimulation focality and depth, with larger coils permitting a deeper but less focal stimulation. As currents induced by circular coils are widely spread under the windings and activate superficial cortical layers, these coils are recommended for stimulating large and superficial areas. Conversely, the figure-of-eight coil provides a more focused stimulation; the electric field is at its maximum under its center (hot spot) where the two rings meet, allowing a more defined area to be stimulated. The electric field of double-cone coils can reach deep cortical layers (Rossi et al., 2021). To establish a standardized measure of stimulation strength across different TMS setups, the output is commonly calibrated based on a discernible and consistent physiological indicator of cortical excitation, like the evoked motor response following the stimulation of the contralateral primary motor cortex. The staircase method is used to provide a discrete index of threshold of the motor response (motor threshold [MT]). Expressing TMS intensity as a percentage of MT allows for a standardized measure of applied stimulation across various coil designs and stimulator models, establishing the basis for standard safety protocols (Rossi et al., 2009; 2021). However, it is essential to note that MT does not offer a direct measure of intrinsic cortical excitability. Some studies have highlighted that MT is significantly influenced by individual variations in the distance between the scalp and the underlying motor cortex (Stokes et al., 2007), and this coil-to-cortex distance is not homogeneous across cortical and cerebellar regions.

Stimulation protocols include "single-pulse" stimulation (spTMS), "paired-pulse" TMS (ppTMS) standard repetitive TMS (rTMS) and patterned rTMS (e.g., theta-burst stimulation [TBS]). Single- or paired-pulse paradigms are generally regarded as inducing an effect that does not last beyond the duration of stimulation. In contrast, rTMS protocols are thought to induce effects that outlast the stimulation period. In particular, long-term facilitative or suppressive effects can be elicited with rTMS applications. Low-frequency rTMS, namely ≤ 1 Hz, is considered primarily as inhibitory (Rossini et al., 2015). On the other hand, high-frequency rTMS (>1 Hz) is expected to have facilitatory effects. Higher-frequency paradigms are commonly applied using short high-frequency bursts interleaved by pauses (e.g., ranging from milliseconds to seconds). The facilitatory or inhibitory effects obtained by rTMS are indeed largely dependent not only on the frequency of stimulation but also on the structure of the train of pulses. While 1-Hz rTMS is generally applied continuously for several minutes, high-frequency rTMS (e.g., 5–25 Hz) is usually applied for a few seconds (1–2 s) followed by a long intertrain interval (20–30 s) without stimulation (Rossi et al., 2009).

A common protocol for patterned rTMS is TBS, in which short bursts of 50-Hz rTMS are repeated at a rate in the theta range (5 Hz) as continuous (cTBS; e.g., 3 pulses of 50 Hz at 5 Hz repeated at intervals of 200 ms for 40 s, for a total of 600 pulses), or intermittent (iTBS; a 2-s train of TBS repeated every 10 s) trains. While cTBS is thought to result in inhibitory after-effects, iTBS results in facilitatory after-effects (Huang et al., 2005).

TMS may be administered "online" or "offline." In the first setting, TMS (single-pulse, paired-pulse, or rTMS) is applied during task execution and participants' performance during stimulation is compared with a sham condition or with performance during the stimulation of a control area that is not expected to be involved in time processing. TMS is then administered to directly quantify its impact on brain function or activity through its direct output or behavioral effect. With "online" paradigms, TMS can induce interference in a specific time window in which the cognitive process of interest takes place, with effects that last up to a few hundred milliseconds. The choice of the timing of TMS stimulation is therefore critical, since different phases of a cognitive task may be characterized by different involvement of the target area (i.e., the "chronometry" of cognition; Pascual-Leone et al., 2000). This is particularly interesting in those tasks in which it is possible to distinguish between the encoding and production phases (time reproduction task) or in which two temporal intervals are compared (time discrimination task). "Online" protocols thus allow the investigation of the role played by a given cortical region and the stage in which this region is involved in the process under study. This relatively high temporal resolution of online TMS makes it a good tool for the study of time processing. In "offline" protocols, on the other hand, the participant's performance can be compared before and after stimulation (e.g., rTMS or TBS) by taking advantage of the long-lasting changes induced by rTMS (i.e., long-term potentiation – or long-term depression-like plasticity), thus resulting in a modulation of cortical excitability (facilitation/inhibition) that may last several minutes to about 1 h after stimulation. It is important to consider that the effects of TMS (but also of tES) delivered on a given brain area can be interpreted in terms of the interaction between the stimulation itself and the ongoing brain activity, which determines how neurons are likely to be activated. Considering that stimulation will have a greater effect on neurons that are not yet depolarized, the activation state of the area at the time of stimulation (i.e., "state dependency"; Silvanto & Pascual-Leone, 2008) must be considered when interpreting the effect obtained from stimulation (Hartwigsen & Silvanto, 2023).

Low-frequency rTMS (1 Hz) – offline procedure

Low-frequency 1-Hz rTMS is the most common suppressive protocol, used offline in most studies (Table 5.1). It is often employed as a single

TABLE 5.1 Studies using low-frequency rTMS (1 Hz)

Author	Exp.	N (gender)	Mean age (SD or range)	Time of stimulation (min)	Intensity	Coil	AREA	Side	Control	Temporal tasks	Temporal intervals (ms)	Modality	Dependent variable	Main findings
Offline procedure														
Koch et al., 2003	1	8 (F = 4)	Range 19–33	10	90% of RMT	70-mm figure-of-eight coil	DLPFC	Bilateral	Pre-stimulation baseline	Time reproduction	5,000 or 15,000	Visual	Reproduced time, SD	Under-estimation after right side stimulation
Koch et al., 2007	1	9 (–)	Range 22–35	10	90% of RMT	70-mm figure-of-eight coil	DLPFC	Right	Pre-stimulation baseline	Time reproduction	Short = 500 (average); long = 2,000 (average)	Visual	Reproduced time, CV	Overestimation long-range intervals; no effect on CV
Correa et al., 2014	1	12 (F = 4)	26.5 (range 22–42)	10	100% of RMT	70-mm figure-of-eight coil	DLPFC	Bilateral	sham	Temporal orienting with symbolic and rhythmic cues[a]	350–1,350	Visual	Mean RT	Reduced RTs for valid symbolic cues as compared to invalid cues after rTMS over left and right DLPFC vs. sham. No effect for the rhythmic cues
Del Olmo et al., 2007	4	14 (F = 4)	28.2 (4.6)	10	90% of RMT	90-mm figure-of-eight coil	SMA	–	–	Paced finger tapping (synchronization-continuation task)	500 (2 Hz)	Auditory	Accuracy, CV	No effect
Giovannelli et al., 2014	1	10 (F = 7)	29.1 (–)	15	Individual resting motor threshold according to recommendations	90-mm figure-of-eight coil	SMA	Bilateral	Oz	Time discrimination	Standard = 1,000; comparison = 800, 1,000, 1,200	Auditory	Percentage of correct responses, RT	No effect on accuracy; general faster RT after stimulation
Théoret et al., 2001	1	7 (F = 1)	Range 28–38	5	90% of RMT	70-mm figure-of-eight coil	M1	Right	–	Paced finger tapping (synchronization-continuation task)	Constant interval (475) and reproduce the tapping	Visual	Mean press, SD	No effect

Study		N	Age		Intensity	Coil	Site	Hemisphere	Control site	Task	Speed/ISI (ms)	Modality	Measure	Result
Jäncke et al., 2004	1	9 (–)	Range 22–34	10	90% of RMT	70-mm figure-of-eight coil	M1	Bilateral	Right brachial plexus	Paced finger tapping task, tapping with maximum speed and tapping with convenient speed	Different speed based on condition	Visual	Mean ITIs, CV	Slowed finger tapping speed after left M1 stimulation
Doumas et al., 2005	1	9 ($F = 2$)	30 (6.5)	25	90% of RMT	70-mm figure-of-eight coil	M1	Left	–	Paced finger tapping (synchronization)	500 (2 Hz)	Auditory	Negative asynchrony and error correction	Decreased asynchrony
Del Olmo et al., 2007	1	14 ($F = 4$)	28.2 (4.6)	10	90% of RMT	90-mm figure-of-eight coil	M1	Right	–	Paced finger tapping (synchronization-continuation task)	2,000 (0.5 Hz), 1,000 (1 Hz), or 500 (2 Hz)	Auditory	Accuracy, CV	No effect
Doumas et al., 2005	1	9 ($F = 2$)	30 (6.5)	25	90% of RMT	70-mm figure-of-eight coil	dPMC	Left	–	Paced finger tapping (synchronization)	500 (2 Hz)	Auditory	Negative asynchrony, error correction	No effect
Del Olmo et al., 2007	4	14 ($F = 4$)	28.2 (4.6)	10	90% of RMT	90-mm figure-of-eight coil	dPMC	Left	–	Paced finger tapping (synchronization-continuation task)	500 (2 Hz)	Auditory	Accuracy, CV	Increase variability in the synchronization phase
Pollok et al., 2008	1	12 ($F = 6$)	29.2 (2.1)	20	90% of RMT	70-mm figure-of-eight coil	dPMC	Bilateral	–	Paced finger-tapping task	800	Auditorily	Asynchrony, variability	Decreased tapping accuracy after left stimulation
Malcolm et al., 2008	1	15 ($F = 11$)	24.6 (3)	15	90% of RMT	70-mm figure-of-eight coil	vPMC	Left	Left STP	Paced finger-tapping task	500 (2 Hz)	Auditorily	Phase synchronization error, absolute period error	Increased synchronization error after STP stimulation
Dormal et al., 2008	1	15 ($F = 3$)	22 (2.7)	15	65% maximum stimulator output	70-mm figure-of-eight coil	IPS	Bilateral	Vertex	Time discrimination	300–900	Visual	RT	No effect
Capizzi et al., 2023a	2	21 ($F = 13$)	25.05 (range 19–36)	20	100% of RMT	70-mm figure-of-eight coil	IPS	Left	Vertex	Temporal orienting	300–1,300	Visual	Mean RT	No effect
Alexander et al., 2005	1	6 ($F = 2$)	30.5 (–)	10	65% of maximum stimulator output	70-mm figure-of-eight coil	IPC (angular gyrus)	Bilateral	Pre-stimulation baseline and vertex	Time discrimination	Standard = 1,200; comparison ± 60, 120, 180, 240	Auditory	RT	Higher reaction time after right stimulation

(*Continued*)

TABLE 5.1 (Continued)

Author	Exp.	N (gender)	Mean age (SD or range)	Time of stimulation (min)	Intensity	Coil	AREA	Side	Control	Temporal tasks	Temporal intervals (ms)	Modality	Dependent variable	Main findings
Oliveri et al., 2009	1	10 (F = 5)	Range 20–30	10	90% of RMT	70-mm figure-of-eight coil	IPC (angular gyrus)	Bilateral	Pre-stimulation baseline	Time reproduction (half reproduction)[b]	Standard 1,600–2,400	Visual	Reproduced time, absolute error	Under-estimation right compared to left side
Oliveri et al., 2009	1	10 (F = 5)	Range 20–30	10	90% of RMT	70-mm figure-of-eight coil	IPC (angular gyrus)	Bilateral	Pre-stimulation baseline	Time reproduction	Standard 1,600–2,400	Visual	Reproduced time, absolute error	No effect
Krause et al., 2012	1	13 (F = 4)	24.08 (0.87)	10	Between 57.88% and 71.92% of stimulator output	59-mm figure-of-eight coil	PPC	Left	Visual cortex	Finger tapping	800	Auditory, visual	Asynchrony, variability	Reduced negative tap-to-pacer asynchronies
Krause et al., 2012	1	13 (F = 4)	24.08 (0.87)	10	Between 57.88% and 71.92% of stimulator output	59-mm figure-of-eight coil	PPC	Right	Visual cortex	Finger tapping	800	Auditory, visual	Asynchrony, variability	No effect
Giovannelli et al., 2014	1	10 (F = 7)	29.1 (–)	15	Individual resting motor threshold according to recommendations	90-mm figure-of-eight coil	PPC	Right	Oz	Time discrimination	Standard = 1,000; comparison = 800, 1,000, 1,200	Auditory	Percentage of correct responses, RT	No effect on accuracy; faster RT after stimulation
Rocha et al., 2019	1	23 (F = 11)	24.2 (–)	15	80% of RMT	70-mm figure-of-eight coil	SPC	Medial	Sham	Time reproduction	1,000, 4,000, 7,000, and 9,000	Visual	Absolute, relative errors	Higher accuracy at 4 and 9 s after stimulation
Giovannelli et al., 2014	1	10 (F = 7)	29.1 (–)	15	Individual resting motor threshold according to recommendations	90-mm figure-of-eight coil	STG	Right	Oz	Time discrimination	Standard = 1,000; comparison = 800, 1,000, 1,200	Auditory	Percentage of correct responses, RT	No effect on accuracy; general faster RT after stimulation
Théoret et al., 2001	1	7 (F = 1)	Range 28–38	5	90% of RMT	70-mm figure-of-eight coil	Cerebellum	Lateral, medial	–	Paced finger tapping (synchronization-continuation task)	Constant interval (475) and reproduce the tapping	Visual	Mean press, SD	Increased Variability only after medial cerebellum stimulation
Jäncke et al., 2004	1	9 (–)	Range 22–34	10	90% of RMT	70-mm figure-of-eight coil	Cerebellum	Bilateral	Right brachial plexus	Paced finger tapping task, tapping with maximum speed and tapping with convenient speed	Different speed based on condition	Visual	Mean ITIs, CV	No effect

Study	Exp	N (F)	Age	n	Coil	Intensity	Target	Laterality	Control	Task	Intervals	Modality	Measure	Results
Fierro et al., 2007	1	10 (F = 4)	23 (4.7)	15	90-mm figure-of-eight coil	90% of RMT	Cerebellum	Bilateral	Pre-stimulation baseline	Time discrimination	Standard = 400; comparison 300–500	Tactile	Percentage of errors	Increased errors; right stimulation
Lee et al., 2007	1	11 (F = 0)	30 (5.1)	8	70-mm figure-of-eight coil	90% of RMT	Cerebellum	Right–left medial	Sham	Time bisection	400–800	Auditory	BP, DL	Overestimation after right and medial stimulation compared to sham
Lee et al., 2007	2	11 (F = 0)	32 (5.2)	8	70-mm figure-of-eight coil	90% of RMT	Cerebellum	Right–left medial	Sham	Time bisection	1,000–2,000	Auditory	BP, DL	No effect compared to sham
Lee et al., 2007	3	18 (F = 0)	27 (7.8)	8	70-mm figure-of-eight coil	90% of RMT	Cerebellum	Medial	Sham	Time bisection	Short = 400–800; long = 1,000–2,000	Auditory	BP, DL	Overestimation after medial stimulation compared to sham for short intervals
Koch et al., 2007	1	9 (–)	22–35	10	70-mm figure-of-eight coil	90% of RMT	Cerebellum	Bilateral	Pre-stimulation baseline	Time reproduction	Short = 500 (average); long = 2,000 (average)	Visual	Reproduced time, CV	Overestimation short range under left site stimulation; no effect on CV
Del Olmo et al., 2007	1	14 (F = 4)	28.2 (4.6)	10	90-mm figure-of-eight coil	90% of RMT	Cerebellum	Bilateral	M1	Paced finger tapping (synchronization-continuation task)	2,000 (0.5 Hz), 1,000 (1 Hz), or 500 (2 Hz)	Auditory	Accuracy, CV	Increased variability for 2-Hz auditory synchronization only for right cerebellar stimulation
Del Olmo et al., 2007	3	14 (F = 4)	28.2 (4.6)	10	90-mm figure-of-eight coil	90% of RMT	Cerebellum	Right	–	Paced finger tapping (synchronization-continuation task)	500 (2 Hz)	Auditory, visual	Accuracy, CV	Increase in variability after rTMS only in the auditory task

(Continued)

TABLE 5.1 (Continued)

Author	Exp.	N (gender)	Mean age (SD or range)	Time of stimulation (min)	Intensity	Coil	AREA	Side	Control	Temporal tasks	Temporal intervals (ms)	Modality	Dependent variable	Main findings
Online procedure														
Verstynen et al., 2006	1	11 (–)	–	During 32 tapping. IPI random between 1 and 2.5 s	38%–65% maximum stimulator output	70-mm figure-of-eight coil	M1	Left	Medial occipital lobe	Paced finger tapping (synchronization-continuation task)	Fast pace (350) and slow pace (550)	Auditory	Variability	Increased variability for longer interval
Verstynen et al., 2006	2	10 (–)	–	During 32 tapping. IPI random between 1.5 and 3 s	Low: range: 29%–40%; medium: range 35%–48%; and high range: 44%–59% of maximum stimulator output	70-mm figure-of-eight coil	M1	Left	10 cm posterior the vertex	Paced finger tapping (synchronization-continuation task)	Pace 450	Auditory	Variability	Increased variability
Levit-Binnun et al., 2007	1	9 (F = 4)	29.3 (8.7)	16 single TMS pulses (each of 200 ms duration) time locked to the metronome click	Low as 50% rMT to as much as 160% rMT	70-mm figure-of-eight coil	M1	–	–	Paced finger tapping (synchronization-continuation task)	2.5 Hz	Auditory	Accuracy	No effect

a Correa et al. (2014) was the only study using an implicit timing task.

b Oliveri et al. (2009) called this task "time bisection"; we renamed it to avoid confusion with the classical time bisection task.

Note: RMT = resting motor threshold; IPI = inter pulse interval; DLPFC = dorsolateral prefrontal cortex; M1 = primary motor area; SMA = supplementary motor area; Left STP = left superior posterior temporal parietal junction; dPMC = dorsal premotor cortex; vPMC = ventral premotor cortex; IPS = intraparietal sulcus; IPC = inferior parietal cortex; PPC = posterior parietal cortex; SPC = superior parietal cortex; STG = superior temporal gyrus; BP = bisection point; CV = coefficient of variation; DL = different limen; RT = reaction time.

stimulation train of one pulse per second lasting 10–20 min, resulting in a suppression of the targeted cortical region. As mentioned, the collective impact of rTMS results in a transient adjustment of cortical excitability in the specific cortical area and its interconnected networks. This adjustment influences task performance after rTMS when compared to the performance level before rTMS or with a control side (Beynel et al., 2019).

Three studies adopted 1-Hz rTMS and targeted frontal areas to investigate their involvement in time processes. Koch et al. (2003) targeted the dorsolateral prefrontal cortex (DLPFC) and observed an under-reproduction of supra-second intervals only following a right, but not a left, DLPFC stimulation. However, Koch et al. (2007; Experiment 1) observed temporal over-reproduction of supra-second temporal intervals when the stimulation was applied over the right DLPFC. Correa et al. (2014) reported a significant effect of offline TMS over right and left DLPFCs (vs. sham) in an implicit temporal orienting task only when symbolic, but not rhythmic, temporal cues were employed.

Giovannelli et al. (2014) and Del Olmo et al. (2007; Experiment 4) targeted the supplementary motor area (SMA) but did not find any effect of stimulation on perceived duration. Giovannelli et al. (2014) found slightly faster reaction times following rTMS but irrespectively of the area of stimulation (the authors also tested parietal and temporal cortices as reported below).

Some studies targeted the premotor cortex (PMC; either dorsal [dPMC] or ventral [vPMC]) using finger-tapping tasks. Doumas et al. (2005) found no alteration of finger-tapping frequency before and after rTMS stimulation. Del Olmo et al. (2007; Experiment 4) observed altered tapping variability when stimulating the ipsilateral cerebellum and contralateral dPMC, but only with sub-second pacing. The authors interpreted this finding as indicating the connectedness of cerebellar and premotor cortices in subsecond motor timing (Del Olmo et al., 2007; Experiment 4). Pollok et al. (2008) tested the different roles of left and right dPMCs and concluded that only the stimulation on the left hemisphere disturbs synchronization accuracy. Malcolm et al. (2008) applied rTMS over left vPMC and left superior temporal–parietal junction (STP) during a tapping task with an audiomotor synchronization paradigm and reported that synchronization errors were marginally increased after STP stimulation.

Other studies have investigated the role of the primary motor cortex (M1) in motor timing paradigms using low-frequency rTMS. The effects have been mixed, as two studies of paced finger tapping (Del Olmo et al., 2007, Experiment 1; Théoret et al., 2001) found no effect of M1 stimulation. However, Doumas et al. (2005) found that offline 1-Hz stimulation decreased tapping asynchrony during auditory synchronization and Jäncke et al. (2004)

showed slowed finger tapping speed after left M1 stimulation. Tapping asynchrony typically demonstrates that participants tap slightly ahead of the occurrence of each beat, indicating a possible anticipation of the beat; a decrease in asynchrony, while making the responses more veridical concerning the stimulus, suggests the elimination of the predictiveness of each response.

Seven studies targeted the parietal cortex, three of them used time discrimination tasks (Alexander et al., 2005; Dormal et al., 2008; Giovannelli et al., 2014), two used time reproduction tasks (Oliveri et al., 2009; Rocha et al., 2019), one a finger tapping task (Krause et al., 2012), and one an implicit temporal orienting task (Capizzi et al., 2023a; Experiment 2). Alexander et al. (2005) and Giovannelli et al. (2014) used auditory stimuli and showed no effect of stimulation on perceived time but increased reaction times when the stimulation was applied over the right cortex; similarly, Dormal et al. (2008) used visual stimuli and showed no effect of stimulation. Oliveri et al. (2009; Experiment 1) showed that the inhibition of the right inferior parietal cortex (IPC; angular gyrus) induced a directional bias with participants under-producing half of a time interval (time reproduction – half reproduction), but no effect was observed when the classical time reproduction task was used. Rocha et al. (2019) showed higher accuracy after rTMS over the superior parietal cortex (SPC), but only at 4 and 9 s. Krause et al. (2012) showed reduced asynchrony in a finger-tapping task after left posterior parietal cortex (PPC) stimulation. As concerns implicit timing, Capizzi et al. (2023a; Experiment 2) found no effect of offline TMS over the left intraparietal sulcus (IPS) in a temporal orienting task.

Only Giovannelli et al. (2014) targeted the temporal cortex, specifically the superior temporal gyrus (STG) using a time discrimination task with auditory stimuli, but failed to find an effect of stimulation on perceived time, showing however increased reaction times after right side stimulation (see also the above-mentioned results from the same study targeting the parietal cortex).

Finally, six studies targeted the cerebellum. Lee et al. (2007) used the time bisection task with auditory stimuli and showed temporal overestimation for short intervals (Experiments 1 and 3), but no effect of stimulation on long intervals (Experiment 2). Fierro et al. (2007) used a time discrimination task and tactile stimuli between 330 and 500 ms and showed an increased number of errors after right cerebellar stimulation. Finally, Koch et al. (2007) selectively disrupted participants' performance for the sub-second but not supra-second time intervals when rTMS was applied over the left side (Experiment 1). Overall, these results suggest that explicit time perception of sub- and supra-second intervals likely depend upon distinct neural systems (Lewis & Miall, 2003) and that the cerebellum is involved in the temporal processing of short temporal intervals. Other studies have used visual stimuli and motor timing tasks (i.e., finger tapping tasks). Théoret et al. (2001) showed increased variability after medial cerebellum stimulation but Jäncke et al. (2004) failed to replicate these results, showing no effect of rTMS stimulation on tapping. Del Olmo et al. (2007; Experiments 1 and 3) showed increased variability in the timing of finger-tapping movements using auditory stimuli. In addition, the authors showed that the effects were limited to the hand ipsilateral to the side of rTMS and were more prominent in high-frequency (2 Hz) tapping. Similar effects after rTMS over the dPMC contralateral to the tapping hand are consistent with the idea that this behavior depends on the cerebellum-premotor circuit.

Low-frequency rTMS (1 Hz) – online procedure

During online rTMS, the stimulation occurs at specific moments while participants are actively performing the timing task. Instead of observing cumulative impacts, this method focuses on evaluating the immediate influence on behavior. This approach offers a more detailed understanding of the timing related neural processing that drives behavior (Beynel et al., 2019).

Only two studies used a low frequency rTMS adopting an online procedure. Verstynen et al. (2006), applied TMS during the unpaced phase with the inter-pulse intervals selected at random from a uniform distribution ranging between 1 and 2.5 s (Experiment 1) and 1.5 and 3 s (Experiment 2). Their results showed that M1 stimulation at random points during the continuation phase of an auditory tapping task led to an increase in inter-tap-interval variability. Moreover, in Experiment 2, the authors also manipulated stimulation intensity and observed a generalized increase in inter-tapping variability and a delay in the subsequent response when the pulse fell within a restricted window before movement onset. However, a later study by Levit-Binnun et al. (2007) failed to replicate the results and showed no effect of stimulation over M1.

Single- and paired-pulse TMS

Single-pulse and paired-pulse stimulations are generally performed online, exploiting the high temporal resolution of these techniques. While spTMS consists of the delivery of single pulses to the target area, in ppTMS two sequential stimuli of varying intensity, separated by preselected interstimulus intervals (ISIs), are applied to one or two cortical sites.

Here, one study used spTMS (Bolognini et al., 2010) and one ppTMS (Salvioni et al., 2013) to interfere with an ongoing temporal processing task (Table 5.2). In particular, Bolognini and colleagues used a time discrimination task with tactile stimuli; the tactile vibration (long or short) was

TABLE 5.2 Studies using single- or paired-pulse TMS

Author	Exp.	N (gender)	Mean age (SD or range)	Time of stimulation	Intensity	Coil	Procedure	AREA	Side	Control	Temporal tasks	Temporal intervals (ms)	Modality	Dependent variable	Main findings
Bolognini et al., 2010	1	13 ($F = 6$)	35	1 single pulse	65% ± 7% of RMT	70-mm figure-of-eight coil	Online (60, 120, or 180 ms after stimulus presentation)	SI	Left	–	Time discrimination	Standard 25	Tactile	Percentage of errors, sensitivity (d') and response criterion	Increased number of errors for ISI = 60 ms compared to other conditions
Bolognini et al., 2010	1	13 ($F = 6$)	35	1 single pulse	65% ± 7% of RMT	70-mm figure-of-eight coil	Online (60, 120, or 180 ms after stimulus presentation)	STG	Left	–	Time discrimination	Standard 25	Tactile	Percentage of errors, sensitivity (d') and response criterion	Increased number of errors for ISI = 180 ms compared to other conditions
Salvioni et al., 2013	1	14 ($F = 9$)	Range 22–28	Paired-pulse (IPI = 35 ms)	55% of the maximum stimulator output	70-mm figure-of-eight coil	Online (stimulation onset standard interval)	V5/MT	Right	Vertex, noTMS	Time discrimination	Standard = 200	Visual	Discrimination threshold	Higher threshold compared at 85- and 120-ms delay
Salvioni et al., 2013	1	14 ($F = 9$)	Range 22–28	Paired-pulse (IPI=35 ms)	55% of the maximum stimulator output	70-mm figure-of-eight coil	Online (stimulation onset standard interval)	V1	Right	Vertex, noTMS	Time discrimination	Standard = 200	Visual	Discrimination threshold	Higher threshold compared at 85 and 120 ms delay
Salvioni et al., 2013	2	14 ($F = 10$)	Range 22–38	Paired-pulse (IPI = 35 ms)	55% of the maximum stimulator output	70-mm figure-of-eight coil	Online (stimulation offset standard interval)	V5/MT	Right	Vertex, noTMS	Time discrimination	Standard = 200	Visual	Discrimination threshold	Higher threshold at 85-ms delay
Salvioni et al., 2013	2	14 ($F = 10$)	Range 22–38	Paired-pulse (IPI = 35 ms)	55% of the maximum stimulator output	70-mm figure-of-eight coil	Online (stimulation offset standard interval)	V1	Right	Vertex, noTMS	Time discrimination	Standard = 200	Visual	Discrimination threshold	Higher threshold at 50-ms delay

Note: **RMT** = resting motor threshold; IPI = inter-pulse interval; SI = primary sensorimotor cortex; V1 = primary visual cortex; STG = superior temporal gyrus; V5/MT = extrastriate visual area; ISI = inter stimulus interval.

followed by a TMS pulse after an ISI of 60, 120, or 180 ms selected randomly. When stimulation was administered to the primary somatosensory cortex (SI) 60 ms after tactile presentation, it disrupted tactile temporal processing, confirming the region's specificity to this sensory modality. Notably, applying TMS over the STG also influenced tactile temporal processing, but with a delay of 180 ms. In both scenarios, the impairment was observed only for touches on the opposite side of stimulation and resulted from decreased perceptual sensitivity. Salvioni et al. (2013) also used a time discrimination task with visual stimuli and applied paired-pulse TMS over the primary visual cortex (V1) and right extrastriate visual area (V5/MT) at three different delays (50–85, 85–120, and 120–155 ms) from the offset of the first flash (i.e., beginning of the first interval) in Experiment 1 and from the offset of the second flash (i.e., end of the first interval and beginning of the retention period) in Experiment 2. Results showed that both V1 and V5/MT are causally involved in encoding and keeping time in memory and that this involvement is independent from low-level visual processing.

High-frequency rTMS (5–20 Hz)

High-frequency rTMS in "online" paradigms can be used to transiently disrupt the functioning of a target cortical region and influence its ongoing cognitive processing (i.e., virtual lesion approach; Beynel et al., 2019). All studies reported here used an online procedure (Table 5.3). However, rTMS using repetition rates of 5 Hz and beyond can be also adopted "offline" to induce long-lasting modulatory effects, considered primarily as facilitatory, thus producing an increase in cortical excitability.

Gironell et al. (2005) used "online" rTMS at 5-Hz stimulation (i.e., trains of 10 s with an inter-stimulus interval of 30 s) targeting the cerebellum and the DLPFC. This study was the only one using a time production task with very long temporal intervals; indeed, participants were asked to produce a 3-min interval by internal counting. No effect of stimulation was observed. The absence of stimulation effects was surprising, particularly when considering the stimulation of DLPFC. Indeed, frontal areas seem to be mainly involved when long temporal intervals are processed (Lewis & Miall, 2003; Mioni et al., 2014) and/or when other cognitive functions are required to perform temporal tasks (Baudouin et al., 2006, 2018; Mioni et al., 2020a). It is important to note that asking participants to internally count may have helped them perform the task and covered a possible effect of stimulation (Grondin et al., 2004; Rattat & Droit-Volet, 2012). Later, Ruspantini et al. (2011) used a finger-tapping task and interfered with the dPMC and vPMC during the task (three pulses at 5 Hz), and results showed reduced synchronization error when TMS was delivered to the vPMC.

TABLE 5.3 Studies using high-frequency rTMS (5–20 Hz)

Author	Exp.	N (gender)	Mean age (SD or range)	TMS freq. (Hz)	Time of stimulation	Intensity	Coil	Procedure	AREA	Control	Side	Temporal tasks	Temporal intervals (ms)	Modality	Dependent variable	Main findings
Gironell et al., 2005	1	16 (–)	26.63 (4.57)	5	Trains of 10 s duration with ISI of 30 s	10% below the motor threshold	70-mm figure-of-eight coil	online	DLPFC	Sham	Bilateral	Time production	180,000	–	Absolute, relative errors	No effect
Ruspantini et al., 2011	1	12 (F = 2)	Range 23–32	5	Three pulses given 100, 300, and 500 ms	90% of RMT	50-mm figure-of-eight coil	Online (three pulses after the third pacing stimulus onset)	vPMC, dPMC	Central sulcus	Left	Paced finger tapping (synchronization-continuation task)	800	Visual	Synchronization error, inter-tap interval, and contact time	Reduced synchronization error after vPMC stimulation
Gironell et al., 2005	1	16 (–)	26.63 (4.57)	5	Trains of 10 s duration with ISI of 30 s	10% below the motor threshold	70-mm figure-of-eight coil	Online	Cerebellum	Sham	Right	Time production	180,000	–	Absolute, relative errors	No effect
Wiener et al., 2010	1	9 (F = 4)	Range 23–36	10	Three pulses (~50-μs pulse, ISI 100 ms)	100% of RMT	70-mm figure-of-eight coil	Online (stimulation onset standard interval)	SMG	Vertex, noTMS	Bilateral	Time discrimination	Standard = 600	Visual	BP, CV	Decreased proportion of long responses after right stimulation
Wiener et al., 2010	2	17 (F = 9)	Range 23–35	10	Three pulses (~50-μs pulse, ISI 100 ms)	100% of RMT	70-mm figure-of-eight coil	Online (stimulation onset comparison interval)	SMG	Vertex, noTMS	Bilateral	Time discrimination	Standard = 600	Visual	BP, CV	Increased proportion of long responses after right stimulation
Wiener et al., 2012	1	19 (F = 9)	25 (range 21–35)	10	Three pulses (~50-μs pulse, ISI 100 ms)	Collected for each subject	70-mm figure-of-eight coil	Online (stimulation onset fixation cross)	SMG	Mid-occipital parietal junction	Right	Time discrimination	Standard = 600	Visual	BP, DL, and CE	Lengthened time
Bueti et al., 2008a	1	8 (F = 1)	28.2 (–)	10	Train of five pulses lasting 500 ms	65% of maximal stimulator output	50-mm figure-of-eight coil	Online (stimulation onset standard interval)	STG	Vertex	Bilateral	Time discrimination	Standard = 600; comparison = ±10, ±20, ±40	Auditory	BP, JND	Lower accuracy and increased variability after right stimulation
Bueti et al., 2008b	1	10 (–)	29.2 (–)	12	500 ms	60% of the stimulator output	70-mm figure-of-eight coil	Online (stimulation onset comparison interval)	IPC (angular gyrus)	Vertex, noTMS	Bilateral	Time discrimination	Standard = 600; comparison = 435–765	Visual moving	BP, WR	Higher variability (WR) after right IPC stimulation

Study																
Bueti et al., 2008b	2	10 (–)	28.2 (–)	12	500 ms	60% of the stimulator output	70-mm figure-of-eight coil	Online (stimulation onset comparison interval)	IPC (angular gyrus)	Bilateral	Vertex, noTMS	Time discrimination	Standard = 600; comparison = 435–765	Visual static	BP, WR	Higher variability (WR) after right IPC stimulation
Bueti et al., 2008b	4	9 (–)	29.6 (–)	12	500 ms	60% of the stimulator output	70-mm figure-of-eight coil	Online (stimulation onset comparison interval)	IPC (angular gyrus)	Bilateral	Vertex, noTMS	Time discrimination	Standard = 600; comparison = 435–765	Auditory BP, WR	Higher variability (WR) after right IPC stimulation
Bueti et al., 2008b	5	5 (–)	28 (–)	12	500 ms	60% of the stimulator output	70-mm figure-of-eight coil	Online (stimulation onset comparison interval)	IPC (angular gyrus)	Right	Vertex, noTMS	Time discrimination	Standard = 600; comparison = 520–680	Auditory BP, WR	Higher variability (WR) after right IPC stimulation
Bueti et al., 2008b	1	10 (–)	29.2 (–)	12	500 ms	60% of the stimulator output	70-mm figure-of-eight coil	Online (stimulation onset comparison interval)	V5/MT	Left	Vertex, noTMS	Time discrimination	Standard = 600; comparison = 435–765	Visual moving BP, WR	Higher WR no effect on BP
Bueti et al., 2008b	2	10 (–)	28.2 (–)	12	500 ms	60% of the stimulator output	70-mm figure-of-eight coil	Online (stimulation onset comparison interval)	V5/MT	Left	Vertex, noTMS	Time discrimination	Standard = 600; comparison = 435–765	Visual static BP, WR	Higher WR no effect on BP
Bueti et al., 2008b	4	9 (–)	29.6 (–)	12	500 ms	60% of the stimulator output	70-mm figure-of-eight coil	Online (stimulation onset comparison interval)	V5/MT	Left	Vertex, noTMS	Time discrimination	Standard = 600; comparison = 435–765	Auditory BP, WR	No effect
Bueti et al., 2008b	5	5 (–)	28 (–)	12	500 ms	60% of the stimulator output	70-mm figure-of-eight coil	Online (stimulation onset comparison interval)	V5/MT	Left	Vertex	Time discrimination	Standard = 600; comparison = 520–680	Auditory BP, WR	No effect

(Continued)

TABLE 5.3 (Continued)

Author	Exp.	N (gender)	Mean age (SD or range)	TMS freq. (Hz)	Time of stimulation	Intensity	Coil	Procedure	AREA	Side	Control	Temporal tasks	Temporal intervals (ms)	Modality	Dependent variable	Main findings
Jones et al., 2004	1	9 (F = 3)	30.6 (6.19)	20	Four pulses	90% of the active leg motor threshold	90-mm figure-of-eight coil	Online (stimulation during encoding or reproduction)	DLPFC	Right	Leg motor area and sham	Time reproduction	Short = 500 (average); long = 2,000 (average)	Visual	Reproduced time	Under-estimation with long intervals (reproduction phase)
Jones et al., 2004	1	9 (F = 3)	30.6 (6.19)	20	Four pulses	90% of the active leg motor threshold	90-mm figure-of-eight coil	Online (stimulation during encoding or reproduction)	SMA	–	Leg motor area and sham	Time reproduction	Short = 500 (average); long = 2,000 (average)	Visual	Reproduced time	No effect
Oliveri et al., 2009	3	7 (F = 3)	Range 20–33	20	Four pulses (200-ms duration)	TMS was given at threshold intensity	70-mm figure-of-eight coil	Online (stimulation onset encoding or reproduction)	IPC (angular gyrus)	Right	Sham	Time reproduction (half reproduction)	Standard 1,600–2,400	Visual	Reproduced time, absolute error	Overestimation compared to sham during reproduction phase
Oliveri et al., 2009	3	7 (F = 3)	Range 20–33	20	Four pulses (200-ms duration)	TMS was given at threshold intensity	70-mm figure-of-eight coil	Online (stimulation onset encoding or reproduction)	IPC (angular gyrus)	Right	Sham	Time reproduction	Standard 1,600–2,400	Visual	Reproduced time, absolute error	No effect
Capizzi et al., 2023a	1	21 (–)	22.84 (Range 19–29)	20	Three pulses (150 ms)	120% of RMT	70-mm figure-of-eight coil	Online (stimulation onset after temporal cue presentation)	IPS	Left	Right IPS, vertex	Temporal orienting	300–1,300	Visual	Mean RT	No effect
Koch et al., 2007	2	8 (–)	Range 22–35	20	Four pulses (stimulation time: 150 ms)	90% of RMT	70-mm figure-of-eight coil	Online (stimulation onset encoding or reproduction)	Cerebellum	Bilateral	Vertex	Time reproduction	Short = 500 (average); long = 2,000 (average)	Visual	Reproduced time, CV	Overestimation short temporal intervals both sides at encoding phase; no effect on CV

Note: RMT = resting motor threshold; DLPFC = dorsolateral prefrontal cortex; dPMC = dorsal premotor cortex; vPMC = ventral premotor cortex; STG = superior temporal gyrus; IPC = inferior parietal cortex; IPS = intraparietal sulcus; V5/MT = extrastriate visual area; BP = bisection point; WR = Weber ratio; CV = coefficient of variation; JND = just noticeable difference; CE = constant error; RT = reaction time.

Three other studies used 10-Hz rTMS stimulation in combination with time discrimination tasks using auditory and/or visual stimuli. Bueti et al. (2008a) showed lower accuracy after right STG stimulation using "online" trains of five pulses lasting 500 ms. Wiener et al. (2010, 2012) stimulated the supramarginal gyrus (SMG) of the inferior parietal lobe and demonstrated that right, but not left, SMG stimulation induced a significant variation in perceived duration. Specifically, the rTMS effect was in the opposite direction in the two experiments presented by Wiener et al. (2010). Whereas rTMS during the standard stimulus in Experiment 1 led to a decrease in the proportion of trials in which participants indicated that the standard (first) stimulus was longer, rTMS during the comparison stimulus led to an increase in the proportion of trials in which participants indicated that the comparison (second) stimulus was longer (Experiment 2). Furthermore, Wiener et al. (2012) replicated this lengthening effect using rTMS over the right SMG. Stimulation of the right SMG seems to selectively increase perceived duration, as this increase occurred during stimulation of either the standard (first) or comparison (second) stimulus. Any disruption of the mechanisms involved in memory or decisional processing would probably not help explain the results.

In line with the idea that the parietal cortex is a multimodal region, Bueti et al. (2008b) stimulated (12 Hz; online procedure) the left and right inferior parietal cortex (IPC, angular gyrus AG) with auditory and visual stimuli and showed a supramodal role of the right AG, with increased variability for both auditory and visual stimuli with stimulation occurring at the comparison stimulus onset.

Finally, four studies used 20-Hz rTMS stimulation with time reproduction tasks and visual stimuli. Jones et al. (2004) showed temporal overestimation associated with the stimulation over the DLPFC for supra-second intervals and when stimulation occurred during reproduction but not during the encoding phase; no effect of stimulation was observed when stimulation was applied over SMA. Koch et al. (2007; Experiment 1) showed that interfering with right DLPFC activity selectively impaired (over-reproduction) temporal processing of long temporal intervals. Finally, Oliveri et al. (2009; Experiment 3) showed that the under-reproduction (time reproduction – half reproduction) following rTMS (trains of four stimuli at 20 Hz and lasting 200 ms) on the right IPC was specifically observed when rTMS trains were delivered online during the retrieval phase. A disruption of the right PPC could cause temporal neglect while participants are retrieving half of a previously encoded time interval. Capizzi et al. (2023a; Experiment 1) employed an online TMS protocol in which they perturbed left IPS activity with three TMS pulses at 20 Hz in an implicit temporal orienting task. Their results showed no modulation of temporal orienting effects after TMS over left IPS as compared to both right IPS and vertex control regions.

Theta-burst stimulation

As mentioned earlier, TBS protocols deliver a pattern of rTMS stimulation that mimics neural oscillatory patterns. Indeed, short bursts of 50 Hz are delivered at a rate in the theta range (5 Hz) as continuous (cTBS) or intermittent (iTBS) trains. While both paradigms involve the same frequency of stimulation within the train, their difference in interstimulus interval between trains is associated with different effects: cTBS generally suppresses neural activity, whereas iTBS facilitates it. The TBS procedure has two advantages: (1) shorter stimulation duration and (2) lower stimulation intensity. These features decrease the likelihood of discomfort from TMS pulses (Fitzgerald et al., 2006; Hong et al., 2015).

Frontal areas were targeted in three studies testing explicit timing: Dusek et al. (2011) used a time reproduction task with visual stimuli and long temporal intervals, Hayashi et al. (2013) also used visual stimuli but a time discrimination task and short temporal intervals, and finally, Méndez et al. (2017) used auditory stimuli and a time generalization task with both short and long temporal intervals (Table 5.4). Hayashi et al. (2013; Experiment 2) showed impaired duration discrimination after cTBS stimulation, while Méndez et al. (2017) showed increased variability after right DLPFC stimulation but only when short temporal intervals were used. Dusek et al. (2011) showed that SMA stimulation with iTBS caused a decrease of variability for 10-s intervals, a result opposite to the one observed after precuneus stimulation and also different from Méndez et al. (2017), who showed increased variability after right SMA stimulation when short temporal intervals were used. Concerning implicit timing, Vallesi et al. (2007) showed a reduction of the foreperiod effect after cTBS over the right DLPFC, whereas no effect was reported after stimulation of the left DLPFC and the right angular gyrus in both simple and choice variable foreperiod (RT) tasks (Experiments 1 and 2, respectively).

Bijsterbosch et al. (2011) targeted left PMC to investigate the temporal control of movement and temporal error correction using a sensorimotor synchronization task in which participants tapped their finger in synchrony with a regular auditory tone train. cTBS, but not iTBS, decreased timing accuracy and increased tapping variability (Experiment 1). Moreover, Experiment 2 showed that error correction changed with practice, regardless of the type of TBS used. The authors concluded that the left PMC is involved in both sensorimotor timing and error correction in both hands.

Three studies targeted the parietal cortex with TBS, using time reproduction (Dusek et al., 2011; Riemer et al. 2016) or time discrimination (Hayashi et al., 2013; Riemer et al., 2016) tasks. Dusek et al. (2011) showed increased variability after precuneus stimulation but only specifically with a 5-s

TABLE 5.4 Studies using theta-burst stimulation

Author	Exp.	N (gender)	Mean age (SD or range)	Theta-burst 50 Hz	Time of stimulation	Intensity	Coil	AREA	Side	Control	Temporal tasks	Temporal intervals (ms)	Modality	Dependent variable	Main findings
Vallesi et al., 2007	1	9 (F = 3)	31 (–)	Continuous	Three pulses, 20 ms among each stimulus, repeated at intervals of 200 ms for 20 s (300 pulses in total)	80% of RMT	50-mm figure-of-eight coil	DLPFC	Right	left DLPFC, right angular gyrus	Variable foreperiod simple-RT task	500, 1,000, and 1,500	Warning signal auditory, foreperiod and target visual	Mean RT	Reduced foreperiod effect after TMS over the right DLPFC (as compared to the pre-TMS baseline)
Vallesi et al., 2007	2	9 (F = 4)	30 (–)	Continuous	Three pulses, 20 ms among each stimulus, repeated at intervals of 200 ms for 20 s (300 pulses in total)	80% of RMT	50-mm figure-of-eight coil	DLPFC	Right	Left DLPFC, right angular gyrus	Variable foreperiod choice-RT task	500, 1,000, and 1,500	Warning signal auditory, foreperiod and target visual	Mean RT	Reduced foreperiod effect after TMS over the right DLPFC (as compared to the pre-TMS baseline)
Méndez et al., 2017	1	15 (F = 5)	24.4 (4.85)	Continuous	Three pulses every 200 ms delivered for 40 s	80% of RMT	70-mm figure-of-eight coil	DLPFC	Right	S1	Temporal categorization task	Short: 200–500; long: 870–1,520	Auditory	Relative threshold, CE	Increased variability; no effect on CE
Hayashi et al., 2013	2	10 (F = 7)	Range 20–30	Continuous	Three pulses every 200 ms delivered for 40 s	40% of stimulator output	70-mm figure-of-eight coil	IFG	Right	Vertex	Time discrimination	450–540, 600–720, and 750–900	Visual	Accuracy	Lower accuracy after IFC stimulation
Dusek et al., 2011	1	19 (F = 11)	25.9 (3)	Intermitted	Three stimuli repeated ten times at intervals of 200 ms; 600 pulses were delivered during the train	100% of RMT	70-mm figure-of-eight coil	SMA	–	–	Time reproduction	5,000, 10,000, and 16,820	Visual	Reproduced time, SD (variability)	No effect on accuracy; decreased variability 10-s interval

(Continued)

TABLE 5.4 (Continued)

Author	Exp.	N (gender)	Mean age (SD or range)	Theta-burst 50 Hz	Time of stimulation	Intensity	Coil	AREA	Side	Control	Temporal tasks	Temporal intervals (ms)	Modality	Dependent variable	Main findings
Méndez et al., 2017	1	15 (F = 5)	24.4 (4.85)	Continuous	Three pulses every 200 ms delivered for 40 s	80% of RMT	90-mm figure-of-eight coil	SMA	Right	S1	Temporal categorization task	Short = 200–500; long = 870–1,520	Auditory	Relative threshold, CE	Increased variability; no effect on CE
Bijsterbosch et al., 2011	1	16 (F = 8)	23.3 (3.9)	Continuous, intermitted	40 s of continuous or 190-s intermitted	90% of RMT	70-mm figure-of-eight coil	PMC	Left	–	Sensorimotor synchronization task and error correction	Paced 600 (regular condition; for error correction, the tone occurred 90 ms earlier or later than expected)	Auditory	Asynchrony, IRI, and SD	Decreased timing accuracy and increased variability after continuous stimulation
Bijsterbosch et al., 2011	2	16 (F = 8)	22.3 (2.5)	Continuous	40 s of continuous or 190-s intermitted	90% of RMT	70-mm figure-of-eight coil	PMC	Left	–	Sensorimotor synchronization task and error correction	Paced 600 (regular condition; for error correction, the tone occurred 90 ms earlier or later than expected)	Auditory	Asynchrony, IRI, and SD	Improved error correction performance after practice
Dusek et al., 2011	1	19 (F = 11)	25.9 (3)	Intermitted	Three pulses repeated ten times at intervals of 200 ms; 600 pulses were delivered during the train	100% of RMT	70-mm figure-of-eight coil	Precuneus	–	–	Time reproduction	5,000, 10,000, and 16,820	Visual	Reproduced time; SD (variability)	No effect on accuracy; increased variability at 5-s interval
Hayashi et al., 2013	2	10 (F = 7)	Range 20–30	Continuous	Three pulses every 200 ms delivered for 40 s	40% of stimulator output	70-mm figure-of-eight coil	IPC	Right	Vertex	Time discrimination	450–540, 600–720, and 750–900	Visual	Accuracy	No effect
Kanai et al., 2011	1	10 (F = 6)	24.3 (2.8)	Continuous	Three pulses every 200 ms delivered for 40 s	40% of stimulator output	70-mm figure-of-eight coil	A1	Right	Pre-stimulation baseline	Time discrimination	Standard = 600	Auditory, visual	Threshold	Increased discrimination threshold independently of the modality

Study		N	Age	TBS	Protocol	Intensity	Coil	Target	Hemisphere	Control	Task	Stimulus	Modality	Measure	Result
Riemer et al., 2016	1	24 ($F=15$)	26.3 (–)	Continuous[a]	Three biphasic pulses for 44 s at 6 Hz; 267 bursts (801 single pulses)	47.4% of the maximal stimulator intensity	75-mm figure-of-eight coil	IPS	Right	Sham	Time reproduction	1,000–5,000	Auditory	Relative error and CV	No effect
Riemer et al., 2016	1	24 ($F=15$)	26.3 (–)	Continuous[a]	Three biphasic pulses for 44 s at 6 Hz; 267 bursts (801 single pulses)	47.4% of the maximal stimulator intensity	75-mm figure-of-eight coil	IPS	Right	Sham	Time discrimination	Standard = 3,000; comparison = 2,500 and 3,500	Auditory	PSE and DL	Reduced uncertainty; no effect on PSE
Kanai et al., 2011	1	10 ($F=6$)	24.3 (2.8)	Continuous	Three pulses every 200 ms delivered for 40 s	40% of stimulator output	70-mm figure-of-eight coil	V1	Left	Pre-stimulation baseline	Time discrimination	Standard = 600	Auditory, visual	Threshold	Increased discrimination threshold in visual modality
Grube et al., 2010	1	12 ($F=0$)	23.5 (4.5)	Continuous	Three pulses every 200 ms delivered for 40 s	80% of RMT	70-mm figure-of-eight coil	Cerebellum	Medial	Sham	Time discrimination	300–600	Auditory	Threshold	Increased discrimination threshold
Méndez et al., 2017	1	15 ($F=5$)	24.4 (4.85)	Continuous	Three pulses every 200 ms delivered for 40 s	80% of RMT	70-mm figure-of-eight coil	Cerebellum	Right	S1	Temporal categorization task	Short = 200–500; long = 870–1,520	Auditory	Relative threshold and CE	Increased variability; no effect on CE

a Riemer et al. (2016) used 30-Hz continuous theta-burst stimulation.

Note: RMT = resting motor threshold; DLPFC = dorsolateral prefrontal cortex; V1 = primary visual area; IPS = intraparietal sulcus; IRI = inter response interval; CE = constant error; CV = coefficient of variation; PSE = point of subjective equality; DL = different limen; RT = reaction time. IFG = inferior frontal gyrus; SMA = supplementary motor area; PMC = premotor cortex; IPC = inferior parietal cortex; S1 = primary somatosensory cortex; A1 = primary auditory area;

interval. Riemer et al. (2016) used short bursts of three pulses at 30 Hz delivered at 6 Hz to inhibit neuronal processes. The authors showed reduced uncertainty (i.e., different limen) for discriminative time judgments after PPC stimulation (inhibitory effect of stimulation). They suggested that the right PPC plays an inhibitory role in time perception, possibly by mediating multisensory integration between temporal stimuli and other quantities (see also Macaluso et al., 2003). While Hayashi et al. (2013) showed no effect of stimulation.

Kanai et al. (2011) aimed to investigate if sensory areas such as the primary visual (V1) or the auditory (A1) cortices were involved in time perception independently of the sensory modality. Results showed a modality-independent role of the auditory cortex, but TMS over the primary visual cortex impaired performance only in visual time discrimination. These asymmetric contributions of the auditory and visual cortices to time perception may be explained by a superiority of the auditory cortex in temporal processing. In this context, the authors postulated that time perception is fundamentally rooted in the auditory system, with visual stimuli being inherently converted into auditory representations during tasks involving temporal discrimination.

Finally, two studies targeted the cerebellum. Méndez et al. (2017) used a cTBS (three pulses every 200 ms delivered for 40 s) in a time generalization task and showed increased variability, while Grube et al. (2010) used cTBS (20-ms intervals at 5 Hz repeated every 200 ms) within a time discrimination task and showed a higher discrimination threshold (lower performance level) in sub-second intervals (300–600 ms), confirming the role of the cerebellum when short temporal intervals are processed.

Transcranial electric stimulation

Transcranial electrical stimulation (tES) is a non-invasive brain stimulation technique that delivers a weak electrical current (normally between 1 and 2 mA) via two or more electrodes to induce a temporary modulation of cortical excitability. Unlike TMS, the electrical current employed in tES techniques is not strong enough to provoke an action potential. Instead, it is kept below the threshold required for action potential generation, aiming solely to influence cortical excitability (Paulus, 2011; Paulus et al., 2016; Reed & Cohen Kadosh, 2018; Woods et al., 2016).

By altering the activity of brain regions involved in a behavior of interest, researchers can observe the resulting behavioral changes and thus establish a causal link between the two. When employing tES, researchers commonly use two conductive rubber electrodes placed in saline-oaked sponges and attach at least one of them to the head with nonconductive elastic straps.

The location of the active electrode (anode) depends on the cortical area to be modulated, while the return electrode (cathode) is usually placed in an area unrelated to the brain processes being examined, such as the forehead or vertex, but could also be placed on an extracephalic location. Although the most common electrode size used is 20–35 cm² (Moreno-Duarte et al. 2014; Reed & Cohen Kadosh, 2018), this can also be modified to suit the needs of the researcher, with a more recent development being the high-definition stimulation, which uses smaller electrodes (e.g., Phi electrodes) or arrays of small electrodes to achieve a more focal area of stimulation (five electrodes montage; Edwards et al., 2013). Precise electrode placement is normally derived from the International electroencephalography (EEG) 10–20 system (Woods et al., 2016).

Three types of tES can be used, depending on the type of current delivered: transcranial direct current stimulation (tDCS), transcranial random noise stimulation (tRNS), and transcranial alternating current stimulation (tACS).

tDCS is the most used of tES techniques; it was re-introduced in the 1990s after having been largely abandoned in the late 1960s (Santarnecchi et al., 2015). A common assumption is that tDCS effects depend on polarity. Cathodal stimulation leads to hyperpolarization and, consequently, to inhibition, while anodal stimulation causes the resting membrane potential to become more positive and, therefore, results in facilitation. Although certain aspects of stimulation parameters may vary, the outcomes have shown a strong correlation with factors such as current density, duration of stimulation, polarity, and the specific location of stimulation (Zaghi et al., 2010). Notably, the polarity of electrodes significantly influences the effects on cortical excitability. For instance, when applying currents up to 1 mA for durations under 20 min, anodal stimulation over the motor cortex enhances the motor evoked potential, whereas a reversal of polarity to cathodal stimulation produces an opposite effect (Paulus, 2011). However, these effects appear to be inconsistent across studies (Costa et al., 2015). Indeed, this distinction is confirmed especially by studies on motor function (Nitsche & Paulus, 2000). Outside the motor system, there is no necessary correspondence between excitation/inhibition and behavioral improvements/impairments, since the timing of stimulation, the excitability status of the cortical area, and the type of task, among others, can influence the outcome (Costa et al., 2015). It is crucial to highlight that modifying stimulation parameters, such as increasing duration or intensity, or alterations in ongoing brain activity, can lead to changes and even a reversal in the direction of excitability modulation (Batsikadze et al., 2013). However, the amount of current that effectively reaches neuronal tissue is influenced by several uncontrollable factors, including the resistance posed by various cephalic structures like the skin, skull, blood vessels, and brain tissue.

tRNS and tACS are highly effective methods for avoiding the directional sensitivity of standard tDCS. For tRNS, the frequency of the current varies in a random manner changing within a spectrum of oscillations ranging from 0.1 to 640 Hz, whereas tACS uses a sinusoidal current that allows the manipulation of intrinsic cortical oscillations (Moreno-Duarte et al., 2014; Reed & Cohen Kadosh, 2018; Woods et al., 2016). tACS is not intended to excite or inhibit cortical activity monotonously as its main goal is to influence brain oscillations. By modulating cortical oscillations, tRNS and tACS enable researchers to investigate the causal link between brain oscillatory activity and specific cognitive processes.

tRNS has clear neuronal and behavioral effects, with 10 min of stimulation increasing motor cortex excitability for ~1 h after stimulation (Terney et al. 2008), although a study by Campana et al. (2016) suggests that low- and high-frequency tRNS can have opposing effects on cortical excitability. tRNS after-effects are intensity-dependent. Stimulation at 1.5 mA leads to an excitability after-effect comparable to what has been observed with anodal tDCS, whereas a lower intensity (0.4 mA) leads to inhibitory after-effect comparable with cathodal tDCS (Paulus et al., 2016). Regarding tACS, entrainment is most effective when tACS is applied at the same frequency as endogenous oscillations. This observation suggests that tACS stimulation is brain state-dependent. For example, tACS modulates alpha activity when cognitive demands are high (Reed & Cohen Kadosh, 2018). Interestingly, fewer sensory sensations are reported during tRNS and tACS, compared to tDCS. Therefore, applying tRNS and tACS might be better suited for placebo-controlled studies.

Studies using transcranial electric stimulation

Transcranial direct current stimulation is the most widely utilized of the tES techniques. The stimulation can result in a polarity-specific change in neural excitability, whereby cathodal tDCS reduces excitability ("inhibitory"-like effect) and anodal tDCS tends to increase excitability ("facilitatory"-like effect) (Nitsche & Paulus, 2000). Table 5.5 reports the studies using tDCS to investigate temporal processing.

In "online" tDCS paradigms, Vicario et al. (2013), Mioni et al. (2016), and Mannarelli et al. (2023) used an extracephalic montage; Oyama et al. (2017) and Yin et al. (2019) positioned the reference electrode over the contralateral supraorbital area, while Tan et al. (2023) over CPz; Javadi et al. (2014) used a bilateral montage (P3/P4); and finally, Johari et al. (2023) were the only authors to use a high-definition montage.

Yin et al. (2019) and Johari et al. (2023) targeted the right DLPFC using a time bisection task and a time generalization task, respectively. Yin et al.

TABLE 5.5 Studies using transcranial direct current stimulation (tDCS)

Author	N (gender)	Mean age (SD or range)	Type of stimulation	Procedure	Intensity (mA)	AREA	Sponges	Montage	Temporal tasks	Temporal intervals (ms)	Modality	Dependent variables	Main findings
Johari et al., 2023	21 (F = 14)	20.23 (range 19–24)	Cathodal	Online 25 min	–	Right DLPFC	5 rings setting	HD	Time generalization	1,500–2,000	Auditory	Accuracy	No effect on accuracy; faster RT during stimulation
Yin et al., 2019	60 (–)	19.8 (1.33)	Anodal, cathodal	Online 20 min	2	Right DLPFC	Active = 5 × 5 cm²; return = 5 × 7 cm²	Supraorbital controlateral	Time bisection	Short block 200–800; long block 1,400–2,600	Visual	Proportion responses, BP and, WR	No effect on short block; long block = temporal overestimation (BP) with anodal and underestimation (BP) with cathodal stimulation
Vicario et al., 2013	15 (F = 9)	25.6 (3.41)	Anodal, cathodal	Online	2	Right PPC	Active = 25 cm²; return = 35 cm²	Extracephalic	Time reproduction	1,500, 1,600, 1,700, 1,800, and 1,900	Visual	Accuracy, CV	Over-reproduction under cathode stimulation; no effect of anodal
Vicario et al., 2013	9 (F = 5)	25.8 (3.88)	Anodal, cathodal	Online	2	Left PPC	Active = 25 cm²; return = 35 cm²	Extracephalic	Time reproduction	1,500, 1,600, 1,700, 1,800, and 1,900	Visual	Accuracy, CV	Reduced variability under cathode stimulation; no effect of anode
Javadi et al., 2014	13 (F = 7)	22.18 (2.18)	Anodal, cathodal	Online, second trial	1.5	Right, left PPC	35 × 35 mm²	Bilaterally	Time discrimination	800, 900, 1,000, 1,100, and 1,200	Visual	Accuracy, SD	Reduced accuracy (anode tDCS to the left PPC and cathode tDCS to the right PPC); increased accuracy (anode tDCS to the right PPC and cathode tDCS to the left PPC)
Oyama et al., 2017	16 (F = 6)	23.7 (1.3)	Anodal, cathodal	Online 20 min	2	Right PPC	Active = 5 × 5 cm²; return = 5 × 7 cm²	Supraorbital controlateral	Time discrimination	Standard 600	Visual	Threshold, SD	Cathode tDCS enhances temporal discrimination

(Continued)

TABLE 5.5 (Continued)

Author	N (gender)	Mean age (SD or range)	Type of stimulation	Procedure	Intensity (mA)	AREA	Sponges	Montage	Temporal tasks	Temporal intervals (ms)	Modality	Dependent variables	Main findings
Tan et al., 2023	16 (F = 8)	23 (1.4)	Anodal	Offline 20 min	2	Right PPC	Active = 38.5 cm²; sham = same intensity applied for 60 s	CPz	Temporal orienting	600–1,400	Visual	Accuracy, mean RT	No effect after anodal or sham tDCS
Mioni et al., 2016	24 (–)	23.85 (1.79)	Anodal, cathodal	Online 20 min	1.5	A1	Active = 25 cm²; return = 35 cm²	Extracephalic	Time bisection	Standard short = 300; standard long = 900	Visual, auditory	BP, WR	Higher variability (WR) under anode compared to sham
Mioni et al., 2016	24 (–)	25.16 (3.34)	Anodal, cathodal	Online 20 min	1.5	V1	Active = 25 cm²; return = 35 cm²	Extracephalic	Time bisection	Standard short = 300; standard long = 900	Visual, auditory	BP, WR	Higher variability (WR) under cathode only in visual modality
Mannarelli et al., 2023	16 (–)	25 (0.8)	Cathodal	Online 20 min	2	Left cerebellum	Active and return = 25 cm²	Extracephalic	Time discrimination	Standard = 1,200; comparison 800, 1,200, 1,600	Auditory	Errors	Higher number of errors and higher RT after stimulation for 800 and 1,200 ms

Note: HD = high definition montage; DLPFC = dorsolateral prefrontal cortex; PPC = posterior parietal cortex; A1 = primary auditory area; V1 = primary visual area; BP = bisection point; WR = Weber ratio; RT = reaction time.

(2019) showed no effect of stimulation when short temporal intervals were used but temporal overestimation under anodal and temporal underestimation under cathodal stimulation with long temporal intervals. However, the results were not replicated in a subsequent study conducted by Johari et al. (2023) where auditory stimuli were used. The results seem to suggest that the right DLPFC is involved in temporal processing when durations longer than 2 s are presented; the involvement of the right DLPFC could be related to the contribution of working memory and executive functions to the estimation of long duration (Friedman & Robbins, 2022; Jones & Graff-Radford, 2021).

Taking advantage of the different effects of anodal and cathodal stimulations, Vicario et al. (2013) and Javadi et al. (2014) investigated the different involvements of left and right PPCs in timing. Vicario et al. (2013) showed that cathodal stimulation over the right PPC affected temporal accuracy by leading participants to over-reproduce time intervals; when the cathodal stimulation was applied to the left PPC, it reduced the variability when reproducing temporal intervals. Javadi et al. (2014) showed that the application of anodal tDCS to the left PPC and cathodal tDCS to the right PPC impaired temporal accuracy in the time discrimination task, while application of anodal tDCS to the right PPC and cathodal tDCS to the left PPC increased accuracy in the duration judgment task. Oyama et al. (2017) only targeted the right PPC and observed improved temporal discrimination after cathodal stimulation. Direct comparisons between these studies are difficult considering the methodological differences. Nevertheless, what can be said is that left and right PPCs seem to play different roles in temporal processing, with more critical involvement of the right PPC side. Concerning implicit timing, Tan et al. (2023) found no significant effect on behavioral performance in an implicit temporal orienting task following anodal tDCS over the right PPC.

Mioni et al. (2016) investigated the modality-independent or modality-specific role in time processing of the primary auditory (A1) and visual (V1) cortices. When tDCS was delivered over the A1, no effect of stimulation was observed on perceived duration, but the authors observed higher temporal variability under anodic stimulation compared to sham and higher variability in the visual compared to the auditory modality. When tDCS was delivered over V1, temporal underestimation and higher variability were observed in the visual compared to the auditory modality. Confirming a modality-independent role of A1 in temporal processing and a modality-specific role of V1 in the processing of temporal intervals in the visual modality, these results were also observed by Kanai et al. (2011) with rTMS.

Finally, Mannarelli et al. (2023) targeted the left cerebellum and showed a higher number of errors and higher RT after cathodal stimulation, pointing to a role of the cerebellum in time perception for sub-second intervals.

Transcranial random noise stimulation

When using tRNS, stimulation frequency is normally distributed between 0.1 and 640 Hz although it can be divided into either low (0.1–100 Hz) or high (101–640) frequencies (Reed & Cohen Kadosh, 2018). All studies reported used high frequency (Table 5.6).

Mioni et al. (2018) used tRNS to test the specific involvement of right posterior parietal (P4) and right frontal (F4) areas on short temporal intervals (<1 s), changing the modality (visual or auditory) used for marking time. Results showed no effect of stimulation when tRNS was applied over F4, probably because frontal areas are involved when long temporal intervals are employed (Lewis & Miall, 2003; Mioni et al., 2014). Indeed, Capizzi et al. (2023b) used longer temporal intervals (480–1,920 ms) and a focal high definition (HD) montage (5-star electrodes montage) over the SMA to investigate the contribution of this area to explicit (time bisection task) and implicit timing (foreperiod task). Results showed temporal overestimation under active stimulation, whereas there was no modulation of implicit timing by HD-tRNS. Mioni et al. (2018) showed temporal overestimation when stimulation was applied over P4 independently of the modality used (visual or auditory), confirming the role of the parietal cortex in multimodal integration (Macaluso et al., 2003). Prete et al. (2021) targeted the parietal cortex, showing that high-frequency tRNS influenced participants' performance compared to the sham condition, and validating the role of the right supramarginal gyrus (SMG; C4/P4) in temporal experience.

Mioni (2020) aimed to replicate with tRNS the results observed with tDCS by Mioni et al. (2016). Temporal overestimation was observed when tRNS was delivered over A1 independently of the modality used (visual or auditory), but when tRNS was delivered over V1, temporal overestimation was observed only for visual stimuli. No effect of stimulation was observed on temporal sensitivity in any condition.

Transcranial alternating current stimulation

Two studies used tACS stimulation with the hypothesis to modulate specific frequency bands to affect subjective time perception (Table 5.6). Wiener et al. (2018) showed that beta stimulation exclusively shifts the perception of time such that stimuli are overestimated, providing evidence for the intrinsic involvement of beta oscillations in the perception of time. In particular, the results pointed to a specific role for beta oscillations in the encoding and retention of memory for temporal intervals. Mioni et al. (2020c) tested the hypothesis that the rate of an internal pacemaker would be driven by neural oscillations in the alpha range (8–12 Hz) (Kononowicz & van Wassenhove, 2016).

TABLE 5.6 Studies using transcranial random noise stimulation (tRNS) and transcranial alternate current stimulation (tACS)

Author	N (gender)	Mean age (SD or range)	Type of stimulation	Procedure	Intensity (mA)	AREA	Sponges	Montage	Temporal tasks	Temporal intervals (ms)	Modality	Dependent variables	Main findings
High-frequency tRNS													
Mioni et al., 2018	40 (F = 30)	22.88 (2.45)	High-frequency tRNS	Online 10 min	1.5	Right frontal (F4)	Active = 16 cm²; return = 60 cm²	Extracephalic	Time bisection	Standard short = 300; standard long = 900	Visual, auditory	BP, WR	No effect
Capizzi et al., 2023a	48 (F = 32)	23.31 (1.63)	High-frequency tRNS	Online 10 min	1	SMA	5 ring setting	Focal high definition	Time bisection, foreperiod task	Standard short = 480; standard long = 1,920	Visual	Bisection task: proportion of long responses; BP; JND, foreperiod task: RTs	Temporal overestimation under active stimulation; no effect on JND nor foreperiod effect
Mioni, 2020	24 (F = 8)	23.88 (3.66)	High-frequency tRNS	Online 15 min	1.5	A1	Active = 25 cm²; return = 35 cm²	Extracephalic	Time bisection	Standard short = 300; standard long = 900	Visual, auditory	BP, WR	Temporal overestimation (BP); no effect on temporal variability (WR)
Mioni et al., 2018	40 (F = 20)	23.04 (1.87)	High-frequency tRNS	Online 10 min	1.5	Right Parietal (P4)	Active = 16 cm²; return = 60 cm²	Extracephalic	Time bisection	Standard short = 300; standard long = 900	Visual, auditory	BP, WR	Temporal overestimation (BP); no effect on temporal variability (WR)
Mioni, 2020	24 (F = 8)	23.22 (2.65)	High-frequency tRNS	Online 15 min	1.5	V1	Active = 25 cm²; return = 35 cm²	Extracephalic	Time bisection	Standard short = 300; standard long = 900	Visual, auditory	BP, WR	Temporal overestimation (BP) only in visual modality; no effect on temporal variability (WR)

(Continued)

TABLE 5.6 (Continued)

Author	N (gender)	Mean age (SD or range)	Type of stimulation	Procedure	Intensity (mA)	AREA	Sponges	Montage	Temporal tasks	Temporal intervals (ms)	Modality	Dependent variables	Main findings
Prete et al., 2021	24 (F = 13)	29.54 ± 1.72 years	High-frequency tRNS	Online 25 min	2	SMG C4/P4	Active = 5 × 5 cm²; return = 5 × 9.5 cm²	Extracephalic	Duration discrimination	Standard = 500; comparison = 350, 450, 550, 650	Visual	Proportion of long responses; BP	hf-tRNS applied on the right SMG increased the aftereffect following duration adaptation
tACS													
Wiener et al., 2018	20 (F = 8)	24 (4)	tACS (alpha and beta)	Online 20 min	1.5	FC1 and FC2	5 × 5 cm²	Bilateral	Time bisection	Standard short = 300; standard long = 900	Visual	BP, CV	BP = overestimation with beta stimulation; no modulation with alpha. No effect on CV
Mioni et al., 2020c	18 (F = 9)	22.89 (4.47)	tACS (alpha)	Online 20 min	1.5	Occipital Oz	Active = 9 cm²; return = 35 cm²	Cz	Time generalization	Standard = 600; comparison short: 300, 400, 500; comparison long: 700, 800, 900	Visual	BP, WR	Temporal modulation of BP based on IAF; no effect on WR

Note. SMA = somatosensory area; A1 = primary auditory area; V1 = primary visual area; SMG = supra marginal gyrus; IAF = individual alpha frequency; BP = bisection point; WR = Weber ratio; CV = coefficient of variation; JND = just noticeable difference; RT = reaction time.

Individual alpha frequencies (IAFs) were recorded at rest before the time generalization task; then, participants performed the timing task while receiving tACS either at their IAFs or at individually accelerated or decelerated rates (IAF ± 2 Hz). Results demonstrated a shift in the psychometric function indicating a modification of perceived duration, such that progressively "faster" alpha stimulation led to longer perceived intervals. These results provide the first evidence that direct manipulations of alpha oscillations can shift perceived time in a manner consistent with a clock speed effect.

Methodological considerations

Focality and intensity of stimulation

TMS has better temporal and spatial resolution than tES. In fact, tES has usually been applied through two large electrodes, with the two of them on the scalp or with one of them placed at an extra-cephalic location (neck or shoulder). Models are being developed that might give some insight into the distribution of the current flow and allow better predictions about the likely sites stimulated. Moreover, smaller electrodes (4 × 1 high-definition tES) have been developed and used with success in other domains. Although TMS can induce distal effects due to brain connectivity, a standard figure-of-eight coil allows a relatively focal direct stimulation. Whereas TMS may induce immediate behavioral effect attesting that the stimulation reached the neural tissue, as in the case for motor cortex stimulation that causes muscle twitches on the opposite side of the body, low-intensity tES produces no similar effects on behavior, so there is no immediate indicator of the success of the stimulation. Although the effect of tES stimulation can be quantified using other neurophysiological techniques, at present, the intensity is simply quantified in terms of the current flow between the two electrodes.

Safety and sensation reported

Differentiating between tolerability and safety is crucial. Tolerability encompasses uncomfortable and unintended effects, such as tingling and itching sensations under the electrodes. Safety, on the other hand, pertains to potential damaging effects. Commonly reported effects using tES include tingling, itching sensations under the electrodes, headaches, and fatigue (Fertonani et al., 2015). Seizures, although rare, pose a potentially serious risk associated with TMS. Common side effects of TMS, such as scalp sensation, headache, and/or neck pain, can significantly impact its tolerability (Peterchev et al., 2017). Additionally, TMS generates a loud clicking noise ranging from 120 to 140 dB, which has the potential to cause acoustic trauma (Dhamne et al., 2014). To prevent this adverse effect, the use of hearing protection is essential. Occupational Safety and Health Administration regulations limit exposure to impulsive noises louder than >140 dB sound pressure level (Tringali et al., 2012). The auditory noise and physical sensation induced by TMS not only discomfort participants but also carry theoretical implications in the study of time processing. Previous studies have shown that the presentation of a rapid sequence of auditory clicks alone can subjectively elongate perceived duration. This effect might be attributed to an increase in arousal, potentially accelerating the pacemaker function (Moss & Maner, 2014; Wearden et al., 2017).

Regarding electrical stimulation aversive effects, the perception of phosphenes, caused by sudden changes in current onset or offset, is prevented by gradually increasing or decreasing current intensity. Erythema (redness) under the electrodes results from tDCS-induced vasodilation and is not considered a safety concern. Unlike rTMS, there have been no reported cases of seizure induction with electric stimulation techniques. Despite the temporal and spatial limitation of tES, no sound is emitted, and fewer sensations are reported during stimulation. Notably, fewer sensory sensations are reported during tRNS, compared to tDCS. Therefore, applying tRNS might be better suited for placebo-controlled studies.

Both TMS and tES are generally safe if used properly. Indeed, it is crucial to properly set safety parameters and screen participants to exclude those with a higher risk of adverse events due to medications, implants (e.g., cardiac pacemakers, brain implants, and hearing aids), or neurological or psychological conditions according to available guidelines (e.g., Rossi et al., 2009; 2021). When designing studies involving non-invasive brain stimulation techniques, we also suggest including questionnaires to control for participants' dominant hand (i.e., Edinburgh Handedness Inventory; Oldfield 1971) and sensation experienced during stimulation (i.e., Fertonani et al., 2015) to control for possible unspecific effect of stimulation.

Conclusions and future research directions

The number of studies using TMS or tES for the investigation of temporal processing has not increased much recently, and most recent studies conducted in the field have used electric stimulation techniques. Probably the most critical issue when it comes to reviewing the effects of brain stimulation is the variety of results observed with the different stimulation techniques. Although it is possible to draw some general conclusions when comparing multiple studies involving the use of non-invasive brain stimulation techniques to study time processing, many methodological differences prevent clear conclusions from being reached.

Through the application of non-invasive brain stimulation techniques, it is possible to identify the direct functional contribution of a particular brain area to the performance of a task, allowing the understanding of where and when a particular process being investigated occurs. However, a large network of areas has been identified as involved in time processing by neuroimaging techniques (for a recent review see Naghibi et al., 2023). When referring to the study of time perception, the adoption of approaches that consider the involvement of the network or connections more thoroughly may be preferable. In this direction, future research should address the combined and complementary use of neuroscientific techniques with different characteristics in terms of spatial and temporal resolution.

Acknowledgments

M. C. is supported by a grant (PID2021-128696NA-I00) funded by MICIU/ AEI/10.13039/501100011033 and ERDF/EU. M. C. also acknowledges support from a María Zambrano Fellowship at the University of Granada from the Spanish Ministry of Universities and the European Union NextGeneration.

References

Alexander, I., Cowey, A., & Walsh, V. (2005). The right parietal cortex and time perception: Back to Critchley and the Zeitraffer phenomenon. *Cognitive Neuropsychology*, *22*(3–4), 306–315. https://doi.org/10.1080/02643290442000356

Baudouin, A., Vanneste, S., Isingrini, M., & Pouthas, V. (2006). Differential involvement of internal clock and working memory in the production and reproduction of duration: A study on older adults. *Acta Psychologica*, *121*(3), 285–296. https://doi.org/10.1016/j.actpsy.2005.07.004

Baudouin, A., Isingrini, M., & Vanneste, S. (2018). Executive functioning and processing speed in age-related differences in time estimation: A comparison of young, old, and very old adults. *Aging, Neuropsychology, & Cognition*, *26*(2), 264–281. https://doi.org/10.1080/13825585.2018.1426715

Batsikadze, G., Moliadze, V., Paulus, W., Kuo, M. F., & Nitsche, M. (2013). Partially non-linear stimulation intensity-dependent effects of direct current stimulation on motor cortex excitability in humans. *The Journal of physiology*, *591*(7), 1987–2000. https://doi.org/10.1113/jphysiol.2012.249730

Bergmann, T. O., & Hartwigsen, G. (2021). Inferring causality from noninvasive brain stimulation in cognitive neuroscience. *Journal of Cognitive Neuroscience*, *33*(2), 195–225. https://doi.org/10.1162/jocn_a_01591

Beynel, L., Appelbaum, L. G., Luber, B., Crowell, C. A., Hilbig, S. A., Lim, W., ... & Deng, Z. D. (2019). Effects of online repetitive transcranial magnetic stimulation (rTMS) on cognitive processing: A meta-analysis and recommendations for future studies. *Neuroscience & Biobehavioral Reviews*, *107*, 47–58. https://doi.org/10.1016/j.neubiorev.2019.08.018

Bijsterbosch, J. D., Lee, K. H., Dyson-Sutton, W., Barker, A. T., & Woodruff, P. W. (2011). Continuous theta burst stimulation over the left pre-motor cortex affects sensorimotor timing accuracy and supraliminal error correction. *Brain Research*, *1410*, 101–111. https://doi.org/10.1016/j.brainres.2011.06.062

Bolognini, N., Papagno, C., Moroni, D., & Maravita, A. (2010). Tactile temporal processing in the auditory cortex. *Journal of Cognitive Neuroscience*, *22*(6), 1201–1211. https://doi.org/10.1162/jocn.2009.21267

Bueti, D., van Dongen, E. V., & Walsh, V. (2008a). The role of superior temporal cortex in auditory timing. *PLoS One, 3*(6), e2481. https://doi.org/10.1371/journal.pone.0002481

Bueti, D., Bahrami, B., & Walsh, V. (2008b). Sensory and association cortex in time perception. *Journal of Cognitive Neuroscience, 20*(6), 1054–1062. https://doi.org/10.1162/jocn.2008.20060

Campana G, Camilleri R, Moret B, Ghin F, Pavan A (2016). Opposite effects of high- and low-frequency transcranial random noise stimulation probed with visual motion adaptation. *Scientific Reports 6*, 38919. https://doi.org/10.1038/srep38919

Capizzi, M., Martín-Signes, M., Coull, J. T., Chica, A. B., & Charras, P. (2023a). A transcranial magnetic stimulation study on the role of the left intraparietal sulcus in temporal orienting of attention. *Neuropsychologia, 184*, 108561. https://doi.org/10.1016/j.neuropsychologia.2023.108561

Capizzi, M., Visalli, A., Wiener, M., & Mioni, G. (2023b). The contribution of the supplementary motor area to explicit and implicit timing: A high-definition transcranial random noise stimulation (HD-tRNS) study. *Behavioural Brain Research, 445*, 114383. https://doi.org/10.1016/j.bbr.2023.114383

Correa, Á., Cona, G., Arbula, S., Vallesi, A., & Bisiacchi, P. (2014). Neural dissociation of automatic and controlled temporal preparation by transcranial magnetic stimulation. *Neuropsychologia, 65*, 131–136. https://psycnet.apa.org/doi/10.1016/j.neuropsychologia.2014.10.023

Costa, T. L., Lapenta, O. M., Boggio, P. S., & Ventura, D. F. (2015). Transcranial direct current stimulation as a tool in the study of sensory-perceptual processing. *Attention, Perception, & Psychophysics, 77*, 1813–1840. https://psycnet.apa.org/doi/10.3758/s13414-015-0932-3

Del Olmo, M. F., Cheeran, B., Koch, G., & Rothwell, J. C. (2007). Role of the cerebellum in externally paced rhythmic finger movements. *Journal of Neurophysiology, 98*(1), 145–152. https://doi.org/10.1152/jn.01088.2006

Deng, Z. D., Lisanby, S. H., & Peterchev, A. V. (2013). Electric field depth–focality tradeoff in transcranial magnetic stimulation: Simulation comparison of 50 coil designs. *Brain Stimulation, 6*(1), 1–13. https://doi.org/10.1016/j.brs.2012.02.005

Dhamne, S. C., Kothare, R. S., Yu, C., Hsieh, T. H., Anastasio, E. M., Oberman, L., ... & Rotenberg, A. (2014). A measure of acoustic noise generated from transcranial magnetic stimulation coils. *Brain Stimulation, 7*(3), 432–434. https://doi.org/10.1016/j.brs.2014.01.056

Dormal, V., Andres, M., & Pesenti, M. (2008). Dissociation of numerosity and duration processing in the left intraparietal sulcus: A transcranial magnetic stimulation study. *Cortex, 44*(4), 462–469. https://psycnet.apa.org/doi/10.1016/j.cortex.2007.08.011

Doumas, M., Praamstra, P., & Wing, A. M. (2005). Low frequency rTMS effects on sensorimotor synchronization. *Experimental Brain Research, 167*, 238–245. https://psycnet.apa.org/doi/10.1007/s00221-005-0029-7

Dusek, P., Jech, R., Havránková, P., Vymazal, J., & Wackermann, J. (2011). Theta-burst transcranial magnetic stimulation over the supplementary motor area decreases variability of temporal estimates. *Neuroendocrinology Letter 32*, 481–486.

Edwards, D., Cortes, M., Datta, A., Minhas, P., Wassermann, E. M., & Bikson, M. (2013). Physiological and modeling evidence for focal transcranial electrical brain stimulation in humans: a basis for high-definition tDCS. *Neuroimage, 74*, 266–275. https://doi.org/10.1016/j.neuroimage.2013.01.042

Farzan, F. (2014). Single-pulse transcranial magnetic stimulation (TMS) protocols and outcome measures. In A. Rotenberg, J. C. Horvath, & A. Pascual-Leone (Eds) *Transcranial magnetic stimulation* (pp. 69–115). Springer.

Fertonani, A., Ferrari, C., & Miniussi, C. (2015). What do you feel if I apply transcranial electric stimulation? Safety, sensations and secondary induced effects. *Clinical Neurophysiology*, *126*, 2181–2188. https://doi.org/10.1016/j.clinph.2015.03.015

Fierro, B., Palermo, A., Puma, A., Francolini, M., Panetta, M. L., Daniele, O., & Brighina, F. (2007). Role of the cerebellum in time perception: A TMS study in normal subjects. *Journal of the Neurological Sciences*, *263*(1–2), 107–112. https://doi.org/10.1016/j.jns.2007.06.033

Fitzgerald, P. B., Fountain, S., & Daskalakis, Z. J. (2006). A comprehensive review of the effects of rTMS on motor cortical excitability and inhibition. *Clinical Neurophysiology*, *117*(12), 2584–2596. https://doi.org/10.1016/j.clinph.2006.06.712

Friedman, N. P., & Robbins, T. W. (2022). The role of prefrontal cortex in cognitive control and executive function. *Neuropsychopharmacology*, *47*(1), 72–89. https://doi.org/10.1038/s41386-021-01132-0

Giovannelli, F., Ragazzoni, A., Battista, D., Tarantino, V., Del Sordo, E., Marzi, T., ... & Cincotta, M. (2014). "... the times they aren't a-changin'..." rTMS does not affect basic mechanisms of temporal discrimination: A pilot study with ERPs. *Neuroscience*, *278*, 302–312. https://doi.org/10.1016/j.neuroscience.2014.08.024

Gironell, A., Rami, L., Kulisevsky, J., & García-Sánchez, C. (2005). Lack of prefrontal repetitive transcranial magnetic stimulation effects in time production processing. *European Journal of Neurology*, *12*(11), 891–896. https://doi.org/10.1111/j.1468-1331.2005.01093.x

Grondin, S., Ouellet, B., & Roussel, M. E. (2004). Benefits and limits of explicit counting for discriminating temporal intervals. *Canadian Journal of Experimental Psychology/Revue Canadienne de Psychologie Expérimentale*, *58*(1), 1. https://doi.org/10.1037/h0087436

Grube, M., Lee, K. H., Griffiths, T. D., Barker, A. T., & Woodruff, P. W. (2010). Transcranial magnetic theta-burst stimulation of the human cerebellum distinguishes absolute, duration-based from relative, beat-based perception of subsecond time intervals. *Frontiers in Psychology*, *1*, 171. https://doi.org/10.3389/fpsyg.2010.00171

Hallett, M. (2000). Transcranial magnetic stimulation and the human brain. *Nature*, *406*(6792), 147–150. https://doi.org/10.1038/35018000

Hayashi, M. J., Kanai, R., Tanabe, H. C., Yoshida, Y., Carlson, S., Walsh, V., & Sadato, N. (2013). Interaction of numerosity and time in prefrontal and parietal cortex. *Journal of Neuroscience*, *33*(3), 883–893. https://doi.org/10.1523/JNEUROSCI.6257-11.2013

Hartwigsen, G., & Silvanto, J. (2023). Noninvasive brain stimulation: Multiple effects on cognition. *The Neuroscientist*, *29*(5), 639–653. https://doi.org/10.1177/10738584221113806

Hong, Y. H., Wu, S. W., Pedapati, E. V., Horn, P. S., Huddleston, D. A., Laue, C. S., & Gilbert, D. L. (2015). Safety and tolerability of theta burst stimulation vs. single and paired pulse transcranial magnetic stimulation: A comparative study of 165 pediatric subjects. *Frontiers in Human Neuroscience*, *9*, 29. https://doi.org/10.3389/fnhum.2015.00029

Huang, Y. Z., Edwards, M. J., Rounis, E., Bhatia, K. P., & Rothwell, J. C. (2005). Theta burst stimulation of the human motor cortex. *Neuron*, *45*(2), 201–206. https://doi.org/10.1016/j.neuron.2004.12.033

Jahanshahi, M., & Rothwell, J. (2000). Transcranial magnetic stimulation studies of cognition: an emerging field. *Experimental Brain Research*, *131*, 1–9. https://doi.org/10.1007/s002219900224

Jäncke, L., Steinmetz, H., Benilow, S., & Ziemann, U. (2004). Slowing fastest finger movements of the dominant hand with low-frequency rTMS of the hand area of the primary motor cortex. *Experimental Brain Research*, *155*, 196–203. https://doi.org/10.1007/s00221-003-1719-7

Javadi, A. H., Brunec, I. K., Walsh, V., Penny, W. D., & Spiers, H. J. (2014). Transcranial electrical brain stimulation modulates neuronal tuning curves in perception of numerosity and duration. *Neuroimage, 102*, 451–457. https://doi.org/10.1016%2Fj.neuroimage.2014.08.016

Johari, K., Tabari, F. & Desai, R.H. (2023). Right frontal HD-tDCS reveals causal involvement of time perception networks in temporal processing of concepts. *Scientific Reports 13*, 16658. https://doi.org/10.1038/s41598-023-43416-z

Jones, D. T., & Graff-Radford, J. (2021). Executive dysfunction and the prefrontal cortex. *CONTINUUM: Lifelong Learning in Neurology, 27*(6), 1586–1601. https://doi.org/10.1212/con.0000000000001009

Jones, C. R., Rosenkranz, K., Rothwell, J. C., & Jahanshahi, M. (2004). The right dorsolateral prefrontal cortex is essential in time reproduction: An investigation with repetitive transcranial magnetic stimulation. *Experimental Brain Research, 158*, 366–372. https://doi.org/10.1007/s00221-004-1912-3

Kanai, R., Lloyd, H., Bueti, D., & Walsh, V. (2011). Modality-independent role of the primary auditory cortex in time estimation. *Experimental Brain Research, 209*, 465–471. https://psycnet.apa.org/doi/10.1007/s00221-011-2577-3

Klomjai, W., Katz, R., & Lackmy-Vallée, A. (2015). Basic principles of transcranial magnetic stimulation (TMS) and repetitive TMS (rTMS). *Annals of Physical and Rehabilitation Medicine, 58*(4), 208–213. https://doi.org/10.1016/j.rehab.2015.05.005

Koch, G., Oliveri, M., Torriero, S., & Caltagirone, C. (2003). Underestimation of time perception after repetitive transcranial magnetic stimulation. *Neurology, 60*(11), 1844–1846. https://doi.org/10.1212/wnl.60.11.1844

Koch, G., Oliveri, M., Torriero, S., Salerno, S., Gerfo, E. L., & Caltagirone, C. (2007). Repetitive TMS of cerebellum interferes with millisecond time processing. *Experimental Brain Research, 179*, 291–299. https://doi.org/10.1007/s00221-006-0791-1

Kononowicz, T. W., & van Wassenhove, V. (2016). In search of oscillatory traces of the internal clock. *Frontiers in Psychology, 7*, 224. https://doi.org/10.3389/fpsyg.2016.00224

Krause, V., Bashir, S., Pollok, B., Caipa, A., Schnitzler, A., & Pascual-Leone, A. (2012). 1 Hz rTMS of the left posterior parietal cortex (PPC) modifies sensorimotor timing. *Neuropsychologia, 50*(14), 3729–3735. https://doi.org/10.1016%2Fj.neuropsychologia.2012.10.020

Lee, K. H., Egleston, P. N., Brown, W. H., Gregory, A. N., Barker, A. T., & Woodruff, P. W. (2007). The role of the cerebellum in subsecond time perception: evidence from repetitive transcranial magnetic stimulation. *Journal of Cognitive Neuroscience, 19*(1), 147–157. https://doi.org/10.1162/jocn.2007.19.1.147

Levit-Binnun, N., Handzy, N. Z., Moses, E., Modai, I., & Peled, A. (2007). Transcranial magnetic stimulation at M1 disrupts cognitive networks in schizophrenia. *Schizophrenia Research, 93*(1–3), 334–344. https://doi.org/10.1016/j.schres.2007.02.019

Lewis, P. A., & Miall, R. C. (2003). Brain activation patterns during measurement of sub-and supra-second intervals. *Neuropsychologia, 41*(12), 1583–1592. https://doi.org/10.1016/s0028-3932(03)00118-0

Liu, A., Vöröslakos, M., Kronberg, G., Henin, S., Krause, M. R., Huang, Y., ... & Buzsáki, G. (2018). Immediate neurophysiological effects of transcranial electrical stimulation. *Nature Communications, 9*(1), 5092. https://doi.org/10.1038/s41467-018-07233-7

Macaluso, E., Driver, J., & Frith, C. D. (2003). Multimodal spatial representations engaged in human parietal cortex during both saccadic and manual spatial orienting. *Current Biology, 13*(12), 990–999. https://doi.org/10.1016/S0960-9822(03)00377-4

Malcolm, M. P., Lavine, A., Kenyon, G., Massie, C., & Thaut, M. (2008). Repetitive transcranial magnetic stimulation interrupts phase synchronization during rhythmic motor entrainment. *Neuroscience Letters, 435*(3), 240–245. https://doi.org/10.1016/j.neulet.2008.02.055

Mannarelli, D., Pauletti, C., Petritis, A., Maffucci, A., Currà, A., Trompetto, C., ... & Fattapposta, F. (2023). The role of cerebellum in timing processing: A contingent negative variation study. *Neuroscience Letters, 808*, 137301. https://doi.org/10.1016/j.neulet.2023.137301

Méndez, J. C., Rocchi, L., Jahanshahi, M., Rothwell, J., & Merchant, H. (2017). Probing the timing network: A continuous theta burst stimulation study of temporal categorization. *Neuroscience, 356*, 167–175. https://doi.org/10.1016/j.neuroscience.2017.05.023

Mioni, G. (2020). Modulating subjective time perception with transcranial random noise stimulation (tRNS). *Journal of Cognitive Enhancement, 4*(1), 71–81. https://link.springer.com/article/10.1007/s41465-019-00128-5

Mioni, G., S. Grondin, and F. Stablum (2014). Temporal dysfunction in traumatic brain injury patients: Primary or secondary impairment? *Frontiers in Human Neuroscience, 8*, 269. https://doi.org/10.3389/fnhum.2014.00269

Mioni, G., Grondin, S., Forgione, M., Fracasso, V., Mapelli, D., & Stablum, F. (2016). The role of primary auditory and visual cortices in temporal processing: A tDCS approach. *Behavioural Brain Research, 313*, 151–157. https://doi.org/10.1016/j.bbr.2016.07.019

Mioni, G., Grondin, S., Mapelli, D., & Stablum, F. (2018). A tRNS investigation of the sensory representation of time. *Scientific Reports, 8*(1), 10364. https://doi.org/10.1038/s41598-018-28673-7

Mioni, G., Grondin, S., Bardi, L., & Stablum, F. (2020b). Understanding time perception through non-invasive brain stimulation techniques: A review of studies. *Behavioural Brain Research, 377*, 112232. https://doi.org/10.1016/j.bbr.2019.112232

Mioni, G., Capizzi, M., & Stablum, F. (2020a). Age-related changes in time production and reproduction tasks: Involvement of attention and working memory processes. *Aging, Neuropsychology, and Cognition, 27*(3), 412–429. https://doi.org/10.1080/13825585.2019.1626799

Mioni, G., Shelp, A., Stanfield-Wiswell, C. T., Gladhill, K. A., Bader, F., & Wiener, M. (2020c). Modulation of individual alpha frequency with tACS shifts time perception. *Cerebral Cortex Communications, 1*(1), tgaa064. https://doi.org/10.1093/texcom/tgaa064

Mondok, C., & Wiener, M. (2023). Selectivity of timing: A meta-analysis of temporal processing in neuroimaging studies using activation likelihood estimation and reverse inference. *Frontiers in Human Neuroscience, 16*, 1000995. https://doi.org/10.3389/fnhum.2022.1000995

Moreno-Duarte, I., Gebodh, N., Schestatsky, P., Guleyupoglu, B., Reato, D., Bikson, M., & Fregni, F. (2014). Transcranial electrical stimulation: Transcranial direct current stimulation (tDCS), transcranial alternating current stimulation (tACS), transcranial pulsed current stimulation (tPCS), and transcranial random noise stimulation (tRNS). In *The stimulated brain* (pp. 35–59). Academic Press.

Moss, J. H., & Maner, J. K. (2014). The clock is ticking: The sound of a ticking clock speeds up women's attitudes on reproductive timing. *Human Nature, 25*, 328–341. https://psycnet.apa.org/doi/10.1007/s12110-014-9210-7

Nani, A., Manuello, J., Liloia, D., Duca, S., Costa, T., & Cauda, F. (2019). The neural correlates of time: A meta-analysis of neuroimaging studies. *Journal of Cognitive Neuroscience, 31*(12), 1796–1826. https://doi.org/10.1162/jocn_a_01459

Naghibi, N., Jahangiri, N., Khosrowabadi, R., Eickhoff, C. R., Eickhoff, S. B., Coull, J. T., & Tahmasian, M. (2023). Embodying time in the brain: A multi-dimensional neuroimaging meta-analysis of 95 duration processing studies. *Neuropsychology Review*, 1–22. https://doi.org/10.1007/s11065-023-09588-1

Nitsche, M. A., & Paulus, W. (2000). Excitability changes induced in the human motor cortex by weak transcranial direct current stimulation. *The Journal of Physiology*, *527*(Pt 3), 633. https://doi.org/10.1111%2Fj.1469-7793.2000.t01-1-00633.x

Niemi, P., & Näätänen, R. (1981). Foreperiod and simple reaction time. *Psychological Bulletin*, *89*(1), 133–162. https://psycnet.apa.org/doi/10.1037/0033-2909.89.1.133

Oliveri, M., Koch, G., Salerno, S., Torriero, S., Gerfo, E. L., & Caltagirone, C. (2009). Representation of time intervals in the right posterior parietal cortex: Implications for a mental timeline. *Neuroimage*, *46*(4), 1173–1179. https://hdl.handle.net/2108/32250

Oldfield, R. C. (1971). The assessment and analysis of handedness: the Edinburgh inventory. *Neuropsychologia*, *9*(1), 97–113. https://doi.org/10.1016/0028-3932(71)90067-4

Oyama, F., Ishibashi, K., & Iwanaga, K. (2017). Cathodal transcranial direct-current stimulation over right posterior parietal cortex enhances human temporal discrimination ability. *Journal of Physiological Anthropology*, *36*(1), 1–10. https://doi.org/10.1186/s40101-017-0157-3

Pascual-Leone, A., Walsh, V., & Rothwell, J. (2000). Transcranial magnetic stimulation in cognitive neuroscience–virtual lesion, chronometry, and functional connectivity. *Current Opinion in Neurobiology*, *10*(2), 232–237. https://doi.org/10.1016/s0959-4388(00)00081-7

Paulus, W. (2011). Transcranial electrical stimulation (tES–tDCS; tRNS, tACS) methods. *Neuropsychological Rehabilitation*, *21*(5), 602–617. https://doi.org/10.1080/09602011.2011.557292

Paulus, W., Nitsche, M. A., & Antal, A. (2016). Application of transcranial electric stimulation (tDCS, tACS, tRNS). *European Psychologist*, *21*(1), 4–14. https://doi.org/10.1027/1016-9040/a000242

Peterchev, A. V., Luber, B., Westin, G. G., & Lisanby, S. H. (2017). Pulse width affects scalp sensation of transcranial magnetic stimulation. *Brain Stimulation*, *10*(1), 99–105. https://doi.org/10.1016%2Fj.brs.2016.09.007

Pollok, B., Rothkegel, H., Schnitzler, A., Paulus, W., & Lang, N. (2008). The effect of rTMS over left and right dorsolateral premotor cortex on movement timing of either hand. *European Journal of Neuroscience*, *27*(3), 757–764. https://doi.org/10.1111/j.1460-9568.2008.06044.x

Prete, G., Lucafò, C., Malatesta, G., & Tommasi, L. (2021). The causal involvement of the right supramarginal gyrus in the subjective experience of time: A hf-tRNS study. *Behavioural Brain Research*, *404*, 113157. https://doi.org/10.1016/j.bbr.2021.113157

Rattat, A. C., & Droit-Volet, S. (2012). What is the best and easiest method of preventing counting in different temporal tasks? *Behavior Research Methods*, *44*(1), 67–80. https://psycnet.apa.org/doi/10.3758/s13428-011-0135-3

Reed, T., & Cohen Kadosh, R. (2018). Transcranial electrical stimulation (tES) mechanisms and its effects on cortical excitability and connectivity. *Journal of Inherited Metabolic Disease*, *41*, 1123–1130. https://doi.org/10.1007%2Fs10545-018-0181-4

Riemer, M., Rhodes, D., & Wolbers, T. (2016). Systematic underreproduction of time is independent of judgment certainty. *Neural Plasticity*, 6890674 https://doi.org/10.1155/2016/6890674

Rocha, K., Marinho, V., Magalhães, F., Ribeiro, J., Oliveira, T., Gupta, D. S., ... & Teixeira, S. (2019). Low-frequency rTMS stimulation over superior parietal cortex medially improves time reproduction and increases the right dorsolateral prefrontal cortex predominance. *International Journal of Neuroscience, 129*(6), 523–533. https://doi.org/10.1080/00207454.2018.1476351

Romero, M. C., Davare, M., Armendariz, M., & Janssen, P. (2019). Neural effects of transcranial magnetic stimulation at the single-cell level. *Nature Communications, 10*(1), 2642. https://doi.org/10.1038/s41467-019-10638-7

Rossi, S., Hallett, M., Rossini, P. M., Pascual-Leone, A., & Safety of TMS Consensus Group. (2009). Safety, ethical considerations, and application guidelines for the use of transcranial magnetic stimulation in clinical practice and research. *Clinical Neurophysiology, 120*(12), 2008–2039. https://doi.org/10.1016/j.clinph.2009.08.016

Rossi, S., Antal, A., Bestmann, S., Bikson, M., Brewer, C., Brockmöller, J., ... & Hallett, M. (2021). Safety and recommendations for TMS use in healthy subjects and patient populations, with updates on training, ethical and regulatory issues: Expert Guidelines. *Clinical Neurophysiology, 132*(1), 269–306. https://doi.org/10.1016/j.clinph.2020.10.003

Rossini, P. M., Burke, D., Chen, R., Cohen, L. G., Daskalakis, Z., Di Iorio, R., ... & Ziemann, U. (2015). Non-invasive electrical and magnetic stimulation of the brain, spinal cord, roots and peripheral nerves: Basic principles and procedures for routine clinical and research application. An updated report from an IFCN Committee. *Clinical Neurophysiology, 126*(6), 1071–1107. https://doi.org/10.1016/j.clinph.2015.02.001

Ruspantini, I., Mäki, H., Korhonen, R., D'Ausilio, A., & Ilmoniemi, R. J. (2011). The functional role of the ventral premotor cortex in a visually paced finger tapping task: A TMS study. *Behavioural Brain Research, 220*(2), 325–330. https://doi.org/10.1016/j.bbr.2011.02.017

Salvioni, P., Murray, M. M., Kalmbach, L., & Bueti, D. (2013). How the visual brain encodes and keeps track of time. *Journal of Neuroscience, 33*(30), 12423–12429. https://doi.org/10.1523/JNEUROSCI.5146-12.2013

Santarnecchi, E., Brem, A. K., Levenbaum, E., Thompson, T., Kadosh, R. C., & Pascual-Leone, A. (2015). Enhancing cognition using transcranial electrical stimulation. *Current Opinion in Behavioral Sciences, 4*, 171–178. https://doi.org/10.1016/j.cobeha.2015.06.003

Siebner, H. R., Funke, K., Aberra, A. S., Antal, A., Bestmann, S., Chen, R., ... & Ugawa, Y. (2022). Transcranial magnetic stimulation of the brain: What is stimulated? A consensus and critical position paper. *Clinical Neurophysiology, 140*, 59–97. https://doi.org/10.1016/j.clinph.2022.04.022

Silvanto, J., & Pascual-Leone, A. (2008). State-dependency of transcranial magnetic stimulation. *Brain Topography, 21*, 1–10. https://doi.org/10.1007%2Fs10548-008-0067-0

Stokes, M. G., Chambers, C. D., Gould, I. C., English, T., McNaught, E., McDonald, O., & Mattingley, J. B. (2007). Distance-adjusted motor threshold for transcranial magnetic stimulation. *Clinical Neurophysiology, 118*(7), 1617–1625. https://doi.org/10.1016/j.clinph.2007.04.004

Tan, B., Liao, Q., Xu, P., Zhang, J., Jin, Z., & Li, L. (2023). Selective enhancement of frontal-posterior functional connectivity by anodal tDCS over the right posterior parietal cortex during temporal attention. *IEEE Journal of Biomedical and Health Informatics.* https://doi.org/10.1109/jbhi.2023.3267063

Teghil, A., Boccia, M., D'Antonio, F., Di Vita, A., de Lena, C., & Guariglia, C. (2019). Neural substrates of internally-based and externally-cued timing: An activation likelihood estimation (ALE) meta-analysis of fMRI studies. *Neuroscience & Biobehavioral Reviews, 96*, 197–209. https://doi.org/10.1016/j.neubiorev.2018.10.003

Terney D, Chaieb L, Moliadze V, Antal A, Paulus W (2008). Increasing human brain excitability by transcranial high-frequency random noise stimulation. *Journal of Neuroscience 28*(52), 14147–14155. https://doi.org/10.1523/JNEUROSCI.4248-08.2008

Théoret, H., Haque, J., & Pascual-Leone, A. (2001). Increased variability of paced finger tapping accuracy following repetitive magnetic stimulation of the cerebellum in humans. *Neuroscience Letters, 306*(1–2), 29–32. https://doi.org/10.1016/s0304-3940(01)01860-2

Tringali, S., Perrot, X., Collet, L., & Moulin, A. (2012). Repetitive transcranial magnetic stimulation: hearing safety considerations. *Brain Stimulation, 5*(3), 354–363. https://doi.org/10.1016/j.brs.2011.06.005

Vallesi, A., Shallice, T., & Walsh, V. (2007). Role of the prefrontal cortex in the foreperiod effect: TMS evidence for dual mechanisms in temporal preparation. *Cerebral Cortex, 17*(2), 466–474. https://doi.org/10.1093/cercor/bhj163

Verstynen, T., Konkle, T., & Ivry, R. B. (2006). Two types of TMS-induced movement variability after stimulation of the primary motor cortex. *Journal of Neurophysiology, 96*(3), 1018–1029. https://doi.org/10.1152/jn.01358.2005

Vicario, C. M., Martino, D., & Koch, G. (2013). Temporal accuracy and variability in the left and right posterior parietal cortex. *Neuroscience, 245*, 121–128. https://doi.org/10.1016/j.neuroscience.2013.04.041

Walsh, V., & Cowey, A. (2000). Transcranial magnetic stimulation and cognitive neuroscience. *Nature Reviews Neuroscience, 1*(1), 73–80. https://doi.org/10.1038/35036239

Wearden, J. H., Williams, E. A., & Jones, L. A. (2017). What speeds up the internal clock? Effects of clicks and flicker on duration judgements and reaction time. *Quarterly Journal of Experimental Psychology, 70*(3), 488–503. https://doi.org/10.1080/17470218.2015.1135971

Wiener, M. (2014). Transcranial magnetic stimulation studies of human time perception: A primer. *Timing & Time Perception, 2*(3), 233–260. https://doi.org/10.1163/22134468-00002022

Wiener, M., Hamilton, R., Turkeltaub, P., Matell, M. S., & Coslett, H. B. (2010). Fast forward: supramarginal gyrus stimulation alters time measurement. *Journal of Cognitive Neuroscience, 22*(1), 23–31. https://doi.org/10.1162/jocn.2009.21191

Wiener, M., Kliot, D., Turkeltaub, P. E., Hamilton, R. H., Wolk, D. A., & Coslett, H. B. (2012). Parietal influence on temporal encoding indexed by simultaneous transcranial magnetic stimulation and electroencephalography. *Journal of Neuroscience, 32*(35), 12258–12267. https://psycnet.apa.org/doi/10.1523/JNEUROSCI.2511-12.2012

Wiener, M., Parikh, A., Krakow, A., & Coslett, H. B. (2018). An intrinsic role of beta oscillations in memory for time estimation. *Scientific Reports, 8*(1), 7992. https://doi.org/10.1038/s41598-018-26385-6

Woods, A. J., Antal, A., Bikson, M., Boggio, P. S., Brunoni, A. R., Celnik, P., … & Nitsche, M. A. (2016). A technical guide to tDCS, and related non-invasive brain stimulation tools. *Clinical Neurophysiology, 127*(2), 1031–1048. https://doi.org/10.1016/j.clinph.2015.11.012

Yin, H. Z., Cheng, M., & Li, D. (2019). The right dorsolateral prefrontal cortex is essential in seconds range timing, but not in milliseconds range timing: An investigation with transcranial direct current stimulation. *Brain and Cognition, 135*, 103568. https://doi.org/10.1016/j.bandc.2019.05.006

Zaghi, S., Acar, M., Hultgren, B., Boggio, P. S., & Fregni, F. (2010). Noninvasive brain stimulation with low-intensity electrical currents: Putative mechanisms of action for direct and alternating current stimulation. *The Neuroscientist, 16*(3), 285–307. https://doi.org/10.1177/1073858409336227

6

PSYCHOPHYSIOLOGICAL CORRELATES OF TIME PERCEPTION

Nicola Cellini and Luigi Micillo

Università di Padova, Padova, Italy

Introduction

Time flies when you're having fun. This very popular idiom, which has a similar translation in several languages, encapsulates a key and well-known element of time processing: our perception of time is consistently modulated (or distorted) by elements of our ongoing experience (see Lake et al., 2016). However, taking this idiom as an example, we may ask: Is it the cognitive aspect of having fun that accelerates the passage of our time? Or may it rather be that time processing is distorted by the physiological activation (arousal) that accompanies our emotional experience? Or is it a combination of cognitive and physiological aspects that drive our ongoing time experience?

Some of these questions were already asked 90 years ago. Indeed, already in 1933, Hoagland proposed the internal chemical-clock hypothesis, which postulated that duration estimates were directly dependent on internal body temperature (see also Hancock, 1993). Years later, Cahoon published a paper called *"Physiological arousal and time estimation"* (1969), in which he manipulated physiological arousal using the threat of shock and measured time using a verbal estimation of a 36-s period between two stimuli, temporal production of a 22-s period, and 1-s finger tapping task. The threat of shock was able to increase heart rate (HR) in the participants, and in this condition, participants who were classified as "chronic arousal" based on an anxiety test showed a positive association between HR and verbal time estimation and finger tapping. The data reported by Cahoon were not easy to interpret, and the author states that "a problem in drawing conclusions from studies of arousal lies in the lack of agreement among investigators of what

DOI: 10.4324/9781003449546-6

constitutes a valid measure of this concept." A few years later, Curton and Lordahl (1974) manipulated physiological and psychological arousal by asking participants to perform a Harward step test or by arousing them with the threat of shock, which was actually never delivered. They manipulated arousal before a 1-s temporal estimation task (they had to connect up to 25 dots, at a rate of 1 dot each second). They found no link between HR changes following the threat of the shock and the temporal performace. However, they observed an unexpected negative correlation between increased HR in the physical condition (i.e., the Horward step test) and temporal estimation. The authors suggested that physical activity may have reduced "mental" arousal or alertness. These two studies published more than 70 years ago, highlight some of the problems we are still facing while investigating the relationship between physiological arousal and temporal processing: lack of a clearly defined concept of both arousal and temporal processing (e.g., which *type* of time), different methodologies, and contrasting results.

In the present chapter, we will not respond to these questions, since no clear-cut answer can be provided at the moment. However, we will try to provide some information to better understand the complex relationship between cognitive and physiological arousal and time processing. We should specify that here we will focus on psychophysiological signals coming from the autonomic nervous system (ANS) since the concept of physiological arousal in time perception is historically linked to temperature and cardiac activity. Also, the role of the central nervous system in time processing is extensively described in other chapters of this book.

Therefore, here we start by briefly describing the ANS and the most used psychophysiological signals in the timing literature. Then we present the timing studies investigating the relationship with psychophysiological signals. Studies are organized according to the design/manipulation used to investigate this relationship. Then we discuss some caveats of using physiological signals to study temporal processing. We conclude by presenting some tools (both open-source and proprietary) that can help researchers analyze their physiological data.

Autonomic nervous system and psychophysiological signals

The ANS is a division of the peripheral nervous system responsible for regulating involuntary bodily functions, including HR, digestion, respiratory rate, and pupillary response. It is called *autonomic* since it operates largely (but not exclusively) unconsciously and automatically to maintain homeostasis within the body. The ANS is divided into two main branches: the sympathetic nervous system (SNS) and the parasympathetic nervous system

(PNS). The SNS is often referred to as the "fight or flight" system since it prepares the body for activity and stress responses. When the SNS increases its activity, it increases HR, redirects blood flow to skeletal muscles, dilates pupil size, and releases stress hormones like adrenaline and noradrenaline. The PNS, often referred to as the "rest and digest" system, promotes relaxation, digestion, and recovery after stress. It slows HR, stimulates digestion, contracts the pupil, and in general promotes the conservation of energy.

Both branches of the ANS work in concert to maintain balance and regulate physiological functions throughout the body (Figure 6.1). Their activities are controlled by complex neural circuits located in the brainstem and higher brain centers, as well as feedback from various sensory receptors and internal organs.

In the next section, we are going to focus on some ANS signals typically used in psychophysiology, such as cardiac and electrodermal activity. To better understand these signals, we are providing a brief introduction to the physiological system behind them.

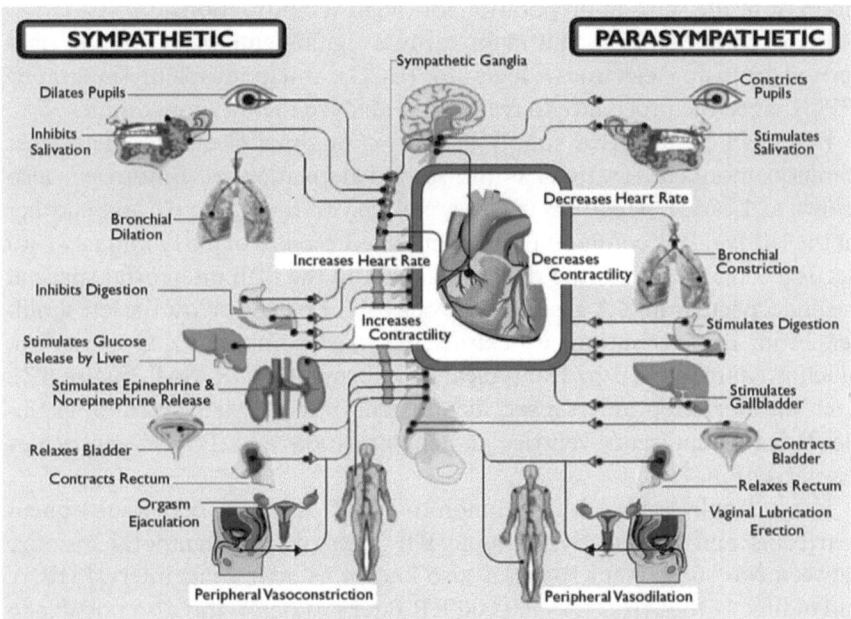

FIGURE 6.1   The two branches of the autonomic nervous system and their main effects on different parts of the body. From Vinik, 2012.

Cardiac activity

With cardiac activity, we refer to the electrical and mechanical processes that occur within the heart to facilitate its function of pumping blood throughout the body. It involves a coordinated series of events, including the generation and propagation of electrical impulses, which lead to the contraction and relaxation of cardiac muscle tissue. Electrical activity in the heart originates from specialized cells within the sinoatrial (SA) node, often referred to as the heart's natural pacemaker. These electrical impulses spread through the atria, causing them to contract and pump blood into the ventricles. The impulses then pass through the atrioventricular node, where they are delayed briefly before continuing down the specialized conduction pathways (the bundle of His, bundle branches, and Purkinje fibers) to the ventricles. This electrical stimulation triggers the ventricles to contract, forcing blood out of the heart and into the arteries. The electrical impulses are mediated by the SNA, with the PNS mainly affecting the SA node, whereas the SNS affects the ventricles. Mechanical activity refers to the physical contraction and relaxation of the heart muscle (myocardium) in response to electrical signals. When the heart contracts (systole), blood is pumped out of the chambers into the circulation. During relaxation (diastole), the heart fills with blood from the veins in preparation for the next contraction.

In psychophysiology, the main cardiac signals employed are the ones derived by using electrocardiography (ECG) and photoplethysmography (PPG), since the procedure to record and analyze them is not complex.

ECG can be recorded simply using two or three electrodes. The most common montage for ECG is the second derivation of Einthoven, also known as Lead II, where electrodes are placed on the right arm and another on the left leg. It is common to use a modified version of this montage, with electrodes placed below the right shoulder and the fifth intercostal space at the midclavicular line. A ground electrode can be placed below the left shoulder. From this derivation, we extract the most common ECG waveform, which is characterized by five typical waveforms (P–Q–R–S–T, Figure 6.2). Each waveform represents a specific moment in the cardiac cycle, for example, R-wave represents ventricular depolarization and T-wave ventricular repolarization.

In psychophysiology, it is common to extract the R-peaks of subsequent heartbeats and to directly compute RR intervals (the temporal distance between two subsequent R-peaks, also known as inter beats interval (IBI)), and indirectly the HR (HR = 60,000/RR intervals (ms)). It is also possible to extract the amplitude of the T-wave, which represents the ventricular repolarization and can be used as an inverse measure of cardiac sympathetic activity (the higher the T-wave amplitude, the lower the sympathetic activity).

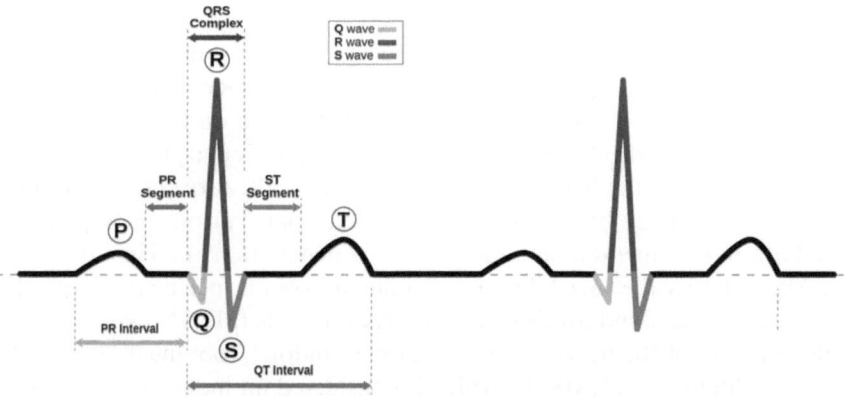

However, for several reasons, including recent challenges on its suitability as an SNS index (Drost et al., 2022), the analysis of the T-wave never became popular, and to date is barely used in psychophysiological studies.

From the RR intervals, we can conduct several types of analysis. The most common are the average RR (or mean HR) in a specific period (e.g., at rest and while performing a task), the phasic cardiac response (PCR; also known as heart rate deceleration [HRD]), and the heart rate variability (HRV).

Mean RR intervals can be used to have an average measure of the cardiac activity of the participant, but alone it does not tell us much information. This is due to the nature of the RR, which is affected by several variables such as age, biological sex, fitness level, and menstrual cycle phase. Instead, it could be more appropriate to collect average RR measures in several time points, for example, before, during, and after a temporal task, or, for instance, dividing a 5-min task into 1-min bins to assess changes in cardiac frequency.

The PCR (aka HRD) is a transient change in HR (or RR intervals) that occurs in response to specific stimuli or events. Conceptually, it is similar to an event-related potential (ERP; see Chapter 3), but the response unfolds on a different timescale (seconds instead of milliseconds). An example of PCR is when a novel visual stimulus occurs on the screen. The novel stimulus elicits an "orienting response" (see Bradley, 2009), and the HR usually shows a deceleration in the 2–4 s after the stimulus onset, which is followed by an acceleration. PCR usually uses 1–2 s before the stimulus as a baseline, and the magnitude of the deceleration depends on several factors. For instance, an unpleasant or a salient stimulus will produce a more marked deceleration.

The HRV is a very popular concept in psychophysiology. It refers to the variation in the time intervals between consecutive heartbeats (IBI or RR), and this variation mainly (but not exclusively) depends on the regulation of the heart by the ANS. Simply speaking, the heart does not beat like a metronome, but it constantly changes its frequency. With HRV, it is possible to assess these variations and to obtain some information on the influence of the PNS on the heart (also known as cardiac "vagal" activity). The analysis of HRV is based on the RR intervals. Using specific software and algorithms (or by hand if you are brave enough), we can calculate various HRV parameters, both in the time domain and in the frequency domain. Time-domain measures quantify the variability of RR intervals over a specific time period. Examples include standard deviation of NN intervals (SDNN, which is the total variability of the heart on a specific time window), root mean square of successive differences (RMSSD, which is considered an index of PNS), and the percentage of successive RR intervals differing by more than 50 ms (pNN50, another PNS measure). In the frequency domain, we analyze the distribution of power within different frequency bands of the HRV spectrum. The two main frequency bands are low frequency (LF-HRV, 0.04–0.15 Hz) and high frequency (HF-HR, 0.15–0.4 Hz). While most of the researchers consider HF-HRV a reliable index of the PNS, the LF-HRV is now considered influenced by both PNS and SNS. It is important to specify that, despite some authors claiming that ultra-short-term HRV measures are feasible and informative (Castaldo et al., 2019), most of the recording and analysis follow the 1996 Task Force Recommendations (Camm et al., 1996). The task force recommends the use of at least 1 min of continuous recording for having a reliable HF-HRV measure, and 2 min for LF-HRV, although the suggested recording for short-term HRV is considered 5 min of recording. These timing constraints need to be considered before planning an HRV study.

Another technique to assess cardiac activity is impedance cardiography (ICG). ICG is a non-invasive technique used to assess cardiac function by measuring changes in thoracic electrical impedance over the cardiac cycle. Usually, it requires the use of four electrodes (either spot or strip electrodes) placed on the neck and thorax to measure transthoracic impedance. It provides information about stroke volume (how much blood is pumped in each heartbeat), cardiac output (how much blood is pumped every minute), and other hemodynamic parameters. Combining ICG with ECG, it is possible to measure the pre-ejection period (PEP; milliseconds), which is the time interval between the onset of ventricular depolarization (the beginning of the QRS complex on the ECG) and the opening of the aortic valve, allowing blood to be ejected from the left ventricle into the aorta during systole. PEP is considered a marker of cardiac contractility, and it is inversely related to cardiac beta-adrenergic SNS activity, that is, the shorter the PEP, the higher the SNS activity in the heart.

Electrodermal activity

Electrodermal activity (EDA), also known as galvanic skin response or skin conductance (SC), refers to changes in the electrical properties of the skin in response to various stimuli (e.g., emotional and physiological). It is based on the activity of eccrine sweat glands, which is controlled by the SNS. Whenever the SNS is activated during emotional arousal, stress, or other stimuli, these glands increase their activity, leading to an increase in sweating, which increases the skin's electrical conductivity due to the presence of electrolytes in sweat. Differently from PEP, EDA is an index of peripheral alpha-adrenergic SNS activity. EDA is typically measured by placing two electrodes on the skin's surface, often on the palms of the hands, the soles of the feet, or the fingers, where the eccrine sweat glands are most present. Since EDA is a measure of conductance and not a voltage, it requires the use of a transducer. EDA is considered a measure of nonspecific arousal and its signal typically consists of two main components: tonic activity and phasic activity. The tonic activity, often referred to as SC level (SCL), represents the baseline level of SC, reflecting the individual's overall arousal state. The phasic activity represents the change of SC in response to specific stimuli or events (SC response (SCR)) or, sometimes, in the absence of any specific event (non-specific SCR). SCRs are characterized by transient increases in SC amplitude that occur 2–4 s after a stimulus presentation, followed by a slow return to the baseline level (Figure 6.3a). It is also possible to assess changes in SCL, which are usually slower changes that occur over a longer period and are associated with sustained changes in arousal or attention (Figure 6.3b).

We recommend that readers interested in using EDA to familiarize themselves with the great work of the *Society for Psychophysiological Research Ad Hoc Committee on Electrodermal Measures* on the recommendations for EDA measurements (Boucsein et al., 2012) in order to familiarize themselves with the principles, use, and constraints of EDA.

Other psychophysiological signals

Although ECG and EDA are the most used physiological signals in the timing literature, mainly due to the ease with which these signals are acquired, other physiological tools are or can be employed. For instance, several studies have focused on the relationship between time perception and temperature, starting from the internal chemical-clock hypothesis (Hoagland, 1933) to more recent studies (e.g., Aschoff, 1998; Mioni et al., 2016). However, it should be noted that measuring core body temperature (e.g., the temperature at the hypothalamus level) can be quite challenging since it is often measured in a discrete and discontinued way (e.g., one assessment every 10 min), for example, using infrared thermometers. Skin temperature is easier to record, also in a continuous fashion, but has a different range of values compared to core temperature, and it is characterized by faster changes than core temperature (Bulcao et al., 2000). These differences need to be considered while planning a study with this variable.

Other useful physiological information can be derived from the use of eye-tracking, which can provide information about pupil size. Indeed, pupillometry is a technique that allows the study of the changes in pupil diameter in

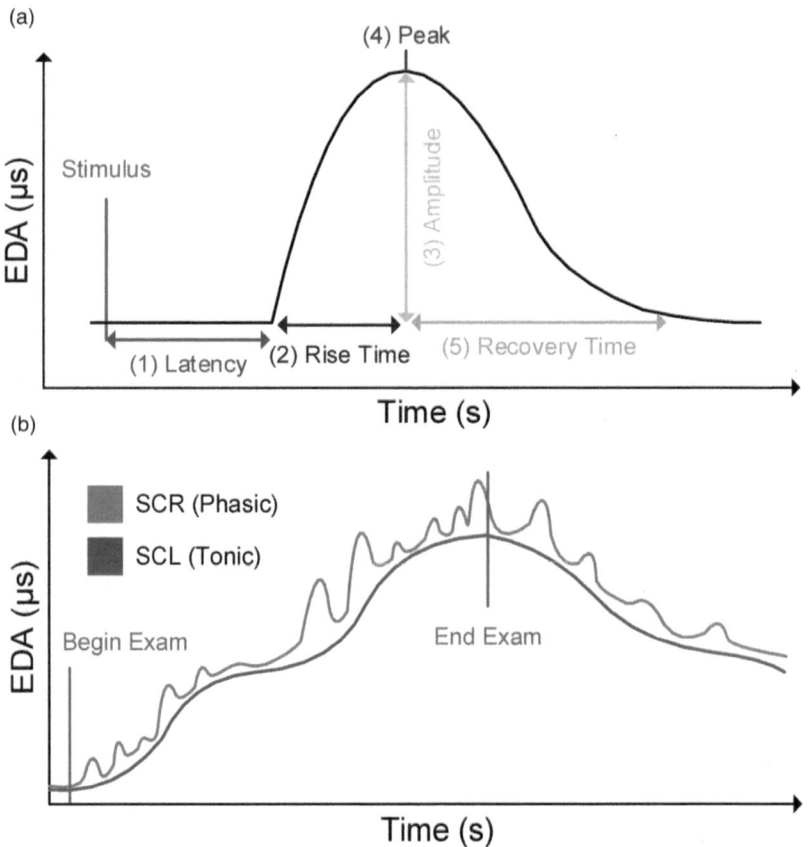

FIGURE 6.3   a) Schematic representation of an SCR after a stimulus. b) Schematic distinction of tonic (SCL) and phasic (SCR) over time from the beginning to the end of an event (e.g., an examination). EDA: Electrodermal Activity; SCR: Skin Conductance Response; SCL: Skin Conductance Level; μS: microsiemens. Adapted from Winter et al., 2020.

response to cognitive processing, emotional arousal, and other stimuli (Sirois & Brisson, 2014). Pupil size is controlled by two iris muscles: the dilator pupillae, which dilates the pupil under sympathetic control, and the sphincter pupillae, which constricts the pupil under parasympathetic control. Therefore, pupillometry can provide information about the SNA and can be used to assess the psychophysiological correlates of temporal processing.

Other physiological indices, such as electromyography or blood pressure, are less useful for investigating the time perception, due to the difficulties of creating research protocols that can reliably discriminate normal, task-unrelated, physiological variations in these signals from variations due to specific manipulations.

Studies investigating the relationship between psychophysiological activity and time processing

In this section, we are providing a list of studies investigating the relationship between psychophysiological activity and temporal processing. The studies are presented based on the manipulation used to modulate physiological activity.

Studies with no direct manipulation of the arousing state

Few studies have investigated the relationship between physiological activity changes during a temporal task and temporal estimation (Table 6.1). For instance, Meissner and Wittmann (2011) investigated the association between participants' performance in a temporal reproduction task (8-, 14-, and 20-s reproduction) and both their interoceptive awareness and ANS response (cardiac, respiratory, and electrodermal) during the task. The RR interval increased (i.e., the HR slow down) across both the encoding and the reproduction phases of the three durations, although this increase was greater during the encoding than the reproduction phase. Instead, SCL showed an initial increase followed by a steady linear decrease across the encoding, and a linear decrease throughout the reproduction phase. The respiration pattern was less consistent and more difficult to interpret. They also reported that the increase of the RR interval during the task was positively associated with accuracy for the longer durations (14 and 20 s), and this relationship was stronger for participants who performed better in the reproduction task. No associations were observed between accuracy and both SCL and respiratory pattern. Another study by the same research group (Otten et al., 2015) replicated the changes in RR interval and SCL activity during a temporal reproduction task (again 8-, 14-, and 20-s reproduction). However, differently from their previous study, this time they found that reproduction

TABLE 6.1 Studies with no direct manipulation of the arousing state

Article Info	Condition	Physiological Activation			Time Perception		
Authors		Recording	Index	Effect	Task	Duration	Effect
Cellini et al. (2015)	-	ECG	PEP	↑	Time Bisection	0.3-0.9s	+
		ECG	HRV	↑	Time Bisection	0.3-0.9s	+
		ECG	TP	↑	Time Bisection	0.3-0.9s	+
		ECG	PEP	↑	Tapping	1s	+
		ECG	HRV	↑	Tapping	1s	+
		ECG	TP	↑	Tapping	1s	+
Fung et al. (2017)	-	ECG	HRV	↑	Time Reproduction	2-5s	-
Meissner & Wittman (2011)	-	ECG	RR interval	↑	Time Reproduction	8-20s	+
		EDA	SCL	↓	Time Reproduction	8-20s	/
Otten et al. (2015)	-	ECG	RR interval	↑	Time Reproduction	8-20s	+
		EDA	SCL	↓	Time Reproduction	8-20s	+
Pollatos et al. (2014)	-	ECG	RMSSD	↑	Time Reproduction	0.5-40s	+

Note: ECG: Electrocardiogram; EDA: Electrodermal Activity; HRV: Heart Rate Variability; RMSSD: Root Mean Square of Successive Differences between normal heartbeats; PEP: Pre-Ejection Period; SCL: Skin Conductance Level; TP: Total Power; +: More Accurate; −: Less Accurate; ↓: Decrease; ↑: Increase; <: Underestimation; >: Overestimation; ↕: No changes; /: No effect.

accuracy for the 8-s interval was positively associated with the RR increase and negatively associated with the SCL decrease across the reproduction phase. No associations were observed for longer durations (14 and 20 s). The authors discussed the discrepancies between the two studies as differences in the characteristics, study sample in terms of age, education, and interoceptive awareness.

A handful of studies have investigated the relationship between physiological activity at rest and temporal processing. For instance, in one of our studies, we investigated the relationship between cardiac activity at rest and the subsequent temporal accuracy in a time bisection task (Cellini et al., 2015). We observed that participants' HF-HRV (i.e., higher PNS activity) at rest was positively associated with accuracy in a time bisection task and a 1-s finger tapping task. Moreover, participants with lower cardiac SNS activity, as indexed by the PEP, had a higher temporal sensitivity in the time bisection task. Of note, differently from Meissner and Wittmann (2011), in our study, we employed sub-second intervals duration. Consistently with our study, a positive relationship between cardiac PNS activity at rest (as indexed by the RMSSD) and time reproduction accuracy of a wide range of durations (0.5, 2, 3, 7, 10, 14, 25, and 40 s) was also found by Pollatos et al. (2014), but not in Fung et al. (2017), who reported a decrease in reproduction accuracy (duration ranging from 2 to 5 s) in participants with high LF-HRV.

Some of the studies have also investigated the relationship between interoceptive awareness and temporal processing. Meissner and Wittmann (2011) showed that interoceptive awareness was positively associated with reproduction accuracy, although they were not able to replicate this result (Otten et al., 2015). The authors explained the lack of replication due to the different levels of interoceptive awareness in the more recent study, which was higher than in the previous one. Also, Pollatos et al. (2014) did not find a significant relationship between interoceptive awareness and temporal accuracy in a reproduction task. In our study (Cellini et al., 2015), we found that interoceptive awareness was negatively associated with the tap rate of a spontaneous finger-tapping task and the coefficient of variation of the 1-s finger-tapping task.

Studies using emotional stimuli to alter physiological arousal

There is extensive literature investigating how emotional information can alter perception (see Droit-Volet & Meck, 2007; Lake et al., 2016 for some reviews on the topic), often capitalizing on emotional stimuli. However, a paucity of these studies has also measured physiological signals as an index of arousal (Table 6.2). Therefore, this paragraph will focus on the studies reporting a physiological measure of arousal.

TABLE 6.2 Studies using emotional stimuli to alter physiological arousal

Article info Authors	Arousal manipulator Type	Condition	Physiological activation Recording	Index	Effect	Time perception Task	Duration	Effect
Angrilli et al. (1997)	IAPS	Low negative	EDA	SCR	↑	Verbal estimation	2–6 s	✓
	IAPS	Low positive	EDA	SCR	↑	Verbal estimation	2–6 s	✓
	IAPS	High negative	EDA	SCR	↑↑	Verbal estimation	2–6 s	✓
	IAPS	High positive	EDA	SCR	↑↑	Verbal estimation	2–6 s	✓
	IAPS	Low negative	ECG	HR	→	Verbal estimation	2–6 s	✓
	IAPS	Low positive	ECG	HR	↑	Verbal estimation	2–6 s	✓
	IAPS	High negative	ECG	HR	→	Verbal estimation	2–6 s	✓
	IAPS	High positive	ECG	HR	↑	Verbal estimation	2–6 s	✓
Appelqvist-Dalton et al. (2022)	Video	Audio + Video	Oculometry	Pupil diameter	↑↑	Retrospective judgment	5–13 s	✓
	Video	Video only	Oculometry	Pupil diameter	↑	Retrospective judgment	5–13 s	✓
	Video	Audio only	Oculometry	Pupil diameter	↑	Retrospective judgment	5–13 s	✓
Gros et al. (2015)	Video vs. Odors	Odors	EDA	EDA	↑	Time bisection	0.4–1.6 s	>
	Video vs. Odors	Video	EDA	EDA	↑↑	Time bisection	0.4–1.6 s	/
Mella et al. (2011)	IADS	High arousal/ Attention time	EDA	SCR	/	Time discrimination	1.8–2.2 s	/
	IADS	Low arousal/ Attention time	EDA	SCR	/	Time discrimination	1.8–2.2 s	/
	IADS	High arousal/ Attention emotion	EDA	SCR	↑	Time discrimination	1.8–2.2 s	>
	IADS	Low arousal/ Attention emotion	EDA	SCR	/	Time discrimination	1.8–2.2 s	/

Study	Stimulus	Condition	Measure		Measure	Task	Duration	
Ogden et al. (20190	IAPS	Low negative	ECG	/	PEP	Verbal estimation	0.2–0.8 s	>
	IAPS	Low positive	ECG	/	PEP	Verbal estimation	0.2–0.8 s	/
	IAPS	High negative	ECG	↑	PEP	Verbal estimation	0.2–0.8 s	>
	IAPS	High positive	ECG	/	PEP	Verbal estimation	0.2–0.8 s	/
Schreuder et al. (2014)	Body posture/ Odor	Unrelaxing chair/ Arousing odor	ECG	↑	HR	Time production	1.33–2.17 m	<
	Body posture/ Odor	Unrelaxing chair/ Arousing odor	EDA	↑	SCR	Time production	1.33–2.17 m	<
	Body posture/ Odor	Unrelaxing chair/ Non-arousing odor	ECG	↑	HR	Time production	1.33–2.17 m	/
	Body posture/ Odor	Unrelaxing chair/ Non-arousing odor	EDA	↑	SCR	Time production	1.33–2.17 m	/
Van Hedger et al. (2017)	TSST + IAPS	Stress positive	ECG	↑	HR	Time reproduction	0.4–4.15 s	>
	TSST + IAPS	Stress negative	ECG	↑	HR	Time reproduction	0.4–4.15 s	>
	TSST + IAPS	Stress neutral	ECG	↑	HR	Time reproduction	0.4–4.15 s	>
	TSST + IAPS	Stress positive	ECG	/	HRV	Time reproduction	0.4–4.15 s	>
	TSST + IAPS	Stress negative	ECG	/	HRV	Time reproduction	0.4–4.15 s	>
	TSST + IAPS	Stress neutral	ECG	/	HRV	Time reproduction	0.4–4.15 s	>
	TSST + IAPS	Stress positive	ECG	↓	PEP	Time reproduction	0.4–4.15 s	>
	TSST + IAPS	Stress negative	ECG	↓	PEP	Time reproduction	0.4–4.15 s	>
	TSST + IAPS	Stress neutral	ECG	↓	PEP	Time reproduction	0.4–4.15 s	>

(Continued)

TABLE 6.2 (Continued)

Article info	Arousal manipulator		Physiological activation			Time perception		
Authors	Type	Condition	Recording	Index	Effect	Task	Duration	Effect
Wöllner et al. (2018)	Video + Speed	Audio + Video (slow)	ECG	HR	↑↑	Retrospective judgment	2.67–40 s	<<
	Video + Speed	Video only (slow)	ECG	HR	↑	Retrospective judgment	2.67–40 s	<<
	Video + Speed	Audio + Video (regular)	ECG	HR	↑↑	Retrospective judgment	2.67–40 s	<
	Video + Speed	Video only (regular)	ECG	HR	↑	Retrospective judgment	2.67–40 s	<
	Video + Speed	Audio + Video (slow)	EDA	SCR	↑↑	Retrospective judgment	2.67–40 s	<<
	Video + Speed	Video only (slow)	EDA	SCR	↑	Retrospective judgment	2.67–40 s	<<
	Video + Speed	Audio + Video (regular)	EDA	SCR	↑↑	Retrospective judgment	2.67–40 s	<
	Video + Speed	Video only (regular)	EDA	SCR	↑	Retrospective judgment	2.67–40 s	<
	Video + Speed	Audio + Video (slow)	Oculometry	Pupil diameter	↑↑	Retrospective judgment	2.67–40 s	<<
	Video + Speed	Video only (slow)	Oculometry	Pupil diameter	↑	Retrospective judgment	2.67–40 s	<<
	Video + Speed	Audio + Video (regular)	Oculometry	Pupil diameter	↑↑	Retrospective judgment	2.67–40 s	<
	Video + Speed	Video only (regular)	Oculometry	Pupil diameter	↑	Retrospective judgment	2.67–40 s	<

Notes: ECG: Electrocardiogram; EDA: Electrodermal Activity; HRV: Heart Rate Variability; IADS: International Affective Digitized Sounds; IAPS: International Affective Picture System; TSST: Trier Social Stress Test; PEP: Pre-Ejection Period; SCL: Skin Conductance Level; SCR: Skin Conductance Response; −: Less Accurate; +: More Accurate; ↓: Decrease; ↑: Increase; <: Underestimation; >: Overestimation; ↕: No changes; /: No effect.

A seminal study on the effect of emotional stimulation and time perception was conducted by Angrilli and colleagues in 1997. In their study, the authors manipulated the participants' emotional states using pleasant, unpleasant, and neutral pictures, taken from the International Affective Picture System (IAPS; Lang et al., 1997), as stimuli for a time reproduction task. Note that these types of stimuli are intrinsically characterized by a valence (e.g., positive-negative and pleasant-unpleasant) and an arousal (low–high) level. At the same time, they recorded EDA and ECG, using SCR and HR as independent indexes of arousal and attention, respectively (Angrilli et al., 1997). Their results showed that highly arousing pleasant pictures were underestimated more than highly arousing unpleasant pictures. The opposite was observed for low-arousing pictures, with the unpleasant ones being more underestimated than the pleasant ones. At the physiological level, pleasant pictures were associated with an HR increase, while the opposite effect (i.e., HR decrease) was observed for unpleasant pictures. Note that HR was assessed only during the first 2 s after picture onset. Arousing stimuli also induced an increase in SCR amplitude between 1 and 4 s after the stimulus onset, regardless of the valence (i.e., pleasant or unpleasant).

Years later, Mella et al. (2011) asked participants to perform a temporal discrimination task with auditory stimuli taken from the IADS (Stevenson & James, 2008) that could be either neutral, high arousing unpleasant, or low arousing pleasant sounds. Moreover, the authors manipulated the task requests, asking participants to either pay attention to the emotional intensity of the stimuli, the duration of the sound, or both features. In the emotional intensity condition, participants perceived the duration of the high-arousal sounds as longer compared to the other conditions. Moreover, in the emotional intensity condition, SCR was higher for high-arousal sounds compared to the sounds.

Ogden et al. (2019) investigated the effect of picture valence (positive, neutral, and negative) and arousal level (high vs. low) on the relationship between physiological arousal and the verbal estimation of the duration of the same pictures. Stimuli could last between 200 and 800 ms. They showed that emotional stimuli were perceived as lasting longer than neutral ones. PNS activity, as assessed by HF-HRV, was not affected by valence and arousal of the stimuli. Instead, cardiac SNS activity, as assessed by the PEP, was higher during the exposure to the negative pictures compared to the neutral ones. Moreover, only for the high arousal negative picture, the increase in SNS activity was associated with longer perceptions of stimuli durations. No association was observed for PNS activity. Interestingly, although positive stimuli induced an increase in SNS activity, for these stimuli, no association was observed between physiological arousal and perceived duration, suggesting that physiological arousal does not influence time processing, at least for positive and low-arousing stimuli. These results were in line with findings from Van Hedger et al. (2017), who increased physiological arousal (HR, HF-HRV, and PEP) of participants using a modified version of the Trier Social Stress Test before and after asking them to perform a reproduction task with IAPS stimuli. The authors reported that emotional stimuli were perceived as lasting longer after the psychophysiological stress compared to neutral ones. Moreover, increased SNS activity (as assessed by PEP) was associated with time reproductions for negative pictures. However, this effect was true only for stimuli lasting 400 ms.

In another study, Ogden et al. (2019) used a similar research design, but this time they also modulated the PNS activity using paced breathing (i.e., asking participants to breath with a frequency of 0.1 Hz, one respiration cycle each 10 s). They showed that paced breathing increased PNS activity (as assessed by HF-HRV) but did not reduce cardiac SNS activity (as assessed by the PEP) compared to normal breathing. After paced breathing, stimuli were perceived as shorter than after normal breathing, but also less accurate. They interpreted these results in line with the clock model, that is, a reduction in arousal, here defined by the increase in PNS activity, can reduce the pacemaker speed and therefore decrease the perceived duration of sub-second intervals.

Another series of studies used video or olfactory stimuli to alter psychophysiological arousal. For instance, Gros et al. (2015) used an auditory temporal bisection task (durations ranging from 400 to 1,600 ms) after emotional video or olfactory stimulation. The authors reported that olfactory stimulation induced a concurrent increase in EDA and a temporal overestimation. However, this result is not in line with a previous study using different types of olfactory stimuli before and after a 15-s production task (Schreuder et al., 2014). Here odors did not affect EDA, HR, and subjective arousal levels, but the most arousing stimuli (rosemary odor) were associated with shorter time productions. Appelqvist-Dalton et al. (2022) tested the effect of subjective and physiological (pupil size) arousal on the temporal estimation of audiovisual, visual-only, and auditory-only clips of various types (e.g., nature documentaries and animation, with durations ranging from 5 to 13 s). They observed that all types of clips were perceived as lasting shorter than the real duration, and greater arousal (assessed with a composite measure of self-reported arousal and pupil size changes) was associated with more accurate estimates of duration. It should be noted that the authors were not able to disentangle the role of physiological arousal from the subjective one. Wöllner et al. (2018) presented participants with clips representing film scenes, sports, or ballet performances in either slow-motion and real-time speed as well as with and without music, and asked participants to rate the perceived duration of each clip. During the task, the respiration rate, HR, SCL, and pupil size were recorded. The results showed that slow-motion decreased pupillary diameters as well as respiration rates. Furthermore, specifically in the slow-motion condition, the presence of music increased cardiac response. From a temporal perspective, participants overestimated more the length of the clips in the slow-motion condition compared to the real-time condition.

Studies manipulating pre-task physiological activity

Physical activity

Another manipulation that can be employed to study the relationship between physiological arousal and time perception, is physical activity, which can be performed before (or between) temporal tasks (Table 6.3). For instance, in 1989, Mihaly and colleagues asked participants to produce 10-s intervals before, during, and after acute physical activity (bike on a cycle ergometer at medium and elevated intensity, as assessed by oxygen volume consumption). They observed that participants were more variable in producing 10-s intervals during the exercise phase, compared to the pre- and post-activity phases. Lambourne (2012) built on this study, and asked participants to cycle or to rest before and after performing a temporal discrimination task with sub-second intervals. They observed an overestimation of sub-second durations after the cycling condition. The authors, who monitored the changes in physiological arousal measuring oxygen volume, suggested that acute exercise can speed up the pacemaker. A similar study was conducted by Dormal et al. (2018), who asked their participants to perform a temporal categorization task (duration from 500 to 900 ms) before and after manipulating their physiological arousal. The manipulation was conducted by asking them to either bike an ergometer cycle, relax (guided muscular relaxation), or play crosswords. They expected that an increase in HR due to cycling would result in a temporal overestimation, whereas a decrease in HR due to the relaxing condition would promote temporal underestimation. They showed that the HR increased in the cycling and decreased in the relaxation condition compared to the others. However, in the post-manipulation temporal task, participants tended to overestimate durations regardless of the condition. The authors hypothesized that the overestimation was due to a combination of interference tasks (the three manipulations) and retention delay between the pre- and post-manipulation tasks. Their second experiment confirmed this hypothesis. They concluded that physical arousal per se does not affect temporal processing.

Schwarz et al. (2013) conducted two experiments to study whether it is the physiological (i.e., HR) or the subjective arousal that influences the pacemaker of the clock model (see Grondin, 2010). They manipulate physiological and subjective arousal by asking participants to perform a muscular

Table 6.3 Studies manipulating pre-task physiological activity

| Article info | Arousal manipulator | Condition | Physiological activation | | | Time perception | | |
Authors	Type		Recording	Index	Effect	Task	Duration	Effect
PHYSICAL ACTIVITY Dormal et al. (2018)	Physical activity	Cycling	ECG	HR	↑	Temporal categorization	0.5–0.9 s	/
	Physical activity	Relax	ECG	HR	↓	Temporal categorization	0.5–0.9 s	/
	Stress	Crossword	ECG	HR	/	Temporal categorization	0.5–0.9 s	/
Lambourne (2012)	Physical activity	Cycling	Respiratory level	Oxygen Peak	↑	Temporal categorization	0.142–1.270 s	>
Mihaly et al. (1988)	Physical activity	Cycling	Oxygen level	–	↑	Temporal generalization	10 s	–
Schwartz et al. (2013)	Physical activity	Muscular exercise	ECG	HR	↑	Time estimation	8 s	/
	Physical activity	Breath hold	ECG	HR	↑	Time estimation	8 s	/
	Physical activity	Muscular exercise	ECG	HR	↑	Time reproduction	8 s	/
	Physical activity	Breath hold	ECG	HR	↑	Time reproduction	8 s	/
PAIN Fayolle et al. (2015)	Pain	Electric shock	EDA	EDA	↑	Time bisection	0.2–8 s	>
Piovesan et al. (2019)	Pain	Electric shock	EDA	SCL	↑	Verbal estimation	0.242–1.296 s	>
	Pain	Thermic shock	EDA	SCL	↑	Verbal estimation	0.242–1.296 s	/
STRESS Cellini et al. (2023)	Stress	PASAT	ECG	HR	↑	Time bisection	0.3–0.9 s	+
	Stress	PASAT	EDA	Global mean	/	Time bisection	0.3–0.9 s	+

Notes: ECG: Electrocardiogram; EDA: Electrodermal Activity; HR: Heart Rate; HRV: Heart Rate Variability; PASAT: Paced Auditory Serial Addition Test; SCL: Skin Conductance Level; +: More Accurate; –: Less Accurate; ↓: Decrease; ↑: Increase; <: Underestimation; >: Overestimation; ‡: No changes; /: No effect.

exercise (ME; which should increase subjective and physiological arousal), a breath-holding (BH; which should decrease physiological while increasing the subjective arousal), and a control condition without any secondary task. During each of these conditions, participants had also to estimate the duration of symbolic figures presented on a screen positioned in front of them, which lasted 8 s each. As expected, the authors found that ME increased HR more than the other two manipulations, whereas both ME and BH induced an increase in perceived arousal compared to the control condition. Participants also perceived the stimuli presented in both the ME and BH conditions as lasting longer, suggesting that subjective, rather than physiological, arousal induces a lengthening of the perceived stimuli duration. They replicated these results in a second experiment, using a time production instead of a time estimation task.

Other studies have tried to reduce physiological arousal before a temporal task by means of meditation (e.g., Kramer et al. 2013; Droit-Volet et al. 2015). Some of these studies showed that extensive or brief meditation periods induced a temporal overestimation in the temporal bisection task. Interestingly, these results seem to contradict the clock model, which would predict an underestimation of duration due to the slowing down of pulse rate induced by the meditation. The authors explained their results suggesting that meditation affects attentional resources (the switch) rather than physiological arousal (the pacemaker). Other studies using meditation or testing meditators showed contrasting patterns (e.g., Droit-Volet & Heros, 2017, Piovesan et al., 2023). However, it should be noted that no physiological measure was assessed during these studies to (dis)confirm these hypotheses. Overall, these studies seem to suggest that modifying physiological arousal (e.g., HR) before a temporal task does not affect temporal estimation.

Shock and pain

Another classic psychophysiological paradigm to modify physiological arousal and recently employed to alter time perception is the use of electrical or thermal shock, which can induce fear and pain. For instance, Fayolle et al. (2015) asked their participants to perform a time bisection task (duration ranging from 0.2 to 8 s), while EDA was concurrently recorded, and each trial could have been followed by an electrical shock or not (probability of 67%). They report a typical literature finding, that is, EDA amplitude was higher in the trials where the shock was presented compared to the ones without the shock delivery. In the shock trials, stimuli were perceived as lasting longer (bisection point shifted to the left), although temporal sensitivity (WR) was not affected by the stimulation.

A similar study was conducted by Piovesan et al. (2019), in which the authors manipulated the arousing state using electro-cutaneous (in

Experiment 1) and thermal stimuli (in Experiment 2), while asking participants to estimate the duration of these stimuli (durations ranging from 242 to 1,296 ms). These experiments used a block design, with each of the three blocks presenting stimuli with different pain intensities (no pain, low pain, and high pain). Experiment 1 showed that the most painful electrocutaneous stimuli induced an increase in SCL and a lengthening of the temporal estimations. Moreover, the increase in SCL during the block of painful stimulations was associated with longer perceptions of the stimuli duration, whereas the increase in HF-HRV in the same block was associated with a shorter perception of stimuli duration. Experiment 2 confirmed the increase of SCL during painful stimulation. However, painful thermal stimulation did not induce a temporal distortion, and the authors did not find any association between physiological changes due to the stimulation and temporal estimation.

Stress induction

Another way to manipulate physiological arousal is to induce stress. We already mentioned the study by Van Hedger et al. (2017), where they induced stress before and after a temporal reproduction task using the Trier Social Stress Test.

We recently tested the effect of pre-task stress in temporal processing (a time bisection task with sub-second intervals, Cellini et al., 2023). Our participants either performed a Paced Auditory Serial Addition Test (PASAT; stress condition), which is a stressful high-cognitive load task where participants have to continuously sum the digits that they hear, or a Paced Auditory Number Reading Task (PANRAT; control condition), which use the same stimuli than the PASAT, but participants have just to repeat aloud the name of each digit they hear. As expected, the PASAT was able to induce greater psychophysiological activation than the PANRAT as assessed by both subjective reports and HR and EDA. Moreover, participants were more accurate after the stress manipulation rather than after the control condition. We explained these results suggesting the psychophysiological stress before a temporal task may increase the "sampling rate" of the clock system, allowing the system to collect more temporal information, which can increase the temporal precision at the decision stage of the clock model.

Methodological considerations

Before starting a study investigating the role of psychophysiological signals in temporal processing, some methodological issues need to be considered.

Temporal resolution. Each physiological signal has its own temporal resolution, and this needs to be carefully considered before running a study.

For instance, EEG has a high temporal resolution (milliseconds), and depending on the setting and the instruments (e.g., amplifiers), it can be used to study processing unfolding in milliseconds (fast oscillations) to seconds (ultraslow oscillations). However, ECG does not have the same resolution. If we think that a typical RR interval lasts between 0.6 and 1.2 s, then in 2 s, we will likely obtain only two values. Therefore, to see a meaningful variation in EGC signals, we usually need several seconds (which can or cannot include a baseline). HRV analyses are even more stringent since we need at least 1 min of recording to get some reliable data. EDA has a temporal resolution in the seconds range, since a typical SCR occurs between 2 and 6 s from the occurrence of a stimulus. It can be analyzed in longer duration (e.g., minutes), but usually EDA tends to decline over time while doing the same task or while resting (see Habituation below). Therefore, researchers may consider segmenting these long recordings into smaller epochs (e.g., 1 min) to assess and account for the expected linear decline in the EDA values.

Habituation. Physiological indices tend to habituation as a function of the repetition of the stimuli (see Bradley, 2009; Walker et al., 2019). It is common to observe a decrease in magnitude response for PCR or SCR/EDA throughout a task, and this habituation is often observed when a test-retest is performed within a short period of time (e.g., a few hours).

Artifacts and confounders (physiological, homeostatic, and circadian). It is important to remember that any type of task response (e.g., manual and vocal) will perturb the ANS to a certain degree. Therefore, there is a need for experimental settings that minimize the vocal or manual response or provide "recovery" periods after the response window to allow a signal to return to a pre-artifact level. Another solution is to match potential artifacts between conditions (e.g., the same type of stimuli and the same rate of vocal response for a stressful or non-stressful condition). Also, the time of the day and circadian preferences may impact the physiological response to specific stimuli, due to the circadian fluctuation of the ANS (Buijs et al., 2013). Therefore, while planning a study using physiological measures, it may be important to test participants at the same time of the day and to control for the participants' chronotype (e.g., Honkalampi et al., 2021). Lastly, physiological measurements are strongly affected by age, biological sex (and for females, menstrual cycle phases), fitness level and body-mass index, and pharmacological treatments (e.g., beta-blockers). We suggest to collect this information when designing a study with physiological measures.

Resources

Physiological data require software, toolbox, or scripts in different programming languages (e.g., R, Python, and MATLAB) to be visualized, processed, and to extract meaningful information. Here we provide a list of software, packages, and toolboxes to process several physiological signals. In the next paragraphs, we are presenting tools specific for either cardiac or electrodermal activity.

EDF browser (https://www.teuniz.net/edfbrowser/) is open-source software designed for viewing, analyzing, and annotating data files in the European data format (EDF), a format typically used to store physiological data. The software provides interactive tools for visualizing and exploring physiological signals in time and frequency domains. Users can view multiple channels of data simultaneously, zoom in/out, scroll through the data, and adjust display settings for optimal visualization. The software offers basic signal processing functionalities, such as filtering, baseline correction, and signal averaging. Users can apply digital filters (e.g., low-pass, high-pass, and band-pass) to remove noise or artifacts from the data and enhance signal quality. It also allows some basic data event detection and export (e.g., R-peaks detections and RR intervals computation). This software is really useful for teaching, as students can easily play with physiological signals even without any knowledge of coding.

NeuroKit2 (https://github.com/neuropsychology/NeuroKit; see also Makowski et al., 2021) is a Python package for neurophysiological signal processing and analysis. It provides a wide range of tools and functionalities for working with various types of physiological data, such as ECG, EDA, EMG, EEG, EOG, and respirations. NeuroKit2 offers a comprehensive set of signal processing tools for filtering, segmenting, and analyzing these signals. It includes functions for baseline correction, artifact removal, peak detection, and feature extraction from several signals. For instance, the package provides tools for computing and analyzing HRV parameters, for example, calculating time-domain, frequency-domain, and nonlinear HRV metrics, as well as performing statistical analysis and visualization of HRV data. For EDA, users can compute SCL, SCRs, and other EDA parameters, as well as visualize and interpret EDA data. It is a very powerful and open-source toolbox, which only requires python environment.

PsPM (from psychophysiological modeling; Bach 2014) is a solid MATLAB-based toolbox for processing, analyzing, and modeling various types of psychophysiological data, including ECG, EDA, EEG, EMG, and pupil size (https://github.com/bachlab/PsPM). PsPM, which requires a MATLAB license to be used, allows users to import raw psychophysiological data from different sources and preprocess it for analysis. This may include artifact detection and removal, baseline correction, filtering, and epoching of data segments relative to experimental events or triggers. Users can compute typical event-related parameters, such as response amplitudes and latencies. The toolbox offers tools for visualizing psychophysiological data and analysis results, including time series plots, ERPs, spectrograms, statistical maps, and model fits. Of note, PsPM also includes SCRalyze, a specific and well-developed tool for the analysis of EDA.

Tools specific for cardiac signals

KUBIOS HRV (Tarvainen et al., 2014) is one of the most widely used software to process, visualize, and interpret ECG/PPG data, in particular HR and HRV. It is not an open-source tool, although they have a free version of the software to analyze data from some wearable devices and simple RR intervals data (https://www.kubios.com/hrv-scientific-lite/). The full version of the software has a user-friendly interface, read data from different formats such as ASCII and EDF, and provides tools for removing artifacts (automatic or manual), interpolating missing data points, detrending, and resampling. It allows the users to analyze data on both the time domain and the frequency domain, and also non-linear HRV analysis like Poincaré plots, approximate entropy, and sample entropy.

ARTiiFACT (Kaufmann et al., 2011) is a MATLAB-based tool, freely available as a stand-alone software, designed to detect and correct artifacts in ECG data (https://github.com/tobias-kaufmann/ARTiiFACT). It also allows the users to detect R-peaks and extract IBI/RR intervals from continuous ECG recordings, to calculate HRV parameters in time and frequency domains, as well as to visualize the raw and analyzed data.

RHRV (Martinez et al., 2017) is an R package for HRV analysis. It provides a comprehensive set of tools and functions for processing, analyzing, and visualizing HRV data in the R ambient (https://rhrv.r-forge.r-project. org/). RHRV can be used to visualize and filter the data and to remove artifacts. Then RHRV can calculate HRV parameters in time and frequency domains, as well as by using nonlinear analysis. The package provides functions for visualizing HRV data and analysis results, including time series plots, spectral plots, scatterplots, histograms, and summary statistics. It also includes functions for statistical analysis of HRV data, such as hypothesis testing, group comparisons, correlation analysis, and regression modeling.

Kardia (Perakakis et al., 2010) is a MATLAB toolbox designed for the analysis of ECG and IBI/RR data (https://github.com/perakakis/KARDIA). Kardia requires MATLAB to be used, and read MAT files. Kardia allows users to conduct both PCR analysis by importing separate files for the cardiac signals and the description and position of any marker/trigger we want to use for the analyses. The toolbox can also be used to conduct time- and frequency-domain HRV analyses.

Tools specific for electrodermal activity

LEDALAB is a very popular MATLAB toolbox specifically designed for the analysis of EDA (http://www.ledalab.de/). The toolbox requires a MATLAB license. LEDALAB reads different types of data and trigger/marker files, and offers various tools for preprocessing raw EDA data. This includes filtering to remove noise and artifacts, linear interpolations to correct artifacts, and baseline correction to adjust for drift. Besides standard EDA analysis (SCR, SCL, etc.), the toolbox provides algorithms for the decomposition of EDA into its constituent components. This typically involves separating the tonic (slow-changing) component from the phasic (rapid) component (Benedek & Kaernbach, 2010a, 2010b).

Breathe Easy EDA (Ksander et al., 2018) is a freely available MATLAB toolbox for EDA data management (e.g., import and pre-processing), cleaning, and analysis (https://github.com/johnksander/BreatheEasyEDA). This toolbox, which requires a MATLAB license, allows to visualize EDA data, identify and correct artifacts (e.g., respiration artifacts), and compute classical EDA statistics (e.g., phasic and tonic EDA, SCR detection, latency and amplitude, and automatic SCR rejections).

EDA MATLAB Toolbox (https://github.com/mateusjoffily/EDA/wiki) is another MATLAB toolbox that can be used to analyze EDA recording from different proprietary file formats, preprocess data (downsampling and filtering), compute standard EDA parameters, manually remove artifacts, and generate descriptive statistics. The analysis can be run using either a graphical user interface or command-line batch scripting.

Conclusion and future research directions

The search for reliable psychophysiological indices of temporal processing is still ongoing. So far, the literature has been providing contrasting and sometimes unclear results. Indeed, there are several issues related to the study of psychophysiological correlates of time perception, such as developing a research paradigm that can fit with the specificity of each physiological signal, the high heterogeneity of temporal tasks and duration employed, and the lack of a clear physiology-based theoretical model of temporal processing. Therefore, to date, clear-cut conclusions on the relationship between physiological activation and time perception cannot be drawn.

In the next years, this area of research may capitalize on the fast developments in physiological data analysis, mainly driven by the increased use of wearable technology. Also, new statistical approaches may help overcome the limitations in terms of temporal resolutions of the different physiological signals. However, it is likely that a big step forward in this literature can be achieved by defining and adopting a standardized approach, and replicating the study over several labs, to define clear and fundamental associations between some physiological and temporal processes.

References

Angrilli, A., Cherubini, P., Pavese, A., & Manfredini, S. (1997). The influence of affective factors on time perception. *Perception & Psychophysics*, *59*, 972–982. https://doi.org/10.3758/BF03205512

Ansari, Y., Mourad, O., Qaraqe, K., & Serpedin, E. (2023). Deep learning for ECG arrhythmia detection and classification: An overview of progress for period 2017–2023. *Frontiers in Physiology*, 14, 1–20. https://doi.org/10.3389/fphys.2023.1246746

Appelqvist-Dalton, M., Wilmott, J. P., He, M., & Simmons, A. M. (2022). Time perception in film is modulated by sensory modality and arousal. *Attention, Perception, & Psychophysics*, *84*(3), 926–942. https://doi.org/10.3758/s13414-022-02464-9

Aschoff, J. (1998). Human perception of short and long time intervals: Its correlation with body temperature and the duration of wake time. *Journal of Biological Rhythms*, *13*(5), 437–442. https://doi.org/10.1177/074873098129000264

Bach, D. R. (2014). A head-to-head comparison of SCRalyze and Ledalab, two model-based methods for skin conductance analysis. *Biological Psychology*, *103*, 63–68. https://doi.org/10.1016/j.biopsycho.2014.08.006

Benedek, M., & Kaernbach, C. (2010a). A continuous measure of phasic electrodermal activity. *Journal of Neuroscience Methods*, *190*(1), 80–91. https://doi.org/10.1016/j.jneumeth.2010.04.028

Benedek, M., & Kaernbach, C. (2010b). Decomposition of skin conductance data by means of nonnegative deconvolution. *Psychophysiology*, *47*(4), 647–658. https://doi.org/10.1111/j.1469-8986.2009.00972.x

Boucsein, W., Fowles, D. C., Grimnes, S., Ben-Shakhar, G., Roth, W. T., Dawson, M.E., & Filion, D. L. (2012). Publication recommendations for electrodermal measurements. *Psychophysiology*, *49*(8), 1017–1034. https://doi.org/10.1111/j.1469-8986.2012.01384.x

Bradley, M. M. (2009). Natural selective attention: Orienting and emotion. *Psychophysiology*, *46*(1), 1–11. https://doi.org/10.1111/j.1469-8986.2008.00702.x

Buijs, R. M., Escobar, C., & Swaab, D. F. (2013). The circadian system and the balance of the autonomic nervous system. *Handbook of Clinical Neurology*, *117*, 173–191. https://doi.org/10.1016/B978-0-444-53491-0.00015-8

Bulcao, C. F., Frank, S. M., Raja, S. N., Tran, K. M., & Goldstein, D. S. (2000). Relative contribution of core and skin temperatures to thermal comfort in humans. *Journal of Thermal Biology*, *25*(1–2), 147–150. https://doi.org/10.1016/S0306-4565(99)00039-X

Cahoon, R. L. (1969). Physiological arousal and time estimation. *Perceptual and motor skills*, *28*(1), 259–268. https://doi.org/10.2466/pms.1969.28.1.259

Camm, A. J., Malik, M., Bigger, J. T., ... Singer, D.H. (1996). Heart rate variability. Standards of measurement, physiological interpretation, and clinical use. *Circulation*, *93*(5), 1043–1065. https://doi.org/10.1161/01.CIR.93.5.1043

Castaldo, R., Montesinos, L., Melillo, P., James, C., & Pecchia, L. (2019). Ultra-short term HRV features as surrogates of short term HRV: A case study on mental stress detection in real life. *BMC Medical Informatics and Decision Making*, *19*(1), 12. https://doi.org/10.1186/s12911-019-0742-y

Cellini, N., Grondin, S., Stablum, F., Sarlo, M., & Mioni, G. (2023). Psychophysiological stress influences temporal accuracy. *Experimental Brain Research*, *241*(9), 2229–2240. https://doi.org/10.1007/s00221-023-06676-9

Cellini, N., Mioni, G., Levorato, I., Grondin, S., Stablum, F., & Sarlo, M. (2015). Heart rate variability helps tracking time more accurately. *Brain and Cognition*, *101*, 57–63. https://doi.org/10.1016/j.bandc.2015.10.003

Curton, E. D., & Lordahl, D. S. (1974). Effects of attentional focus and arousal on time estimation. *Journal of Experimental Psychology*, *103*(5), 861–867. https://doi.org/10.1037/h0037352

Dormal, V., Heeren, A., Pesenti, M., & Maurage, P. (2018). Time perception is not for the faint-hearted? Physiological arousal does not influence duration categorisation. *Cognitive Processing*, *19*, 399–409. https://doi.org/10.1007/s10339-017-0852-3

Droit-Volet, S., Fanget, M., & Dambrun, M. (2015). Mindfulness meditation and relaxation training increases time sensitivity. *Consciousness and Cognition*, *31*, 86–97. https://doi.org/10.1016/j.concog.2014.10.007

Droit-Volet, S., & Heros, J. (2017). Time judgments as a function of mindfulness meditation, anxiety, and mindfulness awareness. *Mindfulness*, *8*(2), 266–275. https://doi.org/10.1007/s12671-016-0597-6

Droit-Volet, S., & Meck, W. H. (2007). How emotions colour our perception of time. *Trends in cognitive sciences*, *11*(12), 504–513.

Drost, L., Finke, J. B., Port, J., & Schächinger, H. (2022). Comparison of TWA and PEP as indices of α2-and ß-adrenergic activation. *Psychopharmacology*, *239*(7), 2277–2288. https://doi.org/10.1007/s00213-022-06114-8

Fayolle, S., Gil, S., & Droit-Volet, S. (2015). Fear and time: Fear speeds up the internal clock. *Behavioural Processes*, *120*, 135–140. https://doi.org/10.1016/j.beproc.2015.09.014

Fung, B. J., Crone, D. L., Bode, S., & Murawski, C. (2017). Cardiac signals are independently associated with temporal discounting and time perception. *Frontiers in Behavioral Neuroscience*, *11*, 1. https://doi.org/10.3389/fnbeh.2017.00001

Grondin, S. (2010). Timing and time perception: A review of recent behavioral and neuroscience findings and theoretical directions. *Attention, Perception, & Psychophysics*, 72, 561–582. https://doi.org/10.3758/APP.72.3.561

Gros, A., Giroud, M., Bejot, Y., Rouaud, O., Guillemin, S., Aboa Eboulé, C., Manera, V., Daumas, A., & Lemesle Martin, M. (2015). A time estimation task as a possible measure of emotions: difference depending on the nature of the stimulus used. *Frontiers in Behavioral Neuroscience*, *9*, 1–11. https://doi.org/10.3389/fnbeh.2015.00143

Hancock, P. A. (1993). Body temperature influence on time perception. *The Journal of General Psychology*, *120*(3), 197–216. https://doi.org/10.1080/00221309.1993.9711144

Hoagland, H. (1933). The physiological control of judgments of duration: Evidence of a chemical clock. *Journal of General Psychology*, *9*, 267–287. https://doi.org/10.1080/00221309.1933.9920937

Honkalampi, K., Järvelin-Pasanen, S., Tarvainen, M. P., Saaranen, T., Vauhkonen, A., Kupari, S., Perkiö-Mäkelä, M., Räsänen, K., & Oksanen, T. (2021). Heart rate variability and chronotype–A systematic review. *Chronobiology International*, *38*(12), 1786–1796. https://doi.org/10.1080/07420528.2021.1939363

Kaufmann, T., Sütterlin, S., Schulz, S. M., & Vögele, C. (2011). ARTiiFACT: A tool for heart rate artifact processing and heart rate variability analysis. *Behavior Research Methods*, *43*, 1161–1170. https://doi.org/10.3758/s13428-011-0107-7

Kramer, R. S., Weger, U. W., & Sharma, D. (2013). The effect of mindfulness meditation on time perception. *Consciousness and Cognition*, *22*(3), 846–852. https://doi.org/10.1016/j.concog.2013.05.008

Ksander, J. C., Kark, S. M., & Madan, C. R. (2018). Breathe Easy EDA: A MATLAB toolbox for psychophysiology data management, cleaning, and analysis. *F1000Research*, *7*, 1–14. https://doi.org/10.12688/f1000research.13849.2

Lake, J. I., LaBar, K. S., & Meck, W. H. (2016). Emotional modulation of interval timing and time perception. *Neuroscience & Biobehavioral Reviews*, *64*, 403–420. https://doi.org/10.1016/j.neubiorev.2016.03.003

Lambourne, K. (2012). The effects of acute exercise on temporal generalization. *Quarterly Journal of Experimental Psychology*, *65*(3), 526–540. https://doi.org/10.1080/17470218.2011.605959

Lang, P. J., Bradley, M. M., & Cuthbert, B. N. (1997). International affective picture system (IAPS): Technical manual and affective ratings. *NIMH Center for the Study of Emotion and Attention*, *1*(39–58), 3.

Makowski, D., Pham, T., Lau, Z. J., Brammer, J. C., Lespinasse, F., Pham, H., Schölzel, C., & Chen, S. A. (2021). NeuroKit2: A Python toolbox for neurophysiological signal processing. *Behavior Research Methods*, *53*, 1689–1696. https://doi.org/10.3758/s13428-020-01516-y

Martínez, C. A. G., Quintana, A. O., Vila, X. A., Touriño, M. J. L., Rodríguez-Liñares, L., Presedo, J. M. R., & Penín, A. J. M. (2017). *Heart rate variability analysis with the R package RHRV*. Springer.

Meissner, K., & Wittmann, M. (2011). Body signals, cardiac awareness, and the perception of time. *Biological Psychology*, *86*(3), 289–297. https://doi.org/10.1016/j.biopsycho.2011.01.001

Mella, N., Conty, L., & Pouthas, V. (2011). The role of physiological arousal in time perception: psychophysiological evidence from an emotion regulation paradigm. *Brain and Cognition*, *75*(2), 182–187. https://doi.org/10.1016/j.bandc.2010.11.012

Mihaly, T., Hancock, P. A., Vercruyssen, M., & Rahimi, M. (1988). Time estimation performance before, during and after physical activity. In *Proceedings of the human factors society annual meeting* (Vol. 32, No. 15, pp. 985–989). SAGE Publications. https://doi.org/10.1518/107118188786761811

Mioni, G., Labonté, K., Cellini, N., & Grondin, S. (2016). Relationship between daily fluctuations of body temperature and the processing of sub-second intervals. *Physiology & Behavior*, *164*, 220–226. https://doi.org/10.1016/j.physbeh.2016.06.008

Ogden, R. S., Henderson, J., McGlone, F., & Richter, M. (2019). Time distortion under threat: Sympathetic arousal predicts time distortion only in the context of negative, highly arousing stimuli. *PloS One*, *14*(5), e0216704. https://doi.org/10.1371/journal.pone.0216704

Otten, S., Schötz, E., Wittmann, M., Kohls, N., Schmidt, S., & Meissner, K. (2015). Psychophysiology of duration estimation in experienced mindfulness meditators and matched controls. *Frontiers in Psychology*, *6*, 1–12. https://doi.org/10.3389/fpsyg.2015.01215

Perakakis, P., Joffily, M., Taylor, M., Guerra, P., & Vila, J. (2010). KARDIA: A MATLAB software for the analysis of cardiac interbeat intervals. *Computer Methods and Programs in Biomedicine*, *98*(1), 83–89. https://doi.org/10.1016/j.cmpb.2009.10.002

Piovesan, A., Mirams, L., Poole, H., Moore, D., & Ogden, R. (2019). The relationship between pain-induced autonomic arousal and perceived duration. *Emotion*, *19*(7), 1148. https://doi.org/10.1037/emo0000512

Piovesan, A., Mirams, L., Poole, H., & Ogden, R. (2023). The effect of mindfulness meditation on the perceived duration of pain. *Journal of Cognitive Psychology*, *35*(2), 233–247. https://doi.org/10.1080/20445911.2022.2154780

Pollatos, O., Yeldesbay, A., Pikovsky, A., & Rosenblum, M. (2014). How much time has passed? Ask your heart. *Frontiers in Neurorobotics, 8,* 1–9. https://doi.org/10.3389/fnbot.2014.00015

Schreuder, E., Hoeksma, M. R., Smeets, M. A., & Semin, G. R. (2014). The effects of odor and body posture on perceived duration. *Frontiers in Neurorobotics, 8,* 1–10. https://doi.org/10.3389/fnbot.2014.00006

Schwarz, M. A., Winkler, I., & Sedlmeier, P. (2013). The heart beat does not make us tick: The impacts of heart rate and arousal on time perception. *Attention, Perception, & Psychophysics, 75,* 182–193. https://doi.org/10.3758/s13414-012-0387-8

Sirois, S., & Brisson, J. (2014). Pupillometry. *Wiley Interdisciplinary Reviews: Cognitive Science, 5*(6), 679–692. https://doi.org/10.1002/wcs.1323

Stevenson, R. A., & James, T. W. (2008). Affective auditory stimuli: Characterization of the international affective digitized sounds (IADS) by discrete emotional categories. *Behavior Research Methods, 40*(1), 315–321. https://doi.org/10.3758/BRM.40.1.315

Tarvainen, M. P., Niskanen, J. P., Lipponen, J. A., Ranta-Aho, P. O., & Karjalainen, P. A. (2014). Kubios HRV–heart rate variability analysis software. *Computer Methods and Programs in Biomedicine, 113*(1), 210–220. https://doi.org/10.1016/j.cmpb.2013.07.024

Van Hedger, K., Necka, E. A., Barakzai, A. K., & Norman, G. J. (2017). The influence of social stress on time perception and psychophysiological reactivity. *Psychophysiology, 54*(5), 706–712. https://doi.org/10.1111/psyp.12836

Vinik, A. I. (2012). The conductor of the autonomic orchestra. *Frontiers in Endocrinology, 3,* 1–13 https://doi.org/10.3389/fendo.2012.00071

Walker, F. R., Thomson, A., Pfingst, K., Vlemincx, E., Aidman, E., & Nalivaiko, E. (2019). Habituation of the electrodermal response–A biological correlate of resilience? *PLoS One, 14*(1), e0210078. https://doi.org/10.1371/journal.pone.0210078

Winter, M., Pryss, R., Probst, T., & Reichert, M. (2020). Towards the applicability of measuring the electrodermal activity in the context of process model comprehension: Feasibility study. *Sensors, 20*(16), 1–23. https://doi.org/10.3390/s20164561

Wöllner, C., Hammerschmidt, D., & Albrecht, H. (2018). Slow motion in films and video clips: Music influences perceived duration and emotion, autonomic physiological activation and pupillary responses. *PloS One, 13*(6), e0199161. https://doi.org/10.1371/journal.pone.0199161

7

TIME AND SENSORY PROCESSING

Insights from deafness and blindness

Maria Bianca Amadeo and Monica Gori
Italian Institute of Technology, Genova, Italy

Nicola Domenici
École normale supérieure, PSL University, CNRS, Paris, France

Introduction

With just the first sentence of this chapter, we would like to challenge the reader with an intriguing question: how long is a second? Certainly, many different answers can be provided to solve this apparently simple quiz, all potentially correct. One can say that 1 s is the time needed to pronounce the word "elephant," hence the counting strategy involving interleaving such animals between digits to achieve consistent timing (saying "one elephant, two elephants, three elephants" should take in fact 3 s, for example. Surprisingly, this strategy is incredibly reliable!). Or, one might decide to time 1 s by repeatedly tapping on a surface at a steady rate. One can even decide to resort to the most technical definition of a second, which is "the duration of 9,9192,631,770 period of the radiation corresponding to the transition between the two hyperfine levels of the ground state of the cae-sium-133 atom" (Essen & Parry, 1955).

With the evident exception of the latter response, it is extremely hard to pinpoint the exact duration of 1 s from a subjective perspective (Grondin et al., 2020). In fact, every time we try to measure 1 s without external, arti-ficial aids, we might provide more or less precise approximations – but we will never reach astounding precision in a consistent fashion. Although most biological processes are inherently stochastic, and noise is one of their essen-tial components (Tsimring, 2014), still it is astonishing to realize how subjec-tive time can be anisomorphic to physical time. Given the crucial role of temporal cues in orienting ourselves within the environment, one would indeed expect our temporal perception to be as accurate and precise as

DOI: 10.4324/9781003449546-7

possible (with higher accuracies indicating a reduction of systematic errors when estimating a given interval, and increased precision suggesting consistencies among repeated estimates). However, our ability to tell time can be easily distorted (Anobile et al., 2020; Burr et al., 2007; Johnston et al., 2006), sometimes in a fashion that changes along the lifespan (Domenici et al., 2021), sometimes due to specific clinical conditions (Amadeo et al., 2022; Pastor et al., 1992).

In this chapter, we proposed a specific framework to understand fluctuations and deficits in temporal perception. Our view does not claim to be exhaustive and is not mutually exclusive with other theoretical implications based on different physiological premises. Mainly, reliable temporal perception is achieved through the activity of an extensive neural circuitry (Merchant et al., 2013), which might be selectively remodeled due to different causes. Our aim is to provide the reader with just but one possible interpretation, centered on the sensory components of time.

In the first part of this chapter, we will link temporal perception with sensory processing, highlighting possible differences in sensitivity related to the sensory channels involved in time estimations. Specifically, we will describe evidence showing how audition can be considered as the most precise sense in coding and decoding temporal information. Then, we will introduce the cross-sensory calibration theory, which will be the focus of our theoretical framework. According to this theory, during development, the most reliable sense (e.g., audition) in perceiving a given feature (e.g., time) calibrates the others, refining perceptual judgments over all sensory modalities (e.g., visual timing). We will then provide evidence of how temporal processing is impaired when hearing experience is lacking, such as in congenital and early deafness, and how auditory information can be exploited to convey compensatory mechanisms when different sensory inputs are missing, such as in blindness.

Time and sensory processing

We constantly process a multitude of sensory information that needs to be integrated, separated and ordered in time to derive a coherent representation of ourselves and the external world. Although the human brain interfaces with a wide variety of temporal scales – from microseconds to daily circadian rhythms – time perception shows some of its most elegant properties within the millisecond-to-second range (Buonomano, 2007; Buonomano & Karmarkar, 2002; Hardy & Buonomano, 2016). It is within such a time span, in fact, that humans can code and decode the complex structure of signals involved in oral communication (Schirmer, 2004; Shannon et al., 1995), process motion information (Johansson, 1976), and develop proper motor control (Diedrichsen et al., 2003; Ivry, 1996).

Given how sub-second timing is crucial to interact with others and the environment, many researchers have sought to explain its elusive nature within the human brain. It should not surprise the reader, then, that in the past century, scientists developed a plethora of different frameworks to conceptualize our experience of time (Gilden & Mezaraups, 2022; Killeen & Grondin, 2022; Buonomano & Maass, 2009; Machado et al., 2009; Treisman et al., 1990; Zakay & Block, 1996). Nonetheless, while theoretical models are extremely insightful, they seldom provide an in-depth explanation of how temporal abilities shape from sensory processing and reshape when one or more sensory modalities are missing. For this very reason, within this chapter, we decided to focus solely on the perceptual implications that might affect our temporal processing (for a detailed review on models of time perception, see Ivry & Schlerf, 2008).

In a compelling work, Shi and Burr (2016) proposed that efficient time perception consolidates from a combination of adaptive recalibration and minimized predictive errors, thus implying that our sense of time emerges from the direct inference of the available sensory signals. In fact, while the world is in an ever-changing state, the brain must update (i.e., recalibrate) its internal representations to anticipate (i.e., predict) all events occurring in the external environment. When sensory signals embody high levels of uncertainty, the brain is forced to use its own internal representation to time its presence in the world. However, when sensory inputs are more reliable, they can be used to improve the brain's predictions in order to better interact with the environment, thus underlying the importance of sensory-specific temporal precisions. On the one hand, this view is supported by the fact that no specific brain region solely dedicated to temporal processing has been identified yet (Rao et al., 2001), and that all sensory signals inherently embody fundamental temporal cues. On the other hand, it entails that asymmetries in performance might emerge when temporal information is processed across different senses, provided that these senses are characterized by different temporal sensitivities. Our perception of time might indeed differ across sensory modalities due to the heterogeneous ways in which each sensory input is encoded and decoded, following a temporal processing hierarchy as some sensory systems are more temporally reliable than others. If that is true, then one would expect that temporal precision will be higher when processing auditory events and lower when processing visual ones, since the auditory system encodes signals in the microseconds range when using interaural differences for source localization (Klumpp & Eady, 2005), while vision relies on neurons with notably lower temporal acuity (Conner, 1982). As we illustrate in the next paragraphs, although specific exceptions might occur, a significant amount of evidence seems to suggest that this might be the case.

Perception of time and timing of perception

Since time is a complex topic, we find it useful to address it by distinguishing two concepts that refer to two distinct aspects related to time: perception of time and timing of perception (Amadeo et al., 2022). The perception of time pertains to the subjective understanding of the passage of time or the duration of events. That is, for instance, the perception of how long a single event is (e.g., "How long did it take to download this chapter?"), or how much time intertwines between two events (e.g., "What time was it when you started reading this chapter?"). Overall, the most direct way to investigate the perception of time is by the use of explicit or implicit duration judgments, obtained through either comparisons or estimations, and interval production and reproduction tasks (see Chapter 1). The timing of perception, instead, relates to temporal resolution whereby we process sensory events. Indeed, every sensory experience is inherently linked to time as perception occurs within specific timeframes (e.g., while you are reading this chapter, your brain is constantly receiving sensory information that needs to be processed). By timing of perception, we thus indicate the low-level temporal resolution of our sensory systems, which determines how finely we can discriminate and perceive the timing of different events or stimuli. To obtain a general measure of such timing, it is often necessary to rely on more indirect measurements such as simultaneity or temporal order judgments (e.g., "Are the sound alarm and the LED indicator on my dashboard synchronized? If not, which one came first?"). In our opinion, keeping in mind the difference between perception of time and timing of perception, regardless of the paradigm type, is important to unfold the intricate relationship between time and sensory processing.

The auditory sense of time

As early as 1963, Paul Fraisse stated: "hearing is the main organ through which we perceive change: it is considered as the 'time sense'" (Fraisse, 1963). Subsequent studies support this idea, showing that the auditory system is the most precise to code and decode temporal information.

Among the firsts to systematically investigate the interconnection of auditory and visual temporal cues, Recanzone (2003) used a sequence discrimination task to highlight differences in temporal sensitivities between vision and audition. In his main experiment, two short streams of stimuli were displayed in quick succession, and participants reported which sequence had the higher temporal rate. Sequences were made of either visual flashes and/or 1-kHz tones and were presented either singularly or in conjunction (i.e., bimodally), thus allowing the quantification of the influence of one sense over the other.

Among the most relevant results, Recanzone found not only that participants showed a higher temporal rate precision when sequences were auditory alone, but also that the presence of auditory sequences significantly impacts on the visual temporal rate, leading to better performance. Surprisingly, the effect was so pronounced that, when comparing two identical visual sequences accompanied by two different auditory ones, participants consistently based their responses on the auditory information.

The phenomenon known as "temporal ventriloquism" is another clear example of audition dominance for perception of time, showing that the presentation of an asynchronous auditory stimulus impacts the perceived timing of a visual stimulus (Aschersleben & Bertelson, 2003; Fendrich & Corballis, 2001). In another study, Burr, Banks, and Morrone (2009) further shed light on audiovisual temporal interactions by using a modified version of the temporal bisection task. Three audiovisual stimuli were sequentially displayed within a fixed interval of 800 ms, with the second stimulus being closer in time either to the first one or to the third one. Then, participants had to report whether the second stimulus appeared earlier or later than the interval's temporal midpoint. Introducing a bi-modal conflict within the task, by simply misaligning in time the auditory and visual components of the first and last stimuli, Burr et al. investigated the influence of each sensory modality on the final temporal estimation. Intriguingly, one of the main results of their study showed that temporal bisection judgments were heavily influenced by the auditory component of the stimuli, so that the final temporal percept was mostly driven by auditory temporal cues, rather than by the temporal occurrence of the visual stimuli. Analogous behavioral results were provided by most of the works investigating the matter (K.-M. Chen & Yeh, 2009; McGovern et al., 2016; Murai & Yotsumoto, 2016; Repp & Penel, 2002), with just few exceptions (Andersen et al., 2004; van Wassenhove et al., 2008).

Returning to the two temporal concepts introduced the aforementioned studies predominantly focus on what we have defined as the perception of time. Nevertheless, additional research into time has uncovered auditory temporal dominance even applying tasks involving timing of perception. For instance, it has been demonstrated that when two beeps are presented simultaneously to a flash, the latter is perceived as two flashes (Shams et al., 2000), and the perceived frequency of flickering lights can be affected by an auditory stimulus presented concurrently at a different rate (Gebhard & Mowbray, 1959; Shipley, 1964).

Additional evidence supporting the auditory temporal dominance was highlighted by transcranial magnetic stimulation (TMS; see Chapter 5), electroencephalography (EEG; see Chapter 3), and neuroimaging studies (see Chapter 2). Obviously, identifying the neural structures directly involved in timing processes is beyond the scope of this chapter, and we acknowledge here evidence supporting the existence of a (or an ensemble of) distributed circuitry (circuitries) recruited during temporal tasks. Specifically, there is strong evidence that different brain substrates, such as the supplementary motor area (Coull et al., 2004; Pouthas et al., 2005), the posterior parietal cortex (Bueti et al., 2008), the basal ganglia (Ferrandez et al., 2003; Nenadic et al., 2003), and the cerebellum (Ivry et al., 2002), are implicated in precise event timing. Notably, neural substrates involved in temporal processing have been pinpointed using a variety of experimental designs, such as sub- and supra-seconds reproduction (Pouthas et al., 2005), and visual (Ferrandez et al., 2003) and auditory temporal discrimination (Bueti et al., 2008). Furthermore, specific brain areas once believed to be mostly unisensory (specifically, auditory), such as the superior temporal gyrus (STG), have been found to be involved in generalized temporal processing as well. Nonetheless, the existence of a generalized timing network implies that time is a-modally represented in the brain, and that temporal information is processed independently from the sensory channels involved. Surprisingly, with TMS, Kanai et al. (2011) showed not only that some degree of sensory specificity is maintained when coding temporal information, but also that auditory areas are crucial to properly process time events in the visual modality. In their work, the authors tested participants with a temporal discrimination task (centered around a 600-ms reference), evaluating in separate blocks both auditory and visual temporal thresholds. Then, they proceeded disrupting primary sensory cortices, aiming at monitoring fluctuations in temporal precision due to the TMS activity. The results of this study showed that the disruption of the primary visual cortex determined an increase in visual temporal thresholds but did not affect performance in the auditory task. Conversely, when disrupting the primary auditory cortex, participants showed lower temporal precision not only in the auditory task (as expected) but also when performing temporal judgments in the visual modality. Similar results were also found by Bolognini et al. (2010), which proved that the disruption of the STG, which is a predominantly acoustic brain area, significantly impaired tactile temporal discrimination within the sub-second range. In fact, through TMS application over the STG, the authors showed how the participants' ability to distinguish between 15 and 25-ms vibrations (delivered at either index fingers) suddenly worsened. However, when the TMS was directed on the somatosensory cortex, no reduction of precision in haptic timing was observed. Functional magnetic resonance imaging (fMRI) studies showed that temporal processing of visual stimuli recruits the STG as well, both within the sub- (Coull et al., 2004; Ferrandez et al., 2003) and supra-second ranges (Lewis & Miall, 2003). Ferrandez et al. (2003), for instance, showed significant activation of the STG during a duration comparison task, in which participants were asked to determine whether the duration of a green LED (lasting either 490, 595, 700, 805, or 910 ms) was equal to the duration of a previously delivered similar standard (lasting 700 ms). Analogous results were obtained by Lewis and Miall (2003), which tested eight participants using a temporal discrimination task centered around either a 600- or 3,000-ms reference, represented by a white line displayed on a gray background. Interestingly, STG activation was more prominent for the sub-second interval, suggesting that sub-second timing might rely to a greater extent on the recruitment of auditory areas compared to supra-second processing.

Moreover, additional studies with EEG showed that areas likely involving the temporal cortices are very early (50–90 ms) recruited during the evaluation and comparison of temporal intervals between stimuli, independently of the sensory modality (Amadeo et al., 2020; Gori et al., 2023).

To sum up, following the intuition brought at the beginning of Section 2 (Shi & Burr, 2016), it is plausible to assume that the sense of time is built using sensory inferences available at a given time The results discussed in this section can be conceptualized considering that differences in time perception across senses emerge due to sensory-specific encodings of temporal cues, which are bounded to the neurophysiological constraints that characterize each perceptual system. On the one hand, in fact, the temporal resolution of the auditory system peaks within the microseconds range, as time differences within this short span are used for sound source localization (Buonomano, 2007). On the other hand, the visual system exhibits a significantly lower temporal resolution (Levitt, Schumer, Sherman, Spear, & Movshon, 2001), peaking at 30 Hz, indicating that we need at least 3.3 ms between two visual events to be able to perceive them as separate. Given such an overwhelming superiority, it is thus no surprise that the auditory system results as the more specialized one in encoding and decoding temporal cues in different tasks. Considering the significant interconnection between time and sensory processing (Shi & Burr, 2016), and the superior role of audition in perceiving time events, an interesting question is thus how early sensory experience determines our ability to feel time, and to what extent early typical/atypical hearing processing influences our perception of time and timing of perception. In the remaining part of this chapter, we will provide the reader with an answer that – hopefully – satisfies both these queries.

The cross-sensory calibration theory

As discussed in the previous section, hearing is the most precise sense in coding and decoding temporal information. However, we still need to clarify how that impacts on overall temporal processing and temporal skills of other sensory modalities. Specifically, what happens during child development, when sensory systems are not mature and perceptual skills are still emerging? To shed light on the matter, in the current paragraph, we will refer to the cross-sensory calibration theory (Burr, 2012; Gori, 2015; Gori et al., 2010), which might be a profitable framework to further conceptualize the link between audition and time. Nonetheless, the reader must be aware that sensory development is a complex topic and that a unified developmental model is still missing (Dekker & Lisi, 2020). Although the cross-sensory calibration hypothesis is one of the dominant explanations for sensory processing during development, it is important to keep in mind that it is not the only one · (see Han et al., 2022; Rohlf et al., 2020).

To better comprehend the relevance of the cross-sensory calibration theory, we must first deepen into how humans integrate sensory information coming from different modalities. In a pioneering work, Ernst and Banks demonstrated that adults combine sensory information coming from different modalities in an optimal fashion (Ernst & Banks, 2002). This occurs as multisensory cues are combined according to a maximum-likelihood integrator, where the cue with the lower variance (i.e., the most reliable) exerts sensory dominance. For example, when localizing the spatial position of audiovisual (Alais & Burr, 2004; Hay et al., 1965) or visuo-tactile (Rock & Victor, 1964) stimuli, vision often dominates the integrated percept (despite both senses providing information for the object's properties estimation). If visual sensory variance increases, however, the overall percept reflects such change and is "pushed away" from the visual cue, so that sensory dominance is reversed (Ernst & Banks, 2002). Surprisingly, the ability to optimally integrate multisensory information is not innate but develops over time, reaching maturation for some skills around late childhood (Ernst, 2008; Gori et al., 2008, 2012; Nardini et al., 2008, 2010; Petrini et al., 2014). Before certain developmental milestones are reached, then young children's perception of a given feature is dominated by a single sensory modality, regardless of the cues' variances. This remarkable dominance of one sense over the others during development seems to be fertile ground for ongoing cross-sensory calibration processes, whereupon the most reliable sense refines the others to preserve perceptual accuracy at the expense of precision, that is, supporting veridical representations of the dominant stimuli while losing typical consistency in estimates fostered by multisensory integration. Turning back to our main focus, stated the auditory dominance in temporal perception (Burr et al., 2009; Chen & Yeh, 2009; McGovern et al., 2016; Murai & Yotsumoto, 2016; Recanzone, 2003; Repp & Penel, 2002), audition would be a powerful temporal calibration system according to the cross-sensory calibration theory. As so, the auditory system could be used during development to calibrate the other sensory channels for temporal perception (Gori, 2015). In support of this hypothesis, during development, it has been observed that audition dominates in an audiovisual temporal bisection task (Gori et al., 2012), the sensitivity to time increases at a faster rate in the auditory modality compared to the visual modality (Zélanti & Droit-Volet, 2012), and time estimates for auditory stimuli are more precise compared to visual ones (Droit-Volet et al., 2007). Moreover, Noel et al. demonstrated a decrease of rapid recalibration (i.e., recalibration based on the single previous trial) and an improvement in multisensory precision when discriminating the simultaneity of audiovisual stimuli from childhood to adulthood (Noel et al., 2016). The latter results have been recently questioned by compelling new evidence proposing that rapid recalibration for audiovisual temporal simultaneity judgment requires the maturation of temporal precision to develop. While these findings would suggest that integration precedes cross-sensory calibration (see also Rohlf et al., 2020), we think that further research is needed to fully understand the development of sensory processing across different modalities. The processes of cross-sensory calibration we refer to are long-lasting processes that may interest the development of more complex perceptual skills, but different mechanisms could underlie rapid recalibration and explain the development of other perceptual skills.

In light of these considerations, if the cross-sensory calibration theory is suitable in some contexts as it may be, then typical/atypical functioning of the calibrating sense should affect the related perceptual skills. In the next paragraphs, we are providing evidence supporting this view by exploring how temporal skills change when the calibrating sense for time (i.e., audition) is missing, such as in deafness, or when it is a crucial source of information about the external environment, such as in blindness.

Developing proper temporal skills without hearing

At this point, a natural question is what happens to the sense of time when the auditory input is missing. Do the perception of time and timing of perception get worse when no auditory signal is properly processed by the brain? Or is our sense of time completely unaffected by deafness?

When one sensory modality is missing, such as in deafness or blindness, thanks to its plasticity the brain undergoes substantial structural and functional reorganization, and the remaining sensory modalities encounter compensatory mechanisms to adapt and face the challenges of interacting with others and the environment without the information coming from the deprived sense (Bola et al., 2017; Kral & Sharma, 2023). Although neural reorganization is often associated with increased abilities in spared sensory modalities (Allman et al., 2009; Barone et al., 2013; Lomber et al., 2010; Strelnikov et al., 2013), this is not always the case and, in line with the cross-sensory calibration theory, a solid body of evidence supports the existence of a generalized temporal impairment when audition is missing (Amadeo et al., 2019; Bolognini et al., 2012; Domenici et al., 2023; Heming & Brown, 2005; Kowalska & Szelag, 2006; Zhang et al., 2020).

Perception of time without hearing

Perception of time intervenes every time we are asked to explicitly or implicitly evaluate temporal durations or intervals. Kowalska and Szelag (2006) investigated temporal perception in 16 congenitally deaf adolescent and 16 age-matched hearing participants. Within their work, the authors involved two different temporal tasks: a production task, in which participants had to actively produce different time intervals (from 1 to 6 s), and a reproduction task, in which participants had to replicate the duration of a series of visual standards (from 1 to 5.5 s) via button keypress. Overall, the temporal performance of deaf adolescents was significantly worse than that of hearing participants in both tasks. In contrast to hearing peers, deaf showed accurate timing only for intervals around 3 s, overestimating intervals shorter than 2 s and underestimating intervals longer than 3 s. In spite of the results, it should also be noted that Kowalska and Szelag involved mostly supra-second timing, which might be influenced by additional, non-specific factors such as counting strategies (Grondin et al., 1999), as well as cognitive processes recruited during the temporal estimation (Brown & Boltz, 2002; Pan & Luo, 2012).

Another compelling study supporting the cross-sensory calibration theory came from Bolognini et al. (2012). In their work, the authors developed both a temporal and a spatial tactile discrimination task, in which they tested 18 and 14 participants equally distributed between hearing and congenitally deaf, respectively (with four deaf participants taking part in both experiments). In the temporal task, stimuli were tactile vibrations delivered at a 100-Hz frequency to the distant phalanx of either index fingers, lasting either 15 or 25 ms, while in the spatial task, the same tactile stimulation was delivered along the entire finger(s), from the distant phalanx to either the intermediate or the proximal one. Participants had then to indicate which stimulus lasted longer, in the temporal discrimination task, or showed greater spatial length, in the spatial discrimination task. While no difference in performance was observed between groups in the spatial task, congenitally deaf individuals were remarkably less precise when discriminating the two stimulations along the temporal dimension. To investigate more deeply the neural substrates potentially involved in this reduced temporal sensitivity, the authors also performed a second experiment involving the TMS, maintaining the same experimental design. Notably, Bolognini et al. delivered single-pulse TMS shortly after the tactile stimulus onset, targeting either the primary somatosensory area (SI) or the STG during both the spatial and temporal procedures. Interestingly, the authors found that, in deaf participants, both spatial and temporal performances decreased after STG stimulation (with the stronger effect peaking when disruption occurred 60 ms after the stimulus onset). However, in hearing participants, STG stimulation determined an impairment of only temporal processing (although at a later stage, as the effect peaked when TMS occurred 180 ms after the stimulus onset), while spatial performance was preserved. Disrupting SI naturally worsened spatial and temporal tactile processing in both groups, but only if TMS were delivered 60 ms after the stimulus onset. Overall, these results showed that STG stimulation in deaf participants determined a significant decrease of performance in the spatial task, in which deaf individuals showed no impairment, as well as in the temporal one, in which performance was already defective. The comparable chronometry between SI and STG observed in deaf participants also advocates in favor of potential cross-modal recalibration mechanisms occurring after early hearing deprivation, further strengthening the link between temporal perception and the ability to process hearing information during early stages of life.

More recently, additional evidence on how the lack of hearing experience impacts on time perception came from Zhang et al. (2020). In their study, the authors tested 16 undergraduate deaf students using a visual temporal bisection task (see Chapter 1) and compared their performance with 16 age-matched hearing participants. At the beginning of the procedure, participants were presented with two different visual standards, lasting 200 and 800 ms, respectively. Then, a set of different visual probes of different durations were displayed, and participants had to indicate whether each probe was more similar to the first standard or to the second standard. Interestingly enough, deaf participants showed not only a strong bias toward the lower interval but also a significant reduction in temporal sensitivity when compared to the hearing ones. Although the deafness' onset was not reported in the study, the authors' findings advocate once again in favor of the relevant interaction between temporal processing and efficient hearing perception.

A variant of the visual temporal bisection task was also used to test deaf individuals (Amadeo et al., 2019; Gori et al., 2022). The task consisted of presenting a sequence of three visual stimuli (2.3° diameter, 75-ms duration) and asking participants to report whether the second one was closer in time to the first or the third one. A bad performance of deaf people was expected for this paradigm as, as mentioned above, a strong auditory dominance has been observed in typical children and adults at the behavioral level, as well as an early (50–90 ms after stimulus onset) recruitment of areas likely involving the temporal cortices was observed with EEG at the neural level. Deaf people not only could not perform it similar to typical hearing individuals, but also they did not show the early recruitment of areas likely involving the temporal cortices observed with EEG in hearing individuals. Interestingly, another subsequent study investigated whether spatial cues could help deaf individuals in performing this specific version of the temporal bisection task. In this case, spatial coordinates of the three consecutive visual stimuli were manipulated together with the temporal delays between them in order to be coherent (i.e., temporal intervals and spatial distances between the three flashes were directly proportional: a longer time delay between the first and second flashes was associated with a longer spatial distance between the two flashes, and the reverse for shorter intervals), or incoherent (i.e., temporal intervals and spatial distances between the three flashes were inversely proportional: a longer time delay between the first and second flashes was associated with a shorter spatial distance between the two flashes, and the reverse for shorter intervals). Notably, spatial cues coherent with the temporal ones helped deaf participants in building the temporal representation necessary to perform the temporal bisection task, suggesting an interaction between the temporal and spatial domains (Amadeo et al., 2019).

In another, recent study, Domenici et al. (2023) tested 15 early and congenitally deaf and 17 hearing participants using a visual oddball-like temporal task (van Wassenhove et al., 2008). In such a task, a train of five visual stimuli was displayed at the center of the screen, all but one sharing the same duration of 500 ms. Differently from the other stimuli, the fourth one (i.e., the target) lasted one of six possible durations (spanning from 380 to 620 ms), and participants had to report whether such stimulus lasted longer compared to the rest of the sequence. The first, relevant result of this study is that deaf observers were not able to systematically discriminate the target's duration, showing both a strong temporal bias and a significant reduction in temporal sensitivity compared to hearing participants. The second, notable result is that a significant gain in temporal precision was observed in deaf when spatially congruent sensory information was added to the target, suggesting that compensation strategies might take place in the brain in order to foster temporal perception after early hearing deprivation.

While conducting a comprehensive review of all studies on temporal perception in deafness exceeds the scope of this chapter, the prevailing findings exploring the perception of time indicate that individuals who are deaf face challenges in estimating temporal intervals and durations. Their performance in these tasks is generally diminished compared to that of individuals with normal hearing.

Timing of perception without hearing

The timing of perception encompasses the resolution at which temporal information is processed. Indeed, it is important to note that every sensory experience is intrinsically intertwined with the concept of time, and as a result, there is an inevitable temporal dimension in which perceived events occur. While much research has agreed on a deficit in the perception of time due to deafness, results are more contrasting when it comes to the timing of perception.

In one of the first studies directly investigating the timing of perception after early hearing loss, Heming and Brown (2005) tested 10 congenitally deaf individuals (with just one of them reporting hearing loss after birth, but before 2 years of age) and 10 hearing participants in a series of visual and tactile temporal tasks. Using pairs of stimulations, the authors asked participants to report whether the two stimuli within each given pair were simultaneously delivered or not. In the tactile condition, two punctate stimulations, delivered to the index and the middle finger of either the same or different hands, were presented in rapid succession, separated by a variable time interval evaluated through an adaptive procedure (i.e., tailored to the participant's performance). In the visual condition, two flashes, displayed at different spatial positions, were again presented in quick succession, separated by the same time interval. Interestingly, deaf participants showed significantly lower temporal sensitivities than hearing participants, as they required longer temporal intervals within pairs to report the stimuli as temporally segregated. Moreover, increases in temporal thresholds were independent of both the sensory modality and the position of the stimulations in space, suggesting a generalized impairment of temporal processing within the deaf group.

Contrarily, other previous works in which temporal resolution in deafness was investigated provide evidence of potential compensatory mechanisms encompassing timing of perception (Nava et al., 2008; Poizner & Tallal, 1987). Poizner and Tallal (1987), for example, performed three different experiments to assess temporal processing abilities in congenitally deaf individuals. In the first and second experiments, the authors evaluated the visual flicker fusion thresholds in 10 congenitally deaf and 12 hearing adults for either a single circle continuously displayed at different frequencies (Experiment 1) or two circles sequentially displayed separated by a variable inter-stimulus interval (Experiment 2). In the third experiment, 13 deaf and 13 hearing participants were recruited and tested with a temporal-order judgment (TOJ) task, which is a standard psychophysical procedure in which participants must report which one of two brief stimuli delivered in sequence, and separated by a given asynchrony, occurred first. In all three experiments, the authors reported that hearing and deaf participants performed in a similar fashion, apparently suggesting that temporal processing might not be necessarily impaired in the visual modality due to the absence of early hearing experience. Nava et al., 2008 further expanded such findings, testing TOJs both in the center and the periphery of the visual field and finding that both the temporal order thresholds (denoting temporal sensitivity) and the points of perceived simultaneity (indicating temporal accuracy) did not differ across congenitally and early deaf and hearing participants. Similar results were obtained for the tactile modality (Moallem et al., 2010). Even though these results, at a first glance, challenge the foundations of the cross-sensory calibration theory, they actually provide evidence that not all temporal skills are affected by lack of audition and temporal resolution intrinsic to our sensory systems is preserved even without auditory stimulation during development (i.e., no impairment in the timing of processing). In a recent work, Whitton et al. (2021) recorded fMRI signals of 10 early deaf and 10 age-matched hearing participants during a visuo-tactile TOJ task, aiming at investigating the cortical substrates potentially recruited in multisensory temporal processing. While no difference in behavior was observed, with deaf showing similar temporal order thresholds, the authors found indicators of plastic reorganization in the deaf brain. In fact, deaf participants showed a significantly larger BOLD activation in the posterior STG, the right inferior parietal lobule, the posterior STG, and the insula cortex at the bilateral level, all areas that have been linked to both multisensory and temporal processing (Cardin et al., 2020). Furthermore, through functional connectivity analysis, the authors highlighted weaker connections between the STG and visual areas in deaf participants while also identifying stronger somato-motor connections. Taken together, these results suggest that the lack of auditory experience impacts on the connectivity between tactile and visual cortical areas while performing temporal multisensory tasks, and potentially justify the absence of difference at the behavioral level. Similar findings were obtained through EEG as well (Scurry et al., 2020).

Evidence from deaf children

Despite the general diffusion of the cross-sensory calibration theory, it is surprising to acknowledge that very few studies on deafness's perceptual consequences involve children. To the best of our knowledge, only one paper specifically addresses temporal processing during development when auditory experience is somehow altered. Gori et al. (2017) investigated the role of audition on temporal discrimination by comparing typically hearing children with deaf children with cochlear implants during an audiovisual temporal bisection task. Specifically, three consecutive auditory, visual, or audiovisual stimuli (each lasting 75 ms) were presented sequentially for a total duration of 1,000 ms. As briefly mentioned above in Section 3, the same authors previously demonstrated that in typical development, audition dominates the final percept during the audiovisual temporal bisection task (Gori et al., 2012). Interestingly, it was observed that deaf children with restored audition did not show the auditory dominance that typical children do but exhibited optimal integration between audition and vision instead. These results agree with the idea that early auditory experience may be fundamental for the temporal calibration of the remaining visual and somatosensory modalities and the lack of it during the first period of life impacts on the development of an auditory-based temporal reference system.

To sum up, although some discrepancies which may depend on different mechanisms underlying different experimental paradigms, research highlighted that lack of audition is often associated with deficits in some temporal tasks, likely involving what we referred to as the perception of time. In line with the assumptions derived from the cross-calibrating theory, when the dominant sensory modality for a specific perceptual skill is missing, its development can be compromised. In the next section, we illustrate a different situation, that is, when vision, which is the dominant sensory modality for space perception (Alais & Burr, 2004; Hay et al., 1965) is missing, and audition becomes one of the main sensory channels for communicating with the external world.

Developing proper temporal skills without vision

Symmetrically to what happens for deafness and time, since vision is the dominant sense for spatial perception, most research on blindness focuses on perceptual consequences in the spatial domain (Gori et al., 2020a) and little is known about the repercussions on the temporal domain. However, as audition and temporal processing are strictly related, one can wonder whether and how temporal skills are impacted by neural reorganization and compensatory mechanisms that interest the auditory system once vision is impaired. One may indeed expect enhanced temporal perception. In this section, we aim to review changes in both the perception of time and timing of perception when the visual system is compromised in adults and during development.

Perception of time without vision

To the best of our knowledge, studies related to the perception of time without vision do not directly address the temporal skills of blind individuals in estimating temporal durations or intervals as the main goal but rather investigate them in control, secondary experiments. For instance, the performance of blind participants in a temporal bisection task was investigated by Gori et al. (2014) in order to compare it with that in a spatial bisection task. In the spatial bisection task, participants were presented with three consecutive sounds of 75 ms delivered from three different spatial positions with 500 ms of delay and had to judge the spatial distances between them (i.e., "Is the second sound closer to the first one or third one in space?"). In the temporal bisection task, participants were presented with the same sounds but delivered from the same spatial position with different delays (max delay < 1 s) and had to judge the temporal intervals between them (i.e., "Is the second sound closer to the first one or third one in time?"). Blind individuals showed a reduced precision in the spatial bisection task, but no differences emerged between the two groups in temporal bisection precision, suggesting an impairment involving the spatial domain only. These results were replicated more recently while investigating with EEG the neural correlates of the spatial bisection deficit caused by blindness (Campus et al., 2019). While demonstrating that early activation in occipital brain areas was affected by lack of vision, this study revealed that the cortical network associated with temporal bisection skills was similar for sighted and blind individuals. Specifically, they found no differences in the early cortical activation involving temporal areas elicited by the processing of the second sound of the temporal bisection task. Based on the preserved temporal skills observed for this kind of paradigm, the same research team subsequently investigated a compelling hypothesis: the preserved temporal representation of events could be used to set impaired auditory spatial representations (Amadeo et al., 2019). This hypothesis turned out to be supported by showing that 17 early blind individuals were attracted by the temporal cues when coherently or incoherently manipulated together with the spatial ones in different versions of the spatial bisection paradigm (the spatial position and the temporal delay of the second stimulus of the bisection task can be manipulated in order to deliver coherent or incoherent spatio-temporal information). In addition, a subsequent EEG study highlighted that also the early neural response of 16 blind people during the spatial bisection task was influenced by the coherent/incoherent temporal features (Gori et al., 2020b), further strengthening the idea that when vision is missing and complex spatial skills are impaired, preserved auditory temporal abilities can guide the construction of spatial representations. In line with this branch of research, Bertonati et al. (2021) investigated the weight ascribed to either spatial or temporal information during an auditory speed evaluation task in 10 early blind and 10 blindfolded adults. They reported that, although both groups preferentially relied on temporal cues to determine the speed of auditory stimuli, this was specifically true for early blind individuals. Taken together, these findings agreed on the auditory system's preference for the time domain, stressing that this can be even increased by the absence of visual experience. Moreover, although they suggested that the perception of time is neither impaired nor improved after blindness, it seems that it comes to the aid when blind people need to deal with complex spatial tasks, they are not able to solve. As we observed for deafness, the temporal and the spatial domains seem to be strictly interrelated when one sense is missing.

Although blindness was not directly assessed, another interesting study implicitly addressing the perception of time in individuals with atypical visual experience is the one conducted by Putzar et al. (2007) involving individuals who had been deprived of vision for at least the first 5 months of their life as a result of congenital binocular cataracts. In the study, 11 age-matched sighted controls and 11 adults being treated for dense congenital binocular cataracts were instructed to identify the specific color from a sequence of rapidly shifting colors that was shown simultaneously with a target flash. A distractor tone, unrelated to the task, was played shortly before or after the target flash. Even though participants were asked to ignore the task-irrelevant sound, when that was presented after the target flash, sighted controls reported that the target flash appeared at a later position, indicating a cross-modal temporal capture effect that suggests auditory dominance. As regards the cataract group, temporal identification of visual stimuli was not impaired by early visual deprivation and was unaffected by task-irrelevant auditory distractor too. In line with the observations in children with restored hearing (Gori et al., 2017), these results suggest that auditory dominance on temporal visual skills can be affected by reduced multisensory interactions during the first months of life and does not easily recover.

Timing of perception without vision

Research on the timing of perception without vision seems to agree with a better temporal resolution of the non-deprived sensory modalities. Among the first to investigate the ability to temporally order sensory events in blindness, Roder et al. (2004) demonstrated with a tactile TOJ task that seven congenitally blind individuals had better tactile temporal resolution compared to a group of age-matched sighted ones, independently of hand posture (i.e., crossed or uncrossed). Similar results have been subsequently replicated by others investigating temporal segregation of tactile (Crollen et al., 2019; Röder et al., 2004; Vanderclausen et al., 2021) and auditory (Stevens & Weaver, 2005) stimuli. In line with that, another study using the TOJ task (Occelli et al., 2008) reported higher ability in 17 early and late blind participants to discriminate which stimulus came first between an auditory and tactile one delivered from different positions in space. This interpretation is also supported by a recent study by Opoku-Baah and Wallace (2020), testing the effect of transient monocular deprivation in judging audiovisual simultaneity (i.e., judging whether two stimuli occurred simultaneously over a range of different stimulus onset asynchronies). They reported that after 90 min of monocular deprivation, audiovisual temporal acuity was enhanced for the deprived eye and reduced for the nondeprived eye, suggesting that even short deprivation windows taking place in adulthood have an impact on temporal acuity. All these findings have been interpreted as a result of compensation mechanisms supported by cross-modal plasticity. In other words, when one sensory modality is deprived, individuals rely heavily on their remaining senses and the brain shows an incredible capacity to reorganize and redistribute neural resources to enhance their processing. Although this seems to be the case for temporal resolution when processing events without vision, it is important to take carefully these conclusions as temporal processing of blind people appears to be sensitive also to spatial features of stimuli and most studies did not take that into account. Notably, similar temporal resolution between sighted and blind adults have been observed when stimuli were delivered from different spatial positions (Crollen et al., 2017; Crollen et al., 2019).

Evidence from visually impaired children

As for deafness, literature about temporal skills in blind children is poor. In a study aiming at exploring spatial skills, a group of seven congenitally blind children aged 9–14 years showed similar performance to blindfolded sighted people in a temporal bisection task conducted as control experiment (Vercillo et al., 2016). However, contrasting results have been recently observed with a different version of the bisection paradigm. In Battistin et al. (2019), 11 blind, 16 visually impaired, 20 sighted, and 16 blindfolded sighted children aged between 6 and 11 years were asked to evaluate the duration of sounds (duration range: 300–900 ms). On the one hand, blind, visually impaired, and blindfolded children were more accurate than sighted participants. On the other hand, blind and visually impaired children were less precise. While the first finding agreed with previous literature showing better performance without vision, authors discussed the second finding as a result of difficulty in maintaining a stable temporal representation in memory rather than in temporal processing.

Although they did not test blind children directly, Chen et al. (2017) investigated the effect of a brief period of monocular or binocular deprivation on the perception of simultaneity of audiovisual and visuo-tactile stimuli during early development. They tested 14 patients treated for binocular congenital cataract aged between 14 and 34 years, and 15 patients treated for monocular congenital cataract aged between 11 and 43 years. Interestingly, there were no differences between patients, either monocularly or binocularly deprived, and age-matched controls for the visuo-tactile condition, as if lack of vision did not impact on temporal segregation of visuo-tactile stimuli. Instead, when auditory stimuli were involved as in the audio-tactile task, monocularly deprived patients were overall less precise regardless of the eye tested, and binocularly deprived patients were less precise only in trials when the visual stimulus came before the auditory one. Authors concluded that the presence of visual input to both eyes and early multisensory interactions is crucial for establishing the neural architecture necessary for the typical development of an audiovisual simultaneity window.

To sum up, not many studies in the past focused on temporal skills following blindness. The few that exist allow us to claim that some aspects of temporal processing seem to benefit from lack of vision as expected based on cross-modal plasticity but not others. Specifically, the ability to evaluate and estimate temporal durations or intervals does not encounter specific improvements due to lack of vision, while the temporal resolution of non-visual sensory modalities is in some contexts higher.

Conclusion

In this chapter, we have delved into the realm of temporal perception from a sensory perspective. Our exploration has focused on the intricate relationship between sensory modalities and the construction of our sense of time. We have observed how our auditory system plays a pivotal role in calibrating some temporal skills across different sensory channels during development, owing to its inherent temporal reliability. By examining the consequences of auditory deprivation, we have discerned that certain temporal skills related to the evaluation of intervals and durations are impaired in the absence of auditory input, while others tied to the intrinsic temporal resolutions of individual sensory systems remain largely unaffected. Furthermore, we explored the context of blindness, where audition assumes a paramount role in conveying crucial information about the external environment. We have uncovered that blind individuals exhibit a remarkable preservation of their ability to estimate time durations and intervals. This suggests that the perception of time itself is largely indifferent to the lack of visual input. However, our exploration has unveiled an intriguing finding: blind individuals demonstrate in some cases an enhanced capacity to temporally segregate incoming sensory events. This improvement in temporal segregation could be attributed to the compensatory mechanisms and cross-modal plasticity that arise when one sensory modality is missing.

In summary, by expanding our understanding of temporal perception, we have witnessed the crucial role of audition in calibrating some temporal skills across sensory modalities, and the differential effects of sensory deprivation on specific temporal abilities. An interesting aspect that warrants further research is the interaction between temporal and spatial domains.

We have briefly touched upon the phenomenon where spatial information aids in decoding visual temporal coordinates when some visual temporal representations are impaired such as in deafness, while temporal information supports complex auditory spatial processing in blindness. How does that happen? These recent findings raise intriguing questions that hold important implications for rehabilitation. Through further investigation and a multidisciplinary approach, we can continue to expand our knowledge of temporal perception and its interactions with spatial processing, opening up new avenues for potential interventions and innovative strategies to improve temporal processing and the overall perceptual experience of individuals with sensory impairments.

References

Alais, D., & Burr, D. (2004). The Ventriloquist effect results from near-optimal bimodal integration. *Current Biology*, *14*(3), 257–262. https://doi.org/10.1016/j.cub.2004.01.029

Allman, B. L., Keniston, L. P., & Meredith, M. A. (2009). Adult deafness induces somatosensory conversion of ferret auditory cortex. *Proceedings of the National Academic of Sciences USA*, *106*(14), 5925–5930. https://doi.org/10.1073/pnas.0809483106

Amadeo, M. B., Campus, C., & Gori, M. (2020). Visual representations of time elicit early responses in human temporal cortex. *NeuroImage*, *217*, 116912. https://doi.org/10.1016/j.neuroimage.2020.116912

Amadeo, M. B., Campus, C., Pavani, F., & Gori, M. (2019). Spatial cues influence time estimations in deaf individuals. *IScience*, *19*, 369–377. https://doi.org/10.1016/j.isci.2019.07.042

Amadeo, M. B., Esposito, D., Escelsior, A., Campus, C., Inuggi, A., Pereira Da Silva, B., Serafini, G., Amore, M., & Gori, M. (2022). Time in schizophrenia: A link between psychopathology, psychophysics and technology. *Translational Psychiatry*, *12*(1), 1. https://doi.org/10.1038/s41398-022-02101-x

Andersen, T. S., Tiippana, K., & Sams, M. (2004). Factors influencing audiovisual fission and fusion illusions. *Brain Research. Cognitive Brain Research*, *21*(3), 301–308. https://doi.org/10.1016/j.cogbrainres.2004.06.004

Anobile, G., Domenici, N., Togoli, I., Burr, D., & Arrighi, R. (2020). Distortions of visual time induced by motor adaptation. *Journal of Experimental Psychology: General*, *149*(7), 1333. https://doi.org/10.1037/xge0000709

Aschersleben, G., & Bertelson, P. (2003). Temporal ventriloquism: Crossmodal interaction on the time dimension. 2. Evidence from sensorimotor synchronization. *International Journal of Psychophysiology: Official Journal of the International Organization of Psychophysiology*, *50*(1–2), 157–163. https://doi.org/10.1016/s0167-8760(03)00131-4

Barone, P., Lacassagne, L., & Kral, A. (2013). Reorganization of the connectivity of cortical field DZ in congenitally deaf cat. *PLoS One*, 8(4), e60093. https://doi.org/10.1371/journal.pone.0060093

Battistin, T., Mioni, G., Schoch, V., & Bisiacchi, P. S. (2019). Comparison of temporal judgments in sighted and visually impaired children. *Research in Developmental Disabilities*, *95*, 103499.

Bertonati, G., Amadeo, M. B., Campus, C., & Gori, M. (2021). Auditory speed processing in sighted and blind individuals. *PLoS One, 16*(9), e0257676. https://doi.org/10.1371/journal.pone.0257676

Bola, Ł., Zimmermann, M., Mostowski, P., Jednoróg, K., Marchewka, A., Rutkowski, P., & Szwed, M. (2017). Task-specific reorganization of the auditory cortex in deaf humans. *Proceedings of the National Academy of Sciences, 114*(4), 4. https://doi.org/10.1073/pnas.1609000114

Bolognini, N., Cecchetto, C., Geraci, C., Maravita, A., Pascual-Leone, A., & Papagno, C. (2012). Hearing shapes our perception of time: Temporal discrimination of tactile stimuli in deaf people. *Journal of Cognitive Neuroscience, 24*(2), 276–286. https://doi.org/10.1162/jocn_a_00135

Bolognini, N., Papagno, C., Moroni, D., & Maravita, A. (2010). Tactile temporal processing in the auditory cortex. *Journal of Cognitive Neuroscience, 22*(6), 1201–1211. https://doi.org/10.1162/jocn.2009.21267

Brown, S. W., & Boltz, M. G. (2002). Attentional processes in time perception: Effects of mental workload and event structure. *Journal of Experimental Psychology: Human Perception and Performance, 28*, 600–615. https://doi.org/10.1037/0096-1523.28.3.600

Bueti, D., Bahrami, B., & Walsh, V. (2008). Sensory and association cortex in time perception. *Journal of Cognitive Neuroscience, 20*(6), 1054–1062. https://doi.org/10.1162/jocn.2008.20060

Buonomano, D. V. (2007). The biology of time across different scales. *Nature Chemical Biology, 3*(10), 594–597. https://doi.org/10.1038/nchembio1007-594

Buonomano, D. V., & Karmarkar, U. R. (2002). How do we tell time? *The Neuroscientist: A Review Journal Bringing Neurobiology, Neurology and Psychiatry, 8*(1), 42–51. https://doi.org/10.1177/107385840200800109

Buonomano, D. V., & Maass, W. (2009). State-dependent computations: Spatiotemporal processing in cortical networks. *Nature Reviews. Neuroscience, 10*(2), 113–125. https://doi.org/10.1038/nrn2558

Burr, D. (2012). Cross-sensory integration and calibration in adults and young children. https://doi.org/10.7551/mitpress/8466.003.0047

Burr, D., Banks, M. S., & Morrone, M. C. (2009). Auditory dominance over vision in the perception of interval duration. *Experimental Brain Research, 198*(1), 49–57. https://doi.org/10.1007/s00221-009-1933-z

Burr, D., Tozzi, A., & Morrone, M. C. (2007). Neural mechanisms for timing visual events are spatially selective in real-world coordinates. *Nature Neuroscience, 10*(4), 423–425. https://doi.org/10.1038/nn1874

Campus, C., Sandini, G., Amadeo, M. B., & Gori, M. (2019). Stronger responses in the visual cortex of sighted compared to blind individuals during auditory space representation. *Scientific Reports, 9*(1), 1. https://doi.org/10.1038/s41598-018-37821-y

Cardin, V., Grin, K., Vinogradova, V., & Manini, B. (2020). Crossmodal reorganisation in deafness: Mechanisms for functional preservation and functional change. *Neuroscience and Biobehavioral Reviews, 113*, 227–237. https://doi.org/10.1016/j.neubiorev.2020.03.019

Chen, K.-M., & Yeh, S.-L. (2009). Asymmetric cross-modal effects in time perception. *Acta Psychologica, 130*(3), 225–234. https://doi.org/10.1016/j.actpsy.2008.12.008

Chen, Y.-C., Lewis, T. L., Shore, D. I., & Maurer, D. (2017). Early binocular input is critical for development of audiovisual but not visuotactile simultaneity perception. *Current Biology, 27*(4), 583–589. https://doi.org/10.1016/j.cub.2017.01.009

Conner, J. D. (1982). The temporal properties of rod vision. *The Journal of Physiology*, *332*, 139–155. https://doi.org/10.1113/jphysiol.1982.sp014406

Coull, J. T., Vidal, F., Nazarian, B., & Macar, F. (2004). Functional anatomy of the attentional modulation of time estimation. *Science*, *303*(5663), 1506–1508. https://doi.org/10.1126/science.1091573

Crollen, V., Lazzouni, L., Rezk, M., Bellemare, A., Lepore, F., & Collignon, O. (2017). Visual experience shapes the neural networks remapping touch into external space. *Journal of Neuroscience*, *37*(42), 10097–10103. https://doi.org/10.1523/JNEUROSCI.1213-17.2017

Crollen, V., Spruyt, T., Mahau, P., Bottini, R., & Collignon, O. (2019). How visual experience and task context modulate the use of internal and external spatial coordinate for perception and action. *Journal of Experimental Psychology: Human Perception and Performance*, *45*(3), 354–362. https://doi.org/10.1037/xhp0000598

Dekker, T., & Lisi, M. (2020). Sensory development: Integration before calibration. *Current Biology*, *30*(9), R409–R412. https://doi.org/10.1016/j.cub.2020.02.060

Diedrichsen, J., Ivry, R. B., & Pressing, J. (2003). Cerebellar and basal ganglia contributions to interval timing. In *Functional and neural mechanisms of interval timing* (pp. 457–483). CRC Press/Routledge/Taylor & Francis Group. https://doi.org/10.1201/9780203009574.ch19

Domenici, N., Tonelli, A., & Gori, M. (2021). Adaptation to high-frequency vibrotactile stimulations fails to affect the clock in young children. *Current Research in Behavioral Sciences*, *2*, 100018. https://doi.org/10.1016/j.crbeha.2021.100018

Domenici, N., Tonelli, A., & Gori, M. (2023). Deaf individuals use compensatory strategies to estimate visual time events. *Brain Research*, *1798*, 148148. https://doi.org/10.1016/j.brainres.2022.148148

Droit-Volet, S., Meck, W. H., & Penney, T. B. (2007). Sensory modality and time perception in children and adults. *Behavioural Processes*, *74*(2), 2. https://doi.org/10.1016/j.beproc.2006.09.012

Ernst, M. O. (2008). Multisensory integration: A late bloomer. *Current Biology*, *18*(12), R519–R521. https://doi.org/10.1016/j.cub.2008.05.002

Ernst, M. O., & Banks, M. S. (2002). Humans integrate visual and haptic information in a statistically optimal fashion. *Nature*, *415*(6870), 6870. https://doi.org/10.1038/415429a

Essen, L., & Parry, J. V. L. (1955). An atomic standard of frequency and time interval: A Cæsium resonator. *Nature*, *176*(4476), 4476. https://doi.org/10.1038/176280a0

Fendrich, R., & Corballis, P. M. (2001). The temporal cross-capture of audition and vision. *Perception & Psychophysics*, *63*(4), 719–725. https://doi.org/10.3758/BF03194432

Ferrandez, A. M., Hugueville, L., Lehéricy, S., Poline, J. B., Marsault, C., & Pouthas, V. (2003). Basal ganglia and supplementary motor area subtend duration perception: An fMRI study. *NeuroImage*, *19*(4), 1532–1544. https://doi.org/10.1016/S1053-8119(03)00159-9

Fraisse, P. (1963). *The psychology of time* (p. 343). Harper & Row.

Gebhard, J. W., & Mowbray, G. H. (1959). On discriminating the rate of visual flicker and auditory flutter. *The American Journal of Psychology*, *72*(4), 521–529.

Gilden, D. L., & Mezaraups, T. M. (2022). Allometric scaling laws for temporal proximity in perceptual organization. *Psychological Review*, *129*(3), 457–483. https://doi.org/10.1037/rev0000307

Gori, M. (2015). Multisensory integration and calibration in children and adults with and without sensory and motor disabilities. *Multisensory Research*, *28*(1–2), 71–99. https://doi.org/10.1163/22134808-00002478

Gori, M., Amadeo, M. B., & Campus, C. (2020a). Spatial metric in blindness: Behavioural and cortical processing. *Neuroscience & Biobehavioral Reviews, 109*, 54–62. https://doi.org/10.1016/j.neubiorev.2019.12.031

Gori, M., Amadeo, M. B., & Campus, C. (2020b). Temporal cues trick the visual and auditory cortices mimicking spatial cues in blind individuals. *Human Brain Mapping, 41*(8), 2077–2091. https://doi.org/10.1002/hbm.2493

Gori, M., Amadeo, M. B., Pavani, F., Valzolgher, C., & Campus, C. (2022). Temporal visual representation elicits early auditory-like responses in hearing but not in deaf individuals. *Scientific Reports, 12*(1), 1. https://doi.org/10.1038/s41598-022-22224-x

Gori, M., Bertonati, G., Campus, C., & Amadeo, M. B. (2023). Multisensory representations of space and time in sensory cortices. *Human Brain Mapping, 44*(2), 656–667. https://doi.org/10.1002/hbm.26090

Gori, M., Chilosi, A., Forli, F., & Burr, D. (2017). Audio-visual temporal perception in children with restored hearing. *Neuropsychologia, 99*, 350–359. https://doi.org/10.1016/j.neuropsychologia.2017.03.025

Gori, M., Del Viva, M., Sandini, G., & Burr, D. C. (2008). Young children do not integrate visual and haptic form information. *Current Biology: CB, 18*(9), 694–698. https://doi.org/10.1016/j.cub.2008.04.036

Gori, M., Sandini, G., & Burr, D. (2012). Development of visuo-auditory integration in space and time. *Frontiers in Integrative Neuroscience, 6*, 77. https://www.frontiersin.org/articles/10.3389/fnint.2012.00077

Gori, M., Sandini, G., Martinoli, C., & Burr, D. (2010). Poor haptic orientation discrimination in nonsighted children may reflect disruption of cross-sensory calibration. *Current Biology, 20*(3), 223–225. https://doi.org/10.1016/j.cub.2009.11.069

Gori, M., Sandini, G., Martinoli, C., & Burr, D. C. (2014). Impairment of auditory spatial localization in congenitally blind human subjects. *Brain: A Journal of Neurology, 137*(Pt 1), 288–293. https://doi.org/10.1093/brain/awt311

Grondin, S., Laflamme, V., & Tetreault, E. (2020). One psychological second does not necessarily last 1000 ms. *Psych Journal, 9*(3), 414–416. https://doi.org/10.1002/pchj.323

Grondin, S., Meilleur-Wells, G., & Lachance, R. (1999). When to start explicit counting in a time-intervals discrimination task: A critical point in the timing process of humans. *Journal of Experimental Psychology: Human Perception and Performance, 25*, 993–1004. https://doi.org/10.1037/0096-1523.25.4.993

Han, S., Chen, Y.-C., Maurer, D., Shore, D. I., Lewis, T. L., Stanley, B. M., & Alais, D. (2022). The development of audio–visual temporal precision precedes its rapid recalibration. *Scientific Reports, 12*(1), 1. https://doi.org/10.1038/s41598-022-25392-y

Hardy, N. F., & Buonomano, D. V. (2016). Neurocomputational models of interval and pattern timing. *Current Opinion in Behavioral Sciences, 8*, 250–257. https://doi.org/10.1016/j.cobeha.2016.01.012

Hay, J. C., Pick, H. L., & Ikeda, K. (1965). Visual capture produced by prism spectacles. *Psychonomic Science, 2*(1), 215–216. https://doi.org/10.3758/BF03343413

Heming, J. E., & Brown, L. N. (2005). Sensory temporal processing in adults with early hearing loss. *Brain and Cognition, 59*, 173–182. https://doi.org/10.1016/j.bandc.2005.05.012

Ivry, R. B. (1996). The representation of temporal information in perception and motor control. *Current Opinion in Neurobiology, 6*(6), 851–857. https://doi.org/10.1016/s0959-4388(96)80037-7

Ivry, R. B., & Schlerf, J. E. (2008). Dedicated and intrinsic models of time perception. *Trends in Cognitive Sciences, 12*(7), 273–280. https://doi.org/10.1016/j.tics.2008.04.002

Ivry, R. B., Spencer, R. M., Zelaznik, H. N., & Diedrichsen, J. (2002). The cerebellum and event timing. *Annals of the New York Academy of Sciences*, *978*(1), 302–317. https://doi.org/10.1111/j.1749-6632.2002.tb07576.x

Johansson, G. (1976). Spatio-temporal differentiation and integration in visual motion perception. *Psychological Research*, *38*(4), 379–393. https://doi.org/10.1007/BF00309043

Johnston, A., Arnold, D. H., & Nishida, S. (2006). Spatially localized distortions of event time. *Current Biology*, *16*(5), 472–479. https://doi.org/10.1016/j.cub.2006.01.032

Kanai, R., Lloyd, H., Bueti, D., & Walsh, V. (2011). Modality-independent role of the primary auditory cortex in time estimation. *Experimental Brain Research*, *209*(3), 465–471. https://doi.org/10.1007/s00221-011-2577-3

Killeen, P. R., & Grondin, S. (2022). A trace theory of time perception. *Psychological Review*, 129(4), 603–639. https://doi.org/10.1037/rev0000308

Klumpp, R. G., & Eady, H. R. (2005). Some measurements of interaural time difference thresholds. *The Journal of the Acoustical Society of America*, *28*(5), 859–860. https://doi.org/10.1121/1.1908493

Kowalska, J., & Szelag, E. (2006). The effect of congenital deafness on duration judgment. *Journal of Child Psychology and Psychiatry, and Allied Disciplines*, *47*(9), 946–953. https://doi.org/10.1111/j.1469-7610.2006.01591.x

Kral, A., & Sharma, A. (2023). Crossmodal plasticity in hearing loss. *Trends in Neurosciences*, *46*(5), 377–393. https://doi.org/10.1016/j.tins.2023.02.004

Levitt, J. B., Schumer, R. A., Sherman, S. M., Spear, P. D., & Movshon, J. A. (2001). Visual response properties of neurons in the LGN of normally reared and visually deprived macaque monkeys. *Journal of Neurophysiology*, *85*(5), 2111–2129. https://doi.org/10.1152/jn.2001.85.5.2111

Lewis, P. A., & Miall, R. C. (2003). Brain activation patterns during measurement of sub- and supra-second intervals. *Neuropsychologia*, *41*(12), 1583–1592. https://doi.org/10.1016/s0028-3932(03)00118-0

Lomber, S. G., Meredith, M. A., & Kral, A. (2010). Cross-modal plasticity in specific auditory cortices underlies visual compensations in the deaf. *Nature Neurosciences*, *13*(11), 1421–1427. https://doi.org/10.1038/nn.2653

Machado, A., Malheiro, M. T., & Erlhagen, W. (2009). Learning to time: A perspective. *Journal of the Experimental Analysis of Behavior*, *92*(3), 423–458. https://doi.org/10.1901/jeab.2009.92-423

McGovern, D. P., Astle, A. T., Clavin, S. L., & Newell, F. N. (2016). Task-specific transfer of perceptual learning across sensory modalities. *Current Biology*, *26*(1), R20–R21. https://doi.org/10.1016/j.cub.2015.11.048

Merchant, H., Harrington, D. L., & Meck, W. H. (2013). Neural basis of the perception and estimation of time. *Annual Review of Neuroscience*, *36*(1), 313–336. https://doi.org/10.1146/annurev-neuro-062012-170349

Moallem, T. M., Reed, C. M., & Braida, L. D. (2010). Measures of tactual detection and temporal order resolution in congenitally deaf and normal-hearing adults. *Journal of the Acoustical Society of America*, *127*(6), 3696–3709. https://doi.org/10.1121/1.3397432

Murai, Y., & Yotsumoto, Y. (2016). Timescale- and sensory modality-dependency of the central tendency of time perception. *PLoS One*, *11*(7), e0158921. https://doi.org/10.1371/journal.pone.0158921

Nardini, M., Bedford, R., & Mareschal, D. (2010). Fusion of visual cues is not mandatory in children. *Proceedings of the National Academy of Sciences*, *107*(39), 17041–17046. https://doi.org/10.1073/pnas.1001699107

Nardini, M., Jones, P., Bedford, R., & Braddick, O. (2008). Development of cue integration in human navigation. *Current Biology: CB, 18*(9), 689–693. https://doi.org/10.1016/j.cub.2008.04.021

Nava, E., Bottari, D., Zampini, M., & Pavani, F. (2008). Visual temporal order judgment in profoundly deaf individuals. *Experimental Brain Research, 190*(2), 179–188. https://doi.org/10.1007/s00221-008-1459-9

Nenadic, I., Gaser, C., Volz, H.-P., Rammsayer, T., Häger, F., & Sauer, H. (2003). Processing of temporal information and the basal ganglia: New evidence from fMRI. *Experimental Brain Research, 148*(2), 238–246. https://doi.org/10.1007/s00221-002-1188-4

Noel, J.-P., Niear, M. D., der Burg, E. V., & Wallace, M. T. (2016). Audiovisual simultaneity judgment and rapid recalibration throughout the lifespan. *PLoS One, 11*(8), e0161698. https://doi.org/10.1371/journal.pone.0161698

Occelli, V., Spence, C., & Zampini, M. (2008). Audiotactile temporal order judgments in sighted and blind individuals. *Neuropsychologia, 46*(11), 2845–2850. https://doi.org/10.1016/j.neuropsychologia.2008.05.023

Opoku-Baah, C., & Wallace, M. T. (2020). Brief period of monocular deprivation drives changes in audiovisual temporal perception. *Journal of Vision, 20*(8), 8. https://doi.org/10.1167/jov.20.8.8

Pan, Y., & Luo, Q.-Y. (2012). Working memory modulates the perception of time. *Psychonomic Bulletin & Review, 19*(1), 46–51. https://doi.org/10.3758/s13423-011-0188-4

Pastor, M. A., Artieda, J., Jahanshahi, M., & Obeso, J. A. (1992). Time estimation and reproduction is abnormal in Parkinson's disease. *Brain: A Journal of Neurology, 115*(Pt 1), 211–225. https://doi.org/10.1093/brain/115.1.211

Petrini, K., Remark, A., Smith, L., & Nardini, M. (2014). When vision is not an option: Children's integration of auditory and haptic information is suboptimal. *Developmental Science, 17*(3), 376–387. https://doi.org/10.1111/desc.12127

Poizner, H., & Tallal, P. (1987). Temporal processing in deaf signers. *Brain and Language, 30*(1), 52–62. https://doi.org/10.1016/0093-934x(87)90027-7

Pouthas, V., George, N., Poline, J.-B., Pfeuty, M., VandeMoorteele, P.-F., Hugueville, L., Ferrandez, A.-M., Lehéricy, S., LeBihan, D., & Renault, B. (2005). Neural network involved in time perception: An fMRI study comparing long and short interval estimation. *Human Brain Mapping, 25*(4), 433–441. https://doi.org/10.1002/hbm.20126

Putzar, L., Goerendt, I., Lange, K., Rösler, F., & Röder, B. (2007). Early visual deprivation impairs multisensory interactions in humans. *Nature Neuroscience, 10*(10), 10. https://doi.org/10.1038/nn1978

Rao, S. M., Mayer, A. R., & Harrington, D. L. (2001). The evolution of brain activation during temporal processing. *Nature Neuroscience, 4*(3), 3. https://doi.org/10.1038/85191

Recanzone, G. H. (2003). Auditory influences on visual temporal rate perception. *Journal of Neurophysiology, 89*(2), 1078–1093. https://doi.org/10.1152/jn.00706.2002

Repp, B. H., & Penel, A. (2002). Auditory dominance in temporal processing: New evidence from synchronization with simultaneous visual and auditory sequences. *Journal of Experimental Psychology. Human Perception and Performance, 28*(5), 1085–1099. https://doi.org/10.1037/0096-1523.28.5.1085

Rock, I., & Victor, J. (1964). Vision and touch: An experimentally created conflict between the two senses. *Science, 143*(3606), 594–596. https://doi.org/10.1126/science.143.3606.594

Röder, B., Rösler, F., & Spence, C. (2004). Early vision impairs tactile perception in the blind. *Current Biology: CB, 14*(2), 121–124. https://doi.org/10.1016/j.cub. 2003.12.054

Roder, B., Rosler, F., Spence, C. (2004). Early vision impairs tactile perception in the blind. *Current Biology; 14*(2), 121–124.

Rohlf, S., Li, L., Bruns, P., & Röder, B. (2020). Multisensory integration develops prior to crossmodal recalibration. *Current Biology: CB, 30*(9), 1726–1732.e7. https://doi.org/10.1016/j.cub.2020.02.048

Schirmer, A. (2004). Timing speech: A review of lesion and neuroimaging findings. *Brain Research. Cognitive Brain Research, 21*(2), 269–287. https://doi.org/10.1016/j. cogbrainres.2004.04.003

Scurry, A. N., Chifamba, K., & Jiang, F. (2020). Electrophysiological dynamics of visual-tactile temporal order perception in early deaf adults. *Frontiers in Neuroscience, 14*, 544472. https://doi.org/10.3389/fnins.2020.544472

Shams, L., Kamitani, Y., & Shimojo, S. (2000). What you see is what you hear. *Nature, 408*(6814), 788.

Shannon, R. V., Zeng, F. G., Kamath, V., Wygonski, J., & Ekelid, M. (1995). Speech recognition with primarily temporal cues. *Science (New York, N. Y.), 270*(5234), 303–304. https://doi.org/10.1126/science.270.5234.303

Shi, Z., & Burr, D. (2016). Predictive coding of multisensory timing. *Current Opinion in Behavioral Sciences, 8*, 200–206. https://doi.org/10.1016/j.cobeha.2016.02.014

Shipley, T. (1964). Auditory flutter-driving of visual flicker. *Science, 145*(3638), 1328–1330.

Stevens, A. A., & Weaver, K. (2005). Auditory perceptual consolidation in early-onset blindness. *Neuropsychologia, 43*(13), 1901–1910. https://doi.org/10.1016/j. neuropsychologia.2005.03.007

Strelnikov, K., Rouger, J., Demonet, J. F., Lagleyre, S., Fraysse, B., Deguine, O., & Barone, P. (2013). Visual activity predicts auditory recovery from deafness after adult cochlear implantation. *Brain, 136*(Pt 12), 3682–3695. https://doi.org/ 10.1093/brain/awt274

Treisman, M., Faulkner, A., Naish, P. L., & Brogan, D. (1990). The internal clock: Evidence for a temporal oscillator underlying time perception with some estimates of its characteristic frequency. *Perception, 19*(6), 705–743. https://doi. org/10.1068/p190705

Tsimring, L. S. (2014). Noise in biology. *Reports on Progress in Physics. Physical Society (Great Britain), 77*(2), 026601. https://doi.org/10.1088/0034-4885/ 77/2/026601

van Wassenhove, V, Buonomano, D. V., Shimojo, S., & Shams, L. (2008). Distortions of subjective time perception within and across senses. *PLoS One, 3*(1), e1437. https://doi.org/10.1371/journal.pone.0001437

Vanderclausen, C., Filbrich, L., De Volder, A., & Legrain, V. (2021). Measuring the sensitivity of tactile temporal order judgments in sighted and blind participants using the adaptive psi method. *Attention, Perception & Psychophysics, 83*(7), 2995–3007. https://doi.org/10.3758/s13414-021-02301-5

Vercillo, T., Burr, D., & Gori, M. (2016). Early visual deprivation severely compromises the auditory sense of space in congenitally blind children. *Developmental Psychology, 52*(6), 847–853. https://doi.org/10.1037/dev0000103

Whitton, S., Kim, J. M., Scurry, A. N., Otto, S., Zhuang, X., Cordes, D., & Jiang, F. (2021). Multisensory temporal processing in early deaf. *Neuropsychologia, 163*, 108069. https://doi.org/10.1016/j.neuropsychologia.2021.108069

Zakay, D., & Block, R. A. (1996). The role of attention in time estimation processes. In *Time, internal clocks and movement* (pp. 143–164). North-Holland/Elsevier Science Publishers. https://doi.org/10.1016/S0166-4115(96)80057-4

Zélanti, P. S., & Droit-Volet, S. (2012). Auditory and visual differences in time perception? An investigation from a developmental perspective with neuropsychological tests. *Journal of Experimental Child Psychology*, *112*(3), 296–311. https://doi.org/10.1016/j.jecp.2012.01.003

Zhang, F., Jin, K., & Zhang, S. (2020). Visual duration bisection in profoundly deaf individuals. *PeerJ*, *8*, e10133. https://doi.org/10.7717/peerj.10133

8

TIME PERCEPTION DEFICITS IN ALZHEIMER'S DISEASE AND MILD COGNITIVE IMPAIRMENT

Martin Riemer

Technical University Berlin, Germany

Introduction

Folk psychology has it that time perception is subject to profound changes across the lifespan. A common conception is that, as we grow older, time seems to run faster (Friedman & Janssen, 2010; Wittmann & Lehnhoff, 2005). One year passes by so quickly, and yet most of us can remember how protracted the same duration appeared to us as a child.

The idea that time perception changes with increasing age is supported by many studies (Block et al., 1998; Gabrian et al., 2017; Gagnon-Harvey et al., 2021; Lustig, 2003; Mioni 2021a), and it has been argued that this effect is driven by altered cognitive processes in aging humans (Balcı et al., 2009; Turgeon et al., 2016; Xu & Church, 2017). Imaging studies have revealed a large network of brain areas being involved in time perception, including parietal and prefrontal cortices (Hayashi et al., 2018), supplemental motor area and insula (Mondok & Wiener, 2023), cerebellum (Mangels et al., 1998), and the basal ganglia (Teki et al., 2012) (see Chapter 2). As all of these regions are affected in later phases of AD progression (Braak & Del Tredici, 2015), it is hard to determine if the timing deficits observed in AD patients are mediated by damage in one specific region. At an early stage of disease progression, however, the brain regions that are mostly affected by neuronal loss lie within the medial temporal lobe, specifically the hippocampus and the entorhinal cortex (Braak & Del Tredici, 2015). Compared to normally aging brains, these areas exhibit significant volume changes, an indicator of neuronal atrophy. Besides their role in long-term memory processes and spatial navigation (O'Keefe & Nadel, 1978), these areas have

DOI: 10.4324/9781003449546-8

recently been shown to play an important role for temporal experience (Melgire et al., 2005; Noulhiane et al., 2007) and the mental construction of temporal order (Hsieh et al., 2014; Long & Kahana, 2019). These findings corroborate the idea that changes in time perception are a common phenomenon of healthy aging, and that a more profound impairment of timing abilities might represent a core aspect of pathological aging (El Haj & Kapogiannis, 2016; Mioni et al., 2021b). In recent years, it has been suggested that the assessment of timing performance could be a useful tool for clinical diagnostics (Carrasco et al., 2000; El Haj & Kapogiannis, 2016; Maaß et al., 2019). Deficits in time perception could be utilized as a behavioral marker for cognitive decline, possibly signalizing the beginning of dementia at an early stage. Hence, standardized timing tasks could provide clinicians with helpful information for the identification of patients at risk for developing dementia. Concerted with other test batteries, these tasks might support the decision for further, more invasive diagnostic tests as PET scans or examinations of cerebrospinal fluid.

In support of this claim, many studies have reported timing deficits in AD patients as compared to healthy, age-matched controls (e.g., Carrasco et al., 2000; Caselli et al., 2009; El Haj et al., 2013; Nichelli et al., 1993; Papagno et al., 2004; Rueda & Schmitter-Edgecombe, 2009), and some studies suggest beginning impairments already in preclinical stages of AD, such as mild cognitive impairment (MCI; e.g., Maaß et al., 2019; Mioni et al., 2016, 2024; but see: Mioni et al., 2019b; Rueda & Schmitter-Edgecombe, 2009). MCI represents a transitive clinical state between age-related cognitive decline within a normal range and pathological cognitive decline (Albert et al., 2011; Petersen & Negash, 2008). Therefore, MCI patients constitute a suitable clinical population to investigate the development of perceptual deficits in AD patients.

The present chapter starts with a brief description of neurodegenerative diseases, focusing on patients diagnosed with AD or MCI, and on individuals presenting with subjective cognitive decline (SCD). I will then provide an overview over the results from studies on timing deficits in these patient groups, discuss the critical interrelation between deficits in interval timing and memory impairments, and I will close this chapter with some methodological aspects that should be considered when assessing timing deficits in older individuals and patients suffering from AD or MCI.

Alzheimer's disease, mild cognitive impairment, and subjective cognitive decline

Alzheimer's disease (AD) is a progressive form of neurodegeneration associated with increasing age (Breijyeh & Karaman, 2020). Neuropathological changes related to the disease consist in an abnormal accumulation of extracellular amyloid-β protein leading to senile plaques, and an abnormal deposit of intracellular tau protein causing neurofibrillary tangles (Braak & Del Tredici, 2015; Breijyeh & Karaman, 2020; He et al., 2018). Recent evidence suggests that an interplay between the deposition of amyloid-β plaques and tau-based neurofibrillary tangles is a main mechanism for the widespread brain changes (Busche et al., 2019; He et al., 2018; Tripathi & Khan, 2020), ultimately leading to detrimental impairments in cognitive functioning. These neuropathological processes occur predominantly in the medial temporal lobe, especially in the hippocampus and the entorhinal cortex (Braak & Del Tredici, 2015; Devanand et al., 2007; Uotani et al., 2006). Braak & Del Tredici (2015) report evidence that the medial temporal lobe is the first region to be affected by neuronal atrophy, already at a very early stage of the disease.

MCI is a transitional state between normal cognitive decline due to healthy aging and pathological cognitive deficits, and is considered a precursor of dementia (Albert et al., 2011; Petersen & Negash, 2008; Whitwell et al., 2007). The amnestic form (aMCI) is primarily marked by memory deficits, as they are characteristic for AD (Petersen et al., 2001). As many of the cortical areas that are affected by neuronal atrophy during AD already exhibit volumetric decrements and reduced cortical thickness in MCI patients (Braak & Del Tredici, 2015; Cheng et al., 2018; Devanand et al., 2007; Fjell et al., 2010; Julkunen et al., 2009), it has been argued that the assessment of elementary cognitive functions associated with those regions could serve as a behavioral marker to support clinical diagnosis (El Haj & Kapogiannis, 2016; Howett et al., 2019). Time perception is one of those cognitive functions, and in recent years, several studies confirmed specific deficits in the timing behavior of AD patients (Carrasco et al., 2000; Caselli et al., 2009; El Haj et al., 2013; Nichelli et al., 1993; Rueda & Schmitter-Edgecombe, 2009). Deficits in timing behavior (although to a smaller degree) have sometimes also been reported in MCI patients (Maaß et al., 2019; Mioni et al., 2021b), and even in healthy subjects performing low in a memory task but not yet fulfilling the diagnostic criteria for MCI (Maaß et al., 2019, 2022).

Another group that recently has received increased attention are patients presenting with SCD. These patients report complaints about cognitive deficits with respect to memory or other cognitive functions, while these deficiencies cannot be detected by standardized neuropsychological test batteries (Jessen et al., 2020; Mitchell et al., 2014). In many studies, it was confirmed that individuals with SCD have an increased risk of developing objectively verifiable cognitive deficits, ultimately fulfilling the diagnostic criteria for MCI (Slot et al., 2019; Visser et al., 2009). Although the SCD group is characterized by a high degree of heterogeneity (ranging from individuals seeking attention to those being over-worried about cognitive decline to those with actual but diagnostically sub-threshold symptoms), and the majority of individuals with SCD will not show a progressive decline of their cognitive abilities (Jessen et al., 2020), it has been argued that SCD should be considered another potential risk factor for developing AD (Warren et al., 2022). This claim is supported by the observation that subjective complaints correlate with amyloid-β deposition, while there was no relationship between subjective complaints and the performance in neuropsychological tests of memory and executive functions (Amariglio et al., 2012).

Deficits in timing and time perception

In recent years, many studies have been conducted to investigate deficits in timing and time perception for AD and MCI patients. In the following section, I will provide an overview about the results of these studies. The overview is structured with respect to four important aspects in time perception research, namely effects on precision versus accuracy of temporal judgments, the length of the tested time intervals, the distinction between prospective and retrospective paradigms, and the distinction between explicit and implicit timing.

Precision and accuracy

When interpreting the performance (and changes in performance) in time perception tasks, it is important to distinguish between the effects on the accuracy and those on the precision of judgments. Mean accuracy refers to the degree to which the judgments are, on average, objectively correct. For example, when we have three trials to produce an interval of 10 s, productions of 9, 10, and 11 s would be equally "accurate" as productions of 5, 10, and 15 s. In contrast, precision refers to the variability of judgments. In the previous example, productions of 15, 15, and 15 s would be extremely precise, though not very accurate. The distinction between accuracy and precision is especially relevant for the study of timing deficits in neurodegenerative diseases, because changes in judgment precision can be explained on the basis of increased noise (possibly caused by memory decay) and do not necessarily imply changes in the underlying timing mechanism itself, whereas changes in judgment accuracy can be interpreted as altered oscillatory processes in terms of internal clock models (Pöppel, 1971; Treisman, 1963). The measure of absolute error score, defined as the mean deviation of temporal estimates from the correct values, but independent of the error's direction (e.g., whether an interval is over- or underestimated), has often been used to describe the performance in timing tasks (e.g., Carrasco et al., 2000; Papagno et al., 2004; Rueda & Schmitter-Edgecombe, 2009). The problem with this measure is that it does not provide information about either mean accuracy or precision. The same absolute error score (e.g., a mean deviation of 5 s in the example of three 10-s-production trials) could be caused by perfect precision and low mean accuracy (productions of 15, 15, and 15 s), as well as by perfect mean accuracy and low precision (productions of 2.5, 10, and 17.5 s).

Precision

With respect to precision, the available data are clearly in favor of a reduced precision in AD patients (Carrasco et al., 2000; Caselli et al., 2009; Nichelli et al., 1993; Pai et al., 2021; Rueda & Schmitter-Edgecombe, 2009). Whether this effect is already present in MCI patients, however, is less clear (Mioni et al., 2021b). Although most of the few studies available do not report differences in the precision of time judgments between MCI patients and healthy controls (Mioni et al., 2019b; Rueda & Schmitter-Edgecombe, 2009), Mioni et al. (2021b) point out that this might be due to the large heterogeneity of the patient group and small sample sizes. In one study examining Parkinson's disease patients with and without MCI, reduced timing precision was found for the MCI group (Mioni et al., 2016). Furthermore, Capizzi et al. (2022) show that timing precision in older adults decreases with lower scores of cognitive functioning as measured with the Mini-Mental State Examination (MMSE), a finding that was recently replicated by Bogon et al. (2024).

Accuracy

In contrast, the data regarding effects on mean accuracy in AD patients are rather inconsistent. Carrasco et al. (2000) asked their participants to produce intervals of different durations and found that the accuracy of AD patients was reduced only for the shortest interval (5 s), whereas longer intervals (10 and 25 s) were produced with an accuracy comparable to healthy, age-matched controls. Using the method of verbal estimation, El Haj et al. (2013) and Ranjbar Pouya et al. (2015) found AD patients' timing accuracy to be impaired, while others, using the same method, did not report such an effect (Nichelli et al., 1993; Pai et al., 2021; Rueda & Schmitter-Edgecombe, 2009). In the study by Nichelli et al. (1993), however, it was found that AD patients (relative to age-matched controls) over-reproduced an interval of 1 s. Caselli et al. (2009) implemented a time bisection task and did not find an effect on accuracy in AD patients. Regarding MCI patients, most studies did not reveal an effect on timing accuracy (Coelho et al., 2016; Heinik & Ayalon, 2010; Mioni et al., 2019b; Rueda & Schmitter-Edgecombe, 2009). The only study providing some evidence for reduced accuracy in MCI patients compared Parkinson's disease patients with and without MCI (Mioni et al., 2016). The authors report that Parkinson patients with MCI perceived temporal intervals on average as shorter than those without MCI. Analyzing the relationship between MMSE scores and timing accuracy, Capizzi et al. (2022) did not find evidence for a reduced accuracy with decreased cognitive functioning. In contrast, Baudouin et al. (2006) showed that accuracy in a reproduction task was correlated with a measure of working memory, in the sense that reduced working memory capacities were associated with an over-reproduction of durations. Also Bogon et al. (2024) reported a correlation between memory capacity and timing accuracy in a non-clinical group of older adults.

Length of the tested time intervals

Although the exact mechanisms of time perception are unknown to date, many scientists have argued that the underlying neuronal processes differ depending on the length of the tested durations (Lewis & Miall, 2003; Paton & Buonomano, 2018; Pouthas et al., 2005; Ulbrich et al., 2007). The processing of different interval lengths has been associated with different brain regions (Lewis & Miall, 2003). By reviewing neuroimaging studies, Lewis and Miall (2003) reported evidence for the dissociation between an automatic timing system dedicated to brief intervals in the sub-second range, and a cognitively controlled timing system for longer intervals. According to this distinction, sub-second intervals are processed mainly in the right cerebellum, the premotor cortex, supplementary motor area, and the left basal ganglia, while intervals in the supra-second range are predominantly processed in prefrontal and parietal cortices. With respect to the current question of timing deficits in neurodegenerative diseases, it is of specific importance that the brain regions activated for short intervals are also associated with the preparation and execution of movements (De Kock et al., 2021), while the areas recruited for longer intervals have been linked to sustained attention and working memory processes (Block et al., 2010; Lustig, 2003), which are characteristic of the cognitive impairments in MCI and AD. The length of durations investigated in various studies range from several milliseconds (as in judgments of temporal order or simultaneity) to several years (as in retrospective judgments on the passage of time).

Temporal order judgments and simultaneity judgments (<300 ms)

The shortest perceivable durations are revealed in judgments of temporal order and simultaneity (e.g., Basharat et al., 2018). In the respective tasks, participants are presented with two successive stimuli and have to judge which one was presented first (temporal order judgments) or whether the stimuli occurred simultaneously or successively (simultaneity judgments). The minimal stimulus onset asynchrony, for which reliable judgments can still be observed, has often been found to lie below 40 ms, although this value varies considerably and depends on the used modality as well as on the manner of presentation (Wackermann, 2007). In several studies, age-related increases of these thresholds were reported (Basharat et al., 2018; Bedard & Barnett-Cowan, 2016; Chan et al., 2014; Craig et al., 2010; Kuehn et al., 2018; Poliakoff et al., 2006; Setti et al., 2011; Szymaszek et al., 2009; Ulbrich et al., 2007) and there is some evidence that these changes are related to the decline of cognitive factors like working memory (Craig et al., 2010; Szymaszek et al., 2009; Ulbrich et al., 2007).

D'Antonio et al. (2019) compared the threshold for perceived simultaneity of two tactile stimuli in patients with AD (mean age: 78 years), MCI (mean age: 71.6 years) and SCD (mean age: 68.4 years) with those of healthy controls. They found that, relative to healthy controls, the tactile simultaneity thresholds of AD and MCI patients were significantly increased (to over 150 ms between stimuli), while individuals with SCD performed at a comparable level with healthy controls (of about 75 ms). Even though the tactile simultaneity threshold of AD patients was numerically larger than the one of MCI patients, this difference did not reach a significant level.

It is important to note, however, that temporal order judgments and simultaneity judgments have been found to be differently affected by increasing age: While older adults perform worse in temporal order judgments, they seem to be unimpaired in judging the simultaneity of stimuli (Basharat et al., 2018; Bedard & Barnett-Cowan, 2016). This finding has been interpreted as reflecting different mechanisms underlying temporal order judgments and simultaneity judgments (Basharat et al., 2018; Wackermann, 2007), but it also suggests that the age-related increase of temporal discrimination thresholds might reflect a deficit in the decision process (i.e., which one of the two stimuli occurred first) rather than a deficit in the perception of temporal simultaneity itself.

The perception of duration (200 ms to 3 s)

According to the classical distinction by Fraisse (1984), direct perception of durations is confined to time intervals of up to 3–5 s, beyond which temporal judgments rely more on abstract memory representations (discussed in the next subsection; see also Grondin 2012). It is important to note that the suggested numerical border between durations that are supposed to be directly perceived and durations that can only be retrospectively estimated varies in the literature, ranging from 2–3 s (e.g., Pöppel, 1997; Ulbrich et al., 2007;

Wittmann, 1999) to values of 7–8 s (Michon, 1978). Moreover, as attention capacities vary between different situations (e.g., Lustig & Meck, 2001) as well as between different individuals (e.g., Mathew et al., 2021), it is plausible to assume similar variations in the range of durations that can form a holistic percept. Consequently, the borderline between durations that can be directly perceived versus those that can only be estimated should be considered a rough approximation. This is especially important for the interpretation of data from patients with AD and MCI whose prominent symptoms are memory and attention impairments (Malhotra, 2019).

Two studies have been conducted to examine timing deficits in AD patients within this range of intervals (Caselli et al., 2009; Nichelli et al., 1993). Testing intervals between 100 ms to 3 s, Caselli et al. (2009) found that AD patients are less precise than a group of age-matched healthy controls. In accordance with this finding, Nichelli et al. (1993) reported a reduced precision in AD patients when they were asked to repeatedly reproduce an interval of 1 s. With respect to MCI patients, similar effects within this duration range were absent. Maaß et al. (2019) did not find differences between aMCI patients and healthy controls in a 1-s-interval production task, and also in the study by Mioni et al. (2019b), no deviating performance was found for MCI patients, neither in a time bisection task testing durations between 400 and 1,600 ms, nor in a 1-s-interval finger tapping task. However, the same group (and using the same bisection task) found that Parkinson's disease patients with MCI performed worse than those without an MCI diagnosis (Mioni et al., 2016), and Capizzi et al. (2022) reported a correlation between cognitive functioning (MMSE scores) and the precision in a time bisection task of a similar duration range (480–1,920 ms). In a recent study by Bogon et al. (2024), the correlation between timing precision and cognitive functioning was specifically found within a group of older (in contrast to younger) participants. In that study, cognitive functioning of older participants was also correlated with the accuracy of timing judgments, mirroring the finding of Mioni et al. (2024) with the same paradigm. Thus, while the few studies conducted on AD patients support the idea of timing deficits for durations below 3 s, the data on MCI patients reveal a more inconsistent picture.

Another performance measure in timing tasks relates to temporal context effects, that is, the degree to which the perceived duration of an interval is influenced by previously experienced durations (Roach et al., 2017; Schlichting et al., 2018). Maaß et al. (2019) found an increased temporal context effect in aMCI patients for durations between 1.2 and 1.7 s. Moreover, the authors found the same effect within the healthy control group, when splitting them up into high- and low-performing individuals (based on their score in a word list recall task; Morris et al., 1989). Low-performing individuals exhibited a more pronounced temporal context effect than high-performing individuals, a finding that was replicated in a later study of the same group (Maaß et al., 2022). Together, these studies suggest that age-related changes in temporal context effects (in contrast to alterations in the perception of single intervals) might be indicative of cognitive decline and hold diagnostic value for dementia at an early, pre-clinical stage.

Within the range of intervals that are assumed to be accessible for direct perception, one prominent distinction is the one between sub-second and supra-second durations, with the idea that the correct perception of intervals below 1 s is pertinent for the coordination of motor behavior, whereas intervals above 1 s are more relevant for judgments at a higher cognitive level (Lewis & Miall, 2003; Paton & Buonomano, 2018; Pouthas et al., 2005; Wiener et al., 2010). It has been suggested that the effect of healthy aging on timing precision is more pronounced when probed with shorter intervals, the largest difference between old and young subjects being observable for sub-second intervals (500–1,000 ms; Lamotte & Droit-Volet, 2017). Using the same paradigm, a time bisection task, the study by Caselli et al. (2009) seems to confirm this pattern for AD patients. The authors found that the generally reduced precision in AD patients (relative to age-matched controls) was more pronounced for sub-second (100–600 ms) compared with supra-second intervals (1–3 s). Also for aMCI patients, Mioni et al. (2024) observed a selective effect on the accuracy for sub-second versus supra-second durations. The precision of judgments was not affected by interval length, but aMCI patients performed significantly worse than healthy controls. Testing a group of patients with lesions in the medial temporal lobe (the anatomical region that shows the first signs of neuronal atrophy in AD; Braak & Del Tredici, 2015), Melgire et al. (2005) also reported a reduced precision in a time bisection task for sub-second (50–200 ms), but not for supra-second intervals (2–8 s). These results promote the use of intervals in the sub-second range to increase a potential diagnostic value of timing performance.

The estimation of duration (>3 s)

Most studies on timing deficits in AD patients focus on rather long intervals of 3 s or longer, for which the involvement of working memory processes can be assumed (Fraisse, 1984; Wittmann, 1999). For durations of up to 2 min, timing performance in AD patients is impaired (Carrasco et al., 2000; El Haj et al., 2013, 2014; Nichelli et al., 1993; Pai et al., 2021; Papagno et al., 2004; Ranjbar Pouya et al., 2015; Rueda & Schmitter-Edgecombe, 2009). Similar deficits were not reported for MCI patients (Coelho et al., 2016; Rueda & Schmitter-Edgecombe, 2009). However, it should be noted that in the study by Coelho et al. (2016), participants were explicitly instructed to count, which should have diminished the influence of a potential timing deficit (Grondin et al., 1999; Riemer et al., 2022). Investigating the effects of age on timing, Perbal et al. (2002) could show that older and younger adults performed at a comparable level when counting was allowed, although significant differences between the groups were observed when participants were distracted from chronometric counting. In fact, chronometric counting is usually considered a compensatory strategy to artificially increase timing performance and conceal existing deficits (Rattat & Droit-Volet, 2012; Riemer et al., 2022; cf. the section on the problem of chronometric counting). Using the method of articulatory suppression to prevent chronometric counting, Baudouin et al. (2006) demonstrated a relationship between working memory capacity and the performance in reproducing intervals of 5 and 14 s.

Passage-of-time judgments

At the very end of the spectrum of interval lengths, some studies implemented passage-of-time judgments (e.g., "How fast did the previous year pass for you?"). The intuitive notion that time seems to run faster as we get older has received some confirmation (e.g., Friedman & Janssen, 2010; Wittmann & Lehnhoff, 2005), but studies investigating systematic changes in the subjective passage of time in AD and MCI patients are scarce. The reason is that the potential differences in these judgments can be easily explained by the probably very different daily routines and lives of the investigated patient groups and healthy controls. Coelho et al. (2016) asked MCI patients to indicate the perceived speed of time during the previous week/month/year, using the questionnaire by Friedman and Janssen (2010) and report that MCI patients perceived the time to be passing slower than healthy controls, but the authors also acknowledge that MCI patients rated their life as being more like a routine compared to healthy controls, suggesting a possible confound influencing the passage-of-time judgments.

Prospective and retrospective timing

An important distinction in the research on human time perception is between prospective and retrospective timing (Block & Zakay, 1997; Mioni et al., 2021b). In prospective paradigms, the participants know in advance that they are expected to judge the duration of temporal intervals, so they can focus their attention on the passage of time, whereas in retrospective paradigms, the time-related nature of the task is only revealed after the duration that is to be judged has elapsed.

Prospective timing

In their meta-analysis, Mioni et al. (2021b) concluded that AD patients show a clear deficit in prospective timing. All studies including AD patients reported a reduced timing performance, both for relatively long intervals above 3 s (Carrasco et al., 2000; El Haj et al., 2013; Nichelli et al., 1993; Pai et al., 2021; Papagno et al., 2004; Rueda & Schmitter-Edgecombe, 2009) and below (Caselli et al., 2009; Nichelli et al., 1993). For MCI patients, the results are less consistent and most studies did not find evidence for a timing deficit (Coelho et al., 2016; Mioni et al., 2019b; Rueda & Schmitter-Edgecombe, 2009; but see Mioni et al., 2016, 2024; Maaß et al., 2019). However, Mioni et al. (2021b) point out that the number of studies on timing deficits in MCI patients are scarce, which is problematic especially in light of the large heterogeneity and small sample sizes that are present in these studies. One should be careful to draw early conclusions, especially considering the timing deficits found in AD patients for which MCI is an important precursor. This caution also seems justified on the basis of a second analysis in Mioni et al. (2021b), in which they used a meta-regression model highlighting the moderating effects of age and type of measure (i.e., whether in the original study the effect was reported as mean score, accuracy or variability of responses). This model revealed a significant effect on timing performance for both AD and MCI patients.

Retrospective timing

The few studies on retrospective timing including AD patients reveal inconsistent results. While El Haj et al. (2013) found that AD patients, relative to healthy controls, tend to under-estimate temporal intervals, Ranjbar Pouya et al., 2015 reported an over-estimation in AD patients. In another study using the reproduction method, El Haj et al. (2014) also reported an under-reproduction of temporal intervals. When directly comparing prospective and retrospective timing, the under-estimation of intervals between 30 and 120 s (that was found in this study) seems to be more pronounced in prospective relative to retrospective conditions, but this effect does not differ between AD patients and healthy controls (El Haj et al., 2013). Heinik (2012) analyzed retrospective judgments about the duration of a clinical interview (with the actual duration ranging between 30 and 93 min) and did not find evidence for a deviating estimates of AD patients. Also in MCI patients, no difference was reported for these duration lengths (Heinik & Ayalon, 2010). It should be noted that due to the one-trial nature of these two studies, it was not possible to calculate the precision of judgments, and that these negative results stand in contrast to the subjective reports of AD patients

(Requena-Komuro et al., 2020). In a semi-structured survey, 79% of AD patients reported difficulties in estimating the time passed since a personal event has occurred or the time until a future event will occur (Requena-Komuro et al., 2020). Also in the study by Coelho et al. (2016), MCI patients performed equal to healthy controls, when they were asked to estimate the time they spent on a clock-drawing task and the duration of the interview.

As Mioni et al. (2021b) point out in their meta-analysis, more studies are needed to answer the question whether AD and MCI patients are impaired in retrospective time judgments, because studies using retrospective timing paradigms are scarce. One reason for this is that retrospective timing studies are limited in the number of trials. After the participants have been asked for a duration judgment in the first trial, the time-related nature of the task is revealed and the participants will most likely adapt their attentional resources accordingly (Matthews & Meck, 2014). This situation reduces the reliability of accuracy measures and prevents the assessment of judgment precision. One possibility to circumvent this constraint consists in the direct presentation of several to-be-judged durations. For example, Riemer et al. (2021) presented their participants with a video containing different events and asked them afterward for comparative judgments with respect to the durations of these events (for a similar approach see Grondin & Plourde, 2007).

Explicit and implicit timing

Another conceptual distinction in the timing literature is between explicit and implicit timing. Implicit timing is often defined as extraction of temporal contingencies between perceived events, whereas explicit timing consists in a deliberate, conscious engagement in the timing process (Herbst et al., 2022; see also Chapter 1). In this vein, implicit timing is often indirectly measured as behavioral facilitation, in contrast to the assessment of overt time judgments in explicit timing tasks. There are several approaches to distinguish explicit from implicit timing processes (e.g., Aufschnaiter et al., 2022; Herbst et al., 2022; Herbst & Obleser, 2019; Salet et al., 2021), and previous research has demonstrated age-related changes in implicit timing tasks (Capizzi et al., 2022; Droit-Volet et al., 2019). For example, Capizzi et al. (2022) used a version of the foreperiod paradigm as an indicator of implicit timing. Their participants were asked to react to an imperative stimulus as quickly as possible. The imperative stimulus itself was preceded by a warning cue, and the temporal interval between the warning cue and the imperative stimulus (i.e., the foreperiod) was varying between 480 and 1,920 ms. Hence, the probability that the imperative stimulus will occur at the relatively longer foreperiods increases with time. Although this task can be solved without attending to the temporal interval between warning cue and the imperative stimulus, shorter reaction times for the longer foreperiods have been frequently observed, indicating that the temporal information is processed (Niemi & Näätänen, 1981). Capizzi et al. (2022) reported that participants with lower cognitive functioning (quantified by MMSE scores) also exhibit a reduced foreperiod effect, suggesting an impairment of implicit timing abilities. It should be noted, however, that the participants in this study also performed an explicit timing task in the same session, using the same intervals and stimuli. While this is advantageous with respect to the direct comparison between the two tasks, it reduces the claim that participants were unaware of the fact that the interval between warning cue and imperative stimulus was of any informative value, because half of the participants (those who performed the explicit task first) were trained to differentiate between the different intervals that were later used as task-irrelevant foreperiods. In other studies, the foreperiod effect has been shown to occur regardless of whether participants are aware of the varying intervals between the warning cue and the imperative stimulus (Kruijne et al., 2022).

Deficits in time perception and memory impairments

The question as to whether time perception is specifically impaired in AD patients, in pre-clinical stages as MCI, or generally in older adults is highly related to the question about age-related memory deficits, because memory impairments are a characteristic symptom in these patient groups. Often it is not entirely clear whether the reduced performance in a time perception task could alternatively be explained by a decrease in mnemonic capabilities (Maaß et al., 2019; Matell et al., 2005; Paraskevoudi et al., 2018; Turgeon et al., 2016). This is a general issue pertaining also to other patient groups with known deficits both in timing performance and working memory. For example, characteristic timing deficits have been described for children with ADHD (Noreika et al., 2013), with some studies providing evidence that these timing deficits are entirely accounted for by impairments in working memory (e.g., Lee & Yang, 2019), while other studies show that the differences in timing abilities between ADHD patients and healthy controls persist after controlling for differences in working memory capacity (e.g., Kerns et al., 2001).

While the perception of very brief intervals as those being tested in temporal order judgments or simultaneity judgments seems to be independent from working memory capacity (Ben-Artzi et al., 2011), the interaction between working memory and timing abilities becomes more relevant when the tested intervals are rather long (e.g., above 3 s), because the longer the time interval to be estimated, the more likely it is that effects of memory come into play. An extreme example might be best suited to illustrate this point: If we ask a participant to reproduce an interval of 1 s, it is rather unlikely that the performance is affected by long-term memory processes. However, if this participant has to reproduce an interval of 1 h, it becomes a matter of prospective memory, and it could be expected that, after the hour has passed, the participant might have forgotten about the task and doesn't respond. Maybe the response will come a bit later, when the participant eventually recalls the timing task. It is obvious that such a deficient performance is not necessarily linked to a problem in the perception of time, but that it might just as well reflect an impairment of time-based prospective memory (Henry et al., 2004; Kliegel et al., 2016), especially as AD and MCI patients exhibit a decrement in prospective memory skills (Costa et al., 2011; Spíndola & Brucki, 2011). The above example also illustrates that memory deficits can interfere with timing performance at two different phases: during the encoding phase, when the to-be-estimated interval is presented, and during the retrieval phase, when a response is required. In a similar vein – and although it might sound trivial – it makes a huge difference whether the over-reproduction of an interval is caused by the reproduced interval being perceived as shorter than the previously presented standard interval, or whether it is simply due to the participant's inability to recall the duration of the presented standard. Misperception and oblivion are not the same.

According to the reasoning that memory-based timing deficits should be more pronounced when relatively long intervals are estimated, while genuine timing deficits should be revealed more in the judgment on relatively short intervals, Nichelli et al. (1993) compared the timing performance for short versus long intervals in AD and amnestic patients. They found that both patient groups were impaired in the estimation of longer time intervals (about 5–40 s). However, for the repeated reproduction of a 1-s interval, only the AD patients showed a deficit, while amnestic patients were unimpaired. This result suggests that (at least for the reproduction of short durations) the decreased timing performance found in AD patients cannot be explained solely on the basis of deficits in working memory.

Of major interest for this issue are timing studies in which the effect of individual differences in working memory capacity was controlled for. Perbal et al. (2002) showed that the influence of working memory depends on the type of the paradigm used for the assessment of timing abilities. The authors found that working memory capacity was correlated with the performance in a time reproduction task, but not with that in a time production task, a finding that was confirmed in other studies (Baudouin et al., 2006, 2019; Mioni et al., 2019a). Thus, in order to answer the question whether age-related timing deficits exist independent from working memory impairments, it is advisable to focus on timing tasks with low demands on working memory (e.g., time production and time estimation tasks) rather than on timing tasks in which the memorization of one or several standard durations is required (e.g., time reproduction or time bisection tasks).

Methodological issues in the study of age-related timing deficits

In this section, I will elaborate on some methodological issues that are pertinent when investigating time perception deficits in older adults and in AD and MCI patients, such as the advantage of repeated testing, motivational factors, and the use of chronometric counting as a compensatory strategy.

Repeated testing

Age-related timing deficits have been proposed as a behavioral marker for neurodegenerative diseases, in order to support clinical diagnostics (El Haj & Kapogiannis, 2016; Maaß et al., 2019; Xu & Church, 2017). Although conventional test batteries for memory impairments (often consisting in the memorization of word lists) are a more direct and intuitive approach to assess pathological cognitive decline in advanced age, the assessment of timing deficits has some specific advantages over those conventional tests and can therefore be a valuable supplement. First, compared to the systematic assessment of memory deficits, timing tasks are easier to conduct. Memory performance is usually quantified by the retrieval rate of a previously learned word list. However, a word list has to be trained before the words can be recalled, while timing tasks are mostly independent of such repeated training sessions.[1] Second, and related to the first point, it is not always clear whether a hypothetically low memorization performance is driven by deficits during the retrieval phase or during the encoding of the target stimuli. For example, a patient with intact mnemonic capacities might have attention-related issues and might just not be capable to focus on the presented words, which ultimately would result in a bad retrieval rate. In some test batteries, this problem is accounted for by having the patients read each word aloud directly after its presentation (e.g., Morris et al., 1989). Third, timing tasks can be presented several times to the same patient, without running into the problem of carry-over effects across repeated testing sessions. Such carry-over effects have been demonstrated for conventional tests for memory impairments and are hardly avoidable when learning standardized word lists (Dikmen et al., 1999), and the construction of parallel test versions involves additional effort and costs. In contrast, the duration stimuli used in timing tasks do not require a specific training and therefore allow for an unlimited series of repeated testing sessions.

The problem of chronometric counting

The assessment of age-related timing deficits is also accompanied with some disadvantages. One of them consists in the obvious compensatory strategy of chronometric counting, that is, instead of focusing on the experience of the duration, the patients might count the seconds in their mind and base their responses on the accumulated number of seconds rather than on a sense of time. In this vein, existing deficits in the perception of time can be concealed, and there are a lot of studies demonstrating the effectiveness of chronometric counting to boost performance in timing tasks (Grondin et al., 1999, 2004; Rattat & Droit-Volet, 2012; Riemer et al., 2022). With respect to the assessment of deficits in older individuals or potential AD or MCI patients, the resort to compensatory strategies to improve performance is of specific relevance, because these groups can be extremely reluctant to admit to cognitive deficits and might rely stronger on compensatory strategies and heuristics to increase their performance. Especially in studies explicitly addressing the effects of age on cognitive functioning, older participants can be extremely motivated to showcase how cognitively fit they still are. This additional incentive can even result in older participants outperforming younger ones (e.g., Riemer et al., 2021).

The capability of chronometric counting to conceal age-related timing deficits is demonstrated in a study by Perbal et al. (2002), where it was shown that performance differences in a timing task between two groups of younger and older participants disappeared when the task instructions allowed for chronometric counting (relative to a situation in which counting was prevented by concurrent reading). To address this issue, there have been various attempts to prevent or reduce the influence of chronometric counting on the performance in timing tasks, including the introduction of a concurrent distracter task (e.g., Perbal et al., 2002), articulatory suppression (e.g., Baudouin et al., 2006), explicit instructions not to count (e.g., Akdoğan & Balcı, 2017), and the design of timing tasks that are more resistant to the effects of chronometric counting (e.g., Riemer et al., 2022). Each one of these approaches, however, comes with some disadvantages (discussed in Rattat & Droit-Volet, 2012, and Riemer et al., 2022). Distracter tasks might also interfere with the timing process. The production of irrelevant speech syllables for articulatory suppression generates countable elements itself. Participants might be unable or unwilling to comply with no-counting instructions. And the modification of a duration discrimination task proposed by Riemer et al. (2022) only reduces the spontaneous tendency to engage in chronometric counting, but not its effectiveness per se. For some research questions, the problem of chronometric counting could be addressed by implementing only sub-second intervals as to-be-judged durations, because it has been shown that the effectiveness of chronometric counting starts for judging durations of about 1.2 s or above (Grondin et al., 1999). However, as the results of many studies suggest that the processing of durations of different lengths is based on different mechanisms (Lewis & Miall, 2003; Paton & Buonomano, 2018; Pouthas et al., 2005; Ulbrich et al., 2007), sticking with brief intervals is no general solution to the problem.

This limitation and potential confound inherent in time perception tasks needs to be considered when assessing age-related timing deficits and when interpreting the results of the respective studies. In this regard, it is important to take into account that in some studies on timing deficits in AD and MCI patients, participants were explicitly instructed to use chronometric counting (Coelho et al., 2016; Pai et al., 2021), which might explain why either no or reduced timing deficits could be observed (Perbal et al., 2002).

Conclusion

AD and its potential precursor MCI are associated with neuronal atrophy in the hippocampus and the entorhinal cortex, already at a very early preclinical state (Braak & Del Tredici, 2015). As both brain regions have been linked to the processing of temporal intervals, many researchers have argued that the clinical assessment of time perception deficits might provide a behavioral marker for AD and an elegant way to identify patients at risk for progressive cognitive decline. Timing deficits in AD have been confirmed in many studies, but seem to be less pronounced in MCI patients (Mioni et al., 2021b). Considering the small number of studies on timing deficits in MCI and the large variability in their experimental design (Coelho et al., 2016, for example, instructed their participants to use chronometric counting to support timing performance), we need more experimental evidence to confirm the usefulness of time perception deficits to detect MCI patients, and hence predict a possible progression to AD.

Note

1 Temporal bisection tasks, in which a small and a large reference interval have to be learned, are an exception.

References

Akdoğan, B., & Balcı, F. (2017). Are you early or late? Temporal error monitoring. *Journal of Experimental Psychology: General*, *146*(3), 347–361. https://doi.org/10.1037/xge0000265

Albert, M. S., DeKosky, S. T., Dickson, D., Dubois, B., Feldman, H. H., Fox, N. C., Gamst, A., Holtzman, D. M., Jagust, W. J., Petersen, R. C., Snyder, P. J., Carrillo, M. C., Thies, B., & Phelps, C. H. (2011). The diagnosis of mild cognitive impairment due to Alzheimer's disease: Recommendations from the National Institute on Aging-Alzheimer's Association workgroups on diagnostic guidelines for Alzheimer's disease. *Alzheimer's & Dementia, 7*(3), 270–279. https://doi.org/10.1016/j.jalz.2011.03.008

Amariglio, R. E., Becker, J. A., Carmasin, J., Wadsworth, L. P., Lorius, N., Sullivan, C., Maye, J. E., Gidicsin, C., Pepin, L. C., Sperling, R. A., Johnson, K. A., & Rentz, D. M. (2012). Subjective cognitive complaints and amyloid burden in cognitively normal older individuals. *Neuropsychologia, 50*(12), 2880–2886. https://doi.org/10.1016/j.neuropsychologia.2012.08.011

Aufschnaiter, S., Zhao, F., Gaschler, R., Kiesel, A., & Thomaschke, R. (2022). Investigating time-based expectancy beyond binary timing scenarios: Evidence from a paradigm employing three predictive pre-target intervals. *Psychological Research, 86*(6), 2007–2020. https://doi.org/10.1007/s00426-021-01606-2

Balcı, F., Meck, W. H., Moore, H., & Brunner, D. (2009). Timing deficits in aging and neuropathology. In J. L. Bizon, & A. Woods (Eds.), *Animal models of human cognitive aging* (pp. 1–41). Humana Press. https://doi.org/10.1007/978-1-59745-422-3_8

Basharat, A., Adams, M. S., Staines, W. R., & Barnett-Cowan, M. (2018). Simultaneity and temporal order judgments are coded differently and change with age: An event-related potential study. *Frontiers in Integrative Neuroscience, 12*, 15. https://doi.org/10.3389/fnint.2018.00015

Baudouin, A., Isingrini, M., & Vanneste, S. (2019). Executive functioning and processing speed in age-related differences in time estimation: A comparison of young, old, and very old adults. *Aging, Neuropsychology, and Cognition, 26*(2), 264–281. https://doi.org/10.1080/13825585.2018.1426715

Baudouin, A., Vanneste, S., Isingrini, M., & Pouthas, V. (2006). Differential involvement of internal clock and working memory in the production and reproduction of duration: A study on older adults. *Acta Psychologica, 121*(3), 285–296. https://doi.org/10.1016/j.actpsy.2005.07.004

Bedard, G., & Barnett-Cowan, M. (2016). Impaired timing of audiovisual events in the elderly. *Experimental Brain Research, 234*(1), 331–340. https://doi.org/10.1007/s00221-015-4466-7

Ben-Artzi, E., Babkoff, H., & Fostick, L. (2011). Auditory temporal processes in the elderly. *Audiology Research, 1*(1), e6. https://doi.org/10.4081/audiores.2011.e6

Block, R. A., Hancock, P. A., & Zakay, D. (2010). How cognitive load affects duration judgments: A meta-analytic review. *Acta Psychologica, 134*(3), 330–343. https://doi.org/10.1016/j.actpsy.2010.03.006

Block, R. A., & Zakay, D. (1997). Prospective and retrospective duration judgments: A meta-analytic review. *Psychonomic Bulletin & Review, 4*(2), 184–197. https://doi.org/10.3758/BF03209393

Block, R. A., Zakay, D., & Hancock, P. A. (1998). Human aging and duration judgments: A meta-analytic review. *Psychology and Aging, 13*(4), 584–596. https://doi.org/10.1037/0882-7974.13.4.584

Bogon, J., Jagorska, C., Steinecker, I., & Riemer, M. (2024). Age-related changes in time perception: Effects of immersive virtual reality and spatial location of stimuli. *Acta Psychologica, 249*, 104460. https://doi.org/10.1016/j.actpsy.2024.104460

Braak, H., & Del Tredici, K. (2015). The preclinical phase of the pathological process underlying sporadic Alzheimer's disease. *Brain, 138*(10), 2814–2833. https://doi.org/10.1093/brain/awv236

Breijyeh, Z., & Karaman, R. (2020). Comprehensive review on Alzheimer's disease: Causes and treatment. *Molecules, 25*(24), 5789. https://doi.org/10.3390/molecules 25245789

Busche, M. A., Wegmann, S., Dujardin, S., Commins, C., Schiantarelli, J., Klickstein, N., Kamath, T. V., Carlson, G. A., Nelken, I., & Hyman, B. T. (2019). Tau impairs neural circuits, dominating amyloid-β effects, in Alzheimer models in vivo. *Nature Neuroscience, 22*(1), 57–64. https://doi.org/10.1038/s41593-018-0289-8

Capizzi, M., Visalli, A., Faralli, A., & Mioni, G. (2022). Explicit and implicit timing in older adults: Dissociable associations with age and cognitive decline. *PloS one, 17*(3), e0264999. https://doi.org/10.1371/journal.pone.0264999

Carrasco, M. C., Guillem, M. J., & Redolat, R. (2000). Estimation of short temporal intervals in Alzheimer's disease. *Experimental Aging Research, 26*(2), 139–151. https://doi.org/10.1080/036107300243605

Caselli, L., Iaboli, L., & Nichelli, P. (2009). Time estimation in mild Alzheimer's disease patients. *Behavioral and Brain Functions, 5*(1), 32. https://doi.org/10.1186/1744-9081-5-32

Chan, Y. M., Pianta, M. J., & McKendrick, A. M. (2014). Older age results in difficulties separating auditory and visual signals in time. *Journal of Vision, 14*(11), 13–13. https://doi.org/10.1167/14.11.13

Cheng, C. P.-W., Cheng, S.-T., Tam, C. W.-C., Chan, W.-C., Chu, W. C.-W., & Lam, L. C.-W. (2018). Relationship between cortical thickness and neuropsychological performance in normal older adults and those with mild cognitive impairment. *Aging and Disease, 9*(6), 1020. https://doi.org/10.14336/AD.2018.0125

Coelho, S., Guerreiro, M., Chester, C., Silva, D., Maroco, J., Coelho, M., Paglieri, F., & de Mendonça, A. (2016). Time perception in mild cognitive impairment: Interval length and subjective passage of time. *Journal of the International Neuropsychological Society, 22*(7), 755–764. https://doi.org/10.1017/S135561 7716000606

Costa, A., Caltagirone, C., & Carlesimo, G. A. (2011). Prospective memory impairment in mild cognitive impairment: An analytical review. *Neuropsychology Review, 21*(4), 390–404. https://doi.org/10.1007/s11065-011-9172-z

Craig, J. C., Rhodes, R. P., Busey, T. A., Kewley-Port, D., & Humes, L. E. (2010). Aging and tactile temporal order. *Attention, Perception, & Psychophysics, 72*(1), 226–235. https://doi.org/10.3758/APP.72.1.226

D'Antonio, F., De Bartolo, M. I., Ferrazzano, G., Trebbastoni, A., Amicarelli, S., Campanelli, A., De Lena, C., Berardelli, A., & Conte, A. (2019). Somatosensory temporal discrimination threshold in patients with cognitive disorders. *Journal of Alzheimer's Disease, 70*(2), 425–432. https://doi.org/10.3233/JAD-190385

De Kock, R., Gladhill, K. A., Ali, M. N., Joiner, W. M., & Wiener, M. (2021). How movements shape the perception of time. *Trends in Cognitive Sciences, 25*(11), 950–963. https://doi.org/10.1016/j.tics.2021.08.002

Devanand, D. P., Pradhaban, G., Liu, X., Khandji, A., De Santi, S., Segal, S., Rusinek, H., Pelton, G. H., Honig, L. S., Mayeux, R., Stern, Y., Tabert, M. H., & De Leon, M. J. (2007). Hippocampal and entorhinal atrophy in mild cognitive impairment: Prediction of Alzheimer disease. *Neurology, 68*(11), 828–836. https://doi.org/10.1212/01.wnl.0000256697.20968.d7

Dikmen, S. S., Heaton, R. K., Grant, I., & Temkin, N. R. (1999). Test–retest reliability and practice effects of expanded Halstead–Reitan neuropsychological test battery. *Journal of the International Neuropsychological Society, 5*(4), 346–356. https://doi.org/10.1017/S1355617799544056

Droit-Volet, S., Lorandi, F., & Coull, J. T. (2019). Explicit and implicit timing in aging. *Acta Psychologica, 193*, 180–189. https://doi.org/10.1016/j.actpsy.2019.01.004

El Haj, M., & Kapogiannis, D. (2016). Time distortions in Alzheimer's disease: A systematic review and theoretical integration. *NPJ Aging and Mechanisms of Disease*, *2*(1), 16016. https://doi.org/10.1038/npjamd.2016.16

El Haj, M., Moroni, C., Samson, S., Fasotti, L., & Allain, P. (2013). Prospective and retrospective time perception are related to mental time travel: Evidence from Alzheimer's disease. *Brain and Cognition*, *83*(1), 45–51. https://doi.org/10.1016/j.bandc.2013.06.008

El Haj, M., Omigie, D., & Moroni, C. (2014). Time reproduction during high and low attentional tasks in Alzheimer's Disease "A watched kettle never boils". *Brain and Cognition*, *88*, 1–5. https://doi.org/10.1016/j.bandc.2014.04.002

Fjell, A. M., Walhovd, K. B., Fennema-Notestine, C., McEvoy, L. K., Hagler, D. J., Holland, D., Brewer, J. B., Dale, A. M., & for the Alzheimer's Disease Neuroimaging Initiative. (2010). CSF biomarkers in prediction of cerebral and clinical change in mild cognitive impairment and Alzheimer's disease. *Journal of Neuroscience*, *30*(6), 2088–2101. https://doi.org/10.1523/JNEUROSCI.3785-09.2010

Fraisse, P. (1984). Perception and estimation of time. *Annual Review of Psychology*, *35*(1), 1–37. https://doi.org/10.1146/annurev.ps.35.020184.000245

Friedman, W. J., & Janssen, S. M. J. (2010). Aging and the speed of time. *Acta Psychologica*, *134*(2), 130–141. https://doi.org/10.1016/j.actpsy.2010.01.004

Gabrian, M., Dutt, A. J., & Wahl, H.-W. (2017). Subjective time perceptions and aging well: A review of concepts and empirical research - A mini-review. *Gerontology*, *63*(4), 350–358. https://doi.org/10.1159/000470906

Gagnon-Harvey, A.-A., McArthur, J., Tétreault, É., Fortin-Guichard, D., & Grondin, S. (2021). Age, Personal Characteristics, and the Speed of Psychological Time. *Timing & Time Perception*, *9*(3), 257–274. https://doi.org/10.1163/22134468-bja10024

Grondin, S. (2012). Violation of the scalar property for time perception between 1 and 2 seconds: Evidence from interval discrimination, reproduction, and categorization. *Journal of Experimental Psychology: Human Perception and Performance*, 38(4), 880–890. https://doi.org/10.1037/a0027188

Grondin, S., Meilleur-Wells, G., & Lachance, R. (1999). When to start explicit counting in a time-intervals discrimination task: A critical point in the timing process of humans. *Journal of Experimental Psychology: Human Perception and Performance*, *25*(4), 993–1004. https://doi.org/10.1037/0096-1523.25.4.993

Grondin, S., Ouellet, B., & Roussel, M.-E. (2004). Benefits and limits of explicit counting for discriminating temporal intervals. *Canadian Journal of Experimental Psychology/Revue Canadienne de Psychologie Expérimentale*, *58*(1), 1–12. https://doi.org/10.1037/h0087436

Grondin, S., & Plourde, M. (2007). Judging multi-minute intervals retrospectively. *Quarterly Journal of Experimental Psychology*, *60*(9), 1303–1312. https://doi.org/10.1080/17470210600988976

Hayashi, M. J., Van Der Zwaag, W., Bueti, D., & Kanai, R. (2018). Representations of time in human frontoparietal cortex. *Communications Biology*, *1*(1), 233. https://doi.org/10.1038/s42003-018-0243-z

He, Z., Guo, J. L., McBride, J. D., Narasimhan, S., Kim, H., Changolkar, L., Zhang, B., Gathagan, R. J., Yue, C., Dengler, C., Stieber, A., Nitla, M., Coulter, D. A., Abel, T., Brunden, K. R., Trojanowski, J. Q., & Lee, V. M.-Y. (2018). Amyloid-β plaques enhance Alzheimer's brain tau-seeded pathologies by facilitating neuritic plaque tau aggregation. *Nature Medicine*, *24*(1), 29–38. https://doi.org/10.1038/nm.4443

Heinik, J. (2012). Accuracy of estimation of time-intervals in psychogeriatric outpatients. *International Psychogeriatrics*, *24*(5), 809–821. https://doi.org/10.1017/S1041610211002596

Heinik, J., & Ayalon, L. (2010). Self-estimation of performance time versus actual performance time in older adults with suspected mild cognitive impairment: A clinical perspective. *The Israel Journal of Psychiatry and Related Sciences*, *47*(4), 291–296.

Henry, J. D., MacLeod, M. S., Phillips, L. H., & Crawford, J. R. (2004). A meta-analytic review of prospective memory and aging. *Psychology and Aging*, *19*(1), 27–39. https://doi.org/10.1037/0882-7974.19.1.27

Herbst, S. K., & Obleser, J. (2019). Implicit temporal predictability enhances pitch discrimination sensitivity and biases the phase of delta oscillations in auditory cortex. *NeuroImage*, *203*, 116198. https://doi.org/10.1016/j.neuroimage.2019.116198

Herbst, S. K., Obleser, J., & Van Wassenhove, V. (2022). Implicit versus explicit timing—Separate or shared mechanisms? *Journal of Cognitive Neuroscience*, 34(8), 1447–1466. https://doi.org/10.1162/jocn_a_01866

Howett, D., Castegnaro, A., Krzywicka, K., Hagman, J., Marchment, D., Henson, R., Rio, M., King, J. A., Burgess, N., & Chan, D. (2019). Differentiation of mild cognitive impairment using an entorhinal cortex-based test of virtual reality navigation. *Brain*, *142*(6), 1751–1766. https://doi.org/10.1093/brain/awz116

Hsieh, L.-T., Gruber, M. J., Jenkins, L. J., & Ranganath, C. (2014). Hippocampal activity patterns carry information about objects in temporal context. *Neuron*, *81*(5), 1165–1178. https://doi.org/10.1016/j.neuron.2014.01.015

Jessen, F., Amariglio, R. E., Buckley, R. F., Van Der Flier, W. M., Han, Y., Molinuevo, J. L., Rabin, L., Rentz, D. M., Rodriguez-Gomez, O., Saykin, A. J., Sikkes, S. A. M., Smart, C. M., Wolfsgruber, S., & Wagner, M. (2020). The characterisation of subjective cognitive decline. *The Lancet Neurology*, *19*(3), 271–278. https://doi.org/10.1016/S1474-4422(19)30368-0

Julkunen, V., Niskanen, E., Muehlboeck, S., Pihlajamäki, M., Könönen, M., Hallikainen, M., Kivipelto, M., Tervo, S., Vanninen, R., Evans, A., & Soininen, H. (2009). Cortical thickness analysis to detect progressive mild cognitive impairment: A reference to Alzheimer's disease. *Dementia and Geriatric Cognitive Disorders*, *28*(5), 389–397. https://doi.org/10.1159/000256274

Kerns, K. A., McInerney, R. J., & Wilde, N. J. (2001). Time reproduction, working memory, and behavioral inhibition in children with ADHD. *Child Neuropsychology*, *7*(1), 21–31. https://doi.org/10.1076/chin.7.1.21.3149

Kliegel, M., Ballhausen, N., Hering, A., Ihle, A., Schnitzspahn, K. M., & Zuber, S. (2016). Prospective memory in older adults: Where we are now and what is next. *Gerontology*, *62*(4), 459–466. https://doi.org/10.1159/000443698

Kruijne, W., Galli, R. M., & Los, S. A. (2022). Implicitly learning when to be ready: From instances to categories. *Psychonomic Bulletin & Review*, *29*(2), 552–562. https://doi.org/10.3758/s13423-021-02004-w

Kuehn, E., Perez-Lopez, M. B., Diersch, N., Döhler, J., Wolbers, T., & Riemer, M. (2018). Embodiment in the aging mind. *Neuroscience & Biobehavioral Reviews*, *86*, 207–225. https://doi.org/10.1016/j.neubiorev.2017.11.016

Lamotte, M., & Droit-Volet, S. (2017). Aging and time perception for short and long durations: A question of attention? *Timing & Time Perception*, *5*(2), 149–167. https://doi.org/10.1163/22134468-00002086

Lee, H.-Y., & Yang, E.-L. (2019). Exploring the effects of working memory on time perception in attention deficit hyperactivity disorder. *Psychological Reports*, *122*(1), 23–35. https://doi.org/10.1177/0033294118755674

Lewis, P. A., & Miall, R. C. (2003). Distinct systems for automatic and cognitively controlled time measurement: Evidence from neuroimaging. *Current Opinion in Neurobiology*, *13*(2), 250–255. https://doi.org/10.1016/S0959-4388(03)00036-9

Long, N. M., & Kahana, M. J. (2019). Hippocampal contributions to serial-order memory. *Hippocampus*, *29*(3), 252–259. https://doi.org/10.1002/hipo.23025

Lustig, C. (2003). Grandfather's clock: Attention and interval timing in older adults. In W. Meck (Ed.), *Functional and neural mechanisms of interval timing* (Vol. 19). CRC Press. https://doi.org/10.1201/9780203009574.ch10

Lustig, C., & Meck, W. H. (2001). Paying attention to time as one gets older. *Psychological Science*, *12*(6), 478–484. https://doi.org/10.1111/1467-9280.00389

Maaß, S. C., Riemer, M., Wolbers, T., & van Rijn, H. (2019). Timing deficiencies in amnestic mild cognitive impairment: Disentangling clock and memory processes. *Behavioural Brain Research*, *373*, 112110. https://doi.org/10.1016/j.bbr.2019.112110

Maaß, S. C., Wolbers, T., van Rijn, H., & Riemer, M. (2022). Temporal context effects are associated with cognitive status in advanced age. *Psychological Research*, *86*(2), 512–521. https://doi.org/10.1007/s00426-021-01502-9

Malhotra, P. A. (2019). Impairments of attention in Alzheimer's disease. *Current Opinion in Psychology*, *29*, 41–48. https://doi.org/10.1016/j.copsyc.2018.11.002

Mangels, J. A., Ivry, R. B., & Shimizu, N. (1998). Dissociable contributions of the prefrontal and neocerebellar cortex to time perception. *Cognitive Brain Research*, *7*(1), 15–39. https://doi.org/10.1016/S0926-6410(98)00005-6

Matell, M. S., Meck, W. H., & Lustig, C. (2005). Not "just" a coincidence: Frontal-striatal interactions in working memory and interval timing. *Memory*, *13*(3–4), 441–448. https://doi.org/10.1080/09658210344000404

Mathew, G. M., Strayer, S. M., Ness, K. M., Schade, M. M., Nahmod, N. G., Buxton, O. M., & Chang, A.-M. (2021). Interindividual differences in attentional vulnerability moderate cognitive performance during sleep restriction and subsequent recovery in healthy young men. *Scientific Reports*, *11*(1), 19147. https://doi.org/10.1038/s41598-021-95884-w

Matthews, W. J., & Meck, W. H. (2014). Time perception: The bad news and the good. *Wiley Interdisciplinary Reviews: Cognitive Science*, *5*(4), 429–446. https://doi.org/10.1002/wcs.1298

Melgire, M., Ragot, R., Samson, S., Penney, T. B., Meck, W. H., & Pouthas, V. (2005). Auditory/visual duration bisection in patients with left or right medial-temporal lobe resection. *Brain and Cognition*, *58*(1), 119–124. https://doi.org/10.1016/j.bandc.2004.09.013

Michon, J. A. (1978). The making of the present: A tutorial review. In *Attention and performance VII* (pp. 89–111). Erlbaum.

Mioni, G., Capizzi, M., & Stablum, F. (2019a). Age-related changes in time production and reproduction tasks: Involvement of attention and working memory processes. *Aging, Neuropsychology, and Cognition*, 1–18. https://doi.org/10.1080/13825585.2019.1626799

Mioni, G., Cardullo, S., Ciavarelli, A., & Stablum, F. (2021a). Age-related changes in time discrimination: The involvement of inhibition, working memory and speed of processing. *Current Psychology*, *40*(5), 2462–2471. https://doi.org/10.1007/s12144-019-00170-8

Mioni, G., Meligrana, L., Grondin, S., Perini, F., Bartolomei, L., & Stablum, F. (2016). Effects of emotional facial expression on time perception in patients with Parkinson's disease. *Journal of the International Neuropsychological Society*, *22*(9), 890–899. https://doi.org/10.1017/S1355617715000612

Mioni, G., Meligrana, L., Perini, F., Marcon, M., & Stablum, F. (2019b). Lack of temporal impairment in patients with mild cognitive impairment. *Frontiers in Integrative Neuroscience, 13*, 42. https://doi.org/10.3389/fnint.2019.00042

Mioni, G., Román-Caballero, R., Clerici, J., & Capizzi, M. (2021b). Prospective and retrospective timing in mild cognitive impairment and Alzheimer's disease patients: A systematic review and meta-analysis. *Behavioural Brain Research, 410*, 113354. https://doi.org/10.1016/j.bbr.2021.113354

Mioni, G., Wolbers, T., & Riemer, M. (2024). Differences between sub-second and supra-second durations for the assessment of timing deficits in amnestic mild cognitive impairment. *Aging Brain, 6*, 100120. https://doi.org/10.1016/j.nbas.2024.100120

Mitchell, A. J., Beaumont, H., Ferguson, D., Yadegarfar, M., & Stubbs, B. (2014). Risk of dementia and mild cognitive impairment in older people with subjective memory complaints: Meta-analysis. *Acta Psychiatrica Scandinavica, 130*(6), 439–451. https://doi.org/10.1111/acps.12336

Mondok, C., & Wiener, M. (2023). Selectivity of timing: A meta-analysis of temporal processing in neuroimaging studies using activation likelihood estimation and reverse inference. *Frontiers in Human Neuroscience, 16*, 1000995. https://doi.org/10.3389/fnhum.2022.1000995

Morris, J. C., Heyman, A., Mohs, R. C., Hughes, J. P., van Belle, G., Fillenbaum, G., Mellits, E. D., & Clark, C. (1989). The consortium to establish a registry for Alzheimer's disease (CERAD). Part I. Clinical and neuropsychological assesment of Alzheimer's disease. *Neurology, 39*(9), 1159–1159. https://doi.org/10.1212/WNL.39.9.1159

Nichelli, P., Venneri, A., Molinari, M., Tavani, F., & Grafman, J. (1993). Precision and accuracy of subjective time estimation in different memory disorders. *Cognitive Brain Research, 1*(2), 87–93. https://doi.org/10.1016/0926-6410(93)90014-V

Niemi, P., & Näätänen, R. (1981). Foreperiod and simple reaction time. *Psychological Bulletin, 89*(1), 133–162. https://doi.org/10.1037/0033-2909.89.1.133

Noreika, V., Falter, C. M., & Rubia, K. (2013). Timing deficits in attention-deficit/hyperactivity disorder (ADHD): Evidence from neurocognitive and neuroimaging studies. *Neuropsychologia, 51*(2), 235–266. https://doi.org/10.1016/j.neuropsychologia.2012.09.036

Noulhiane, M., Pouthas, V., Hasboun, D., Baulac, M., & Samson, S. (2007). Role of the medial temporal lobe in time estimation in the range of minutes. *NeuroReport, 18*(10), 1035–1038. https://doi.org/10.1097/WNR.0b013e3281668be1

O'Keefe, J., & Nadel, L. (1978). *The hippocampus as a cognitive map*. Oxford University Press.

Pai, M.-C., Yang, C.-J., & Fan, S.-Y. (2021). Time perception in prodromal Alzheimer's dementia and in prodromal dementia with lewy bodies. *Frontiers in Psychiatry, 12*, 728344. https://doi.org/10.3389/fpsyt.2021.728344

Papagno, C., Allegra, A., & Cardaci, M. (2004). Time estimation in Alzheimer's disease and the role of the central executive. *Brain and Cognition, 54*(1), 18–23. https://doi.org/10.1016/S0278-2626(03)00237-9

Paraskevoudi, N., Balcı, F., & Vatakis, A. (2018). "Walking" through the sensory, cognitive, and temporal degradations of healthy aging: Multisensory processing in aging. *Annals of the New York Academy of Sciences, 1426*(1), 72–92. https://doi.org/10.1111/nyas.13734

Paton, J. J., & Buonomano, D. V. (2018). The neural basis of timing: Distributed mechanisms for diverse functions. *Neuron, 98*(4), 687–705. https://doi.org/10.1016/j.neuron.2018.03.045

Perbal, S., Droit-Volet, S., Isingrini, M., & Pouthas, V. (2002). Relationships between age-related changes in time estimation and age-related changes in processing speed, attention, and memory. *Aging, Neuropsychology, and Cognition*, 9(3), 201–216. https://doi.org/10.1076/anec.9.3.201.9609

Petersen, R. C., Doody, R., Kurz, A., Mohs, R. C., Morris, J. C., Rabins, P. V., Ritchie, K., Rossor, M., Thal, L., & Winblad, B. (2001). Current concepts in mild cognitive impairment. *Archives of Neurology*, 58(12), 1985. https://doi.org/10.1001/archneur.58.12.1985

Petersen, R. C., & Negash, S. (2008). Mild cognitive impairment: An overview. *CNS Spectrums*, 13(1), 45–53. https://doi.org/10.1017/S1092852900016151

Poliakoff, E., Shore, D. I., Lowe, C., & Spence, C. (2006). Visuotactile temporal order judgments in ageing. *Neuroscience Letters*, 396(3), 207–211. https://doi.org/10.1016/j.neulet.2005.11.034

Pöppel, E. (1971). Oscillations as possible basis for time perception. *Studium Generale; Zeitschrift Fur Die Einheit Der Wissenschaften Im Zusammenhang Ihrer Begriffsbildungen Und Forschungsmethoden*, 24(1), 85–107.

Pöppel, E. (1997). A hierarchical model of temporal perception. *Trends in Cognitive Sciences*, 1(2), 56–61. https://doi.org/10.1016/S1364-6613(97)01008-5

Pouthas, V., George, N., Poline, J.-B., Pfeuty, M., VandeMoorteele, P.-F., Hugueville, L., Ferrandez, A.-M., Lehéricy, S., LeBihan, D., & Renault, B. (2005). Neural network involved in time perception: An fMRI study comparing long and short interval estimation. *Human Brain Mapping*, 25(4), 433–441. https://doi.org/10.1002/hbm.20126

Ranjbar Pouya, O., Kelly, D. M., & Moussavi, Z. (2015). Tendency to overestimate the explicit time interval in relation to aging and cognitive decline. In *2015 37th annual international conference of the IEEE engineering in medicine and biology society (EMBC)* (pp. 4692–4695). https://doi.org/10.1109/EMBC.2015.7319441

Rattat, A.-C., & Droit-Volet, S. (2012). What is the best and easiest method of preventing counting in different temporal tasks? *Behavior Research Methods*, 44(1), 67–80. https://doi.org/10.3758/s13428-011-0135-3

Requena-Komuro, M.-C., Marshall, C. R., Bond, R. L., Russell, L. L., Greaves, C., Moore, K. M., Agustus, J. L., Benhamou, E., Sivasathiaseelan, H., Hardy, C. J. D., Rohrer, J. D., & Warren, J. D. (2020). Altered time awareness in dementia. *Frontiers in Neurology*, 11, 291. https://doi.org/10.3389/fneur.2020.00291

Riemer, M., Vieweg, P., Van Rijn, H., & Wolbers, T. (2022). Reducing the tendency for chronometric counting in duration discrimination tasks. *Attention, Perception, & Psychophysics*, 84(8), 2641–2654. https://doi.org/10.3758/s13414-022-02523-1

Riemer, M., Wolbers, T., & van Rijn, H. (2021). Age-related changes in time perception: The impact of naturalistic environments and retrospective judgements on timing performance. *Quarterly Journal of Experimental Psychology*, 74(11), 2002–2012. https://doi.org/10.1177/17470218211023362

Roach, N. W., McGraw, P. V., Whitaker, D. J., & Heron, J. (2017). Generalization of prior information for rapid Bayesian time estimation. *Proceedings of the National Academy of Sciences*, 114(2), 412–417. https://doi.org/10.1073/pnas.1610706114

Rueda, A. D., & Schmitter-Edgecombe, M. (2009). Time estimation abilities in mild cognitive impairment and Alzheimer's disease. *Neuropsychology*, 23(2), 178–188. https://doi.org/10.1037/a0014289

Salet, J. M., Kruijne, W., & Van Rijn, H. (2021). Implicit learning of temporal behavior in complex dynamic environments. *Psychonomic Bulletin & Review*, 28(4), 1270–1280. https://doi.org/10.3758/s13423-020-01873-x

Schlichting, N., Damsma, A., Aksoy, E. E., Wächter, M., Asfour, T., & van Rijn, H. (2018). Temporal context influences the perceived duration of everyday actions:

Assessing the ecological validity of lab-based timing phenomena. *Journal of Cognition*, *2*(1), 1. https://doi.org/10.5334/joc.4

Setti, A., Finnigan, S., Sobolewski, R., McLaren, L., Robertson, I. H., Reilly, R. B., Anne Kenny, R., & Newell, F. N. (2011). Audiovisual temporal discrimination is less efficient with aging: An event-related potential study. *NeuroReport*, *22*(11), 554–558. https://doi.org/10.1097/WNR.0b013e328348c731

Slot, R. E. R., Sikkes, S. A. M., Berkhof, J., Brodaty, H., Buckley, R., Cavedo, E., Dardiotis, E., Guillo-Benarous, F., Hampel, H., Kochan, N. A., Lista, S., Luck, T., Maruff, P., Molinuevo, J. L., Kornhuber, J., Reisberg, B., Riedel-Heller, S. G., Risacher, S. L., Roehr, S., … van der Flier, W. M. (2019). Subjective cognitive decline and rates of incident Alzheimer's disease and non-Alzheimer's disease dementia. *Alzheimer's & Dementia*, *15*(3), 465–476. https://doi.org/10.1016/j.jalz.2018.10.003

Spíndola, L., & Brucki, S. M. D. (2011). Prospective memory in Alzheimer's disease and mild cognitive impairment. *Dementia & Neuropsychologia*, *5*(2), 64–68. https://doi.org/10.1590/S1980-57642011DN05020002

Szymaszek, A., Sereda, M., Pöppel, E., & Szelag, E. (2009). Individual differences in the perception of temporal order: The effect of age and cognition. *Cognitive Neuropsychology*, *26*(2), 135–147. https://doi.org/10.1080/02643290802504742

Teki, S., Grube, M., & Griffiths, T. D. (2012). A unified model of time perception accounts for duration-based and beat-based timing mechanisms. *Frontiers in Integrative Neuroscience*, *5*. https://doi.org/10.3389/fnint.2011.00090

Treisman, M. (1963). Temporal discrimination and the indifference interval: Implications for a model of the 'internal clock'. *Psychological Monographs: General and Applied*, *77*(13), 1–31. https://doi.org/10.1037/h0093864

Tripathi, T., & Khan, H. (2020). Direct interaction between the β-amyloid core and tau facilitates cross-seeding: A novel target for therapeutic intervention. *Biochemistry*, *59*(4), 341–342. https://doi.org/10.1021/acs.biochem.9b01087

Turgeon, M., Lustig, C., & Meck, W. H. (2016). Cognitive aging and time perception: Roles of Bayesian optimization and degeneracy. *Frontiers in Aging Neuroscience*, *8*. https://doi.org/10.3389/fnagi.2016.00102

Ulbrich, P., Churan, J., Fink, M., & Wittmann, M. (2007). Temporal reproduction: Further evidence for two processes. *Acta Psychologica*, *125*(1), 51–65. https://doi.org/10.1016/j.actpsy.2006.06.004

Uotani, C., Sugimori, K., & Kobayashi, K. (2006). Association of minimal thickness of the medial temporal lobe with hippocampal volume, maximal and minimal hippocampal length: Volumetric approach with horizontal magnetic resonance imaging scans for evaluation of a diagnostic marker for neuroimaging of Alzheimer's disease. *Psychiatry and Clinical Neurosciences*, *60*(3), 319–326. https://doi.org/10.1111/j.1440-1819.2006.01508.x

Visser, P. J., Verhey, F., Knol, D. L., Scheltens, P., Wahlund, L.-O., Freund-Levi, Y., Tsolaki, M., Minthon, L., Wallin, Å. K., Hampel, H., Bürger, K., Pirttila, T., Soininen, H., Rikkert, M. O., Verbeek, M. M., Spiru, L., & Blennow, K. (2009). Prevalence and prognostic value of CSF markers of Alzheimer's disease pathology in patients with subjective cognitive impairment or mild cognitive impairment in the DESCRIPA study: A prospective cohort study. *The Lancet Neurology*, *8*(7), 619–627. https://doi.org/10.1016/S1474-4422(09)70139-5

Wackermann, J. (2007). Inner and outer horizons of time experience. *The Spanish Journal of Psychology*, *10*(1), 20–32. https://doi.org/10.1017/S1138741600006284

Warren, S. L., Reid, E., Whitfield, P., & Moustafa, A. A. (2022). Subjective memory complaints as a predictor of mild cognitive impairment and Alzheimer's disease. *Discover Psychology*, *2*(1), 13. https://doi.org/10.1007/s44202-022-00031-9

Whitwell, J. L., Przybelski, S. A., Weigand, S. D., Knopman, D. S., Boeve, B. F., Petersen, R. C., & Jack, C. R. (2007). 3D maps from multiple MRI illustrate changing atrophy patterns as subjects progress from mild cognitive impairment to Alzheimer's disease. *Brain, 130*(7), 1777–1786. https://doi.org/10.1093/brain/awm112

Wiener, M., Turkeltaub, P., & Coslett, H. B. (2010). The image of time: A voxel-wise meta-analysis. *NeuroImage, 49*(2), 1728–1740. https://doi.org/10.1016/j.neuroimage.2009.09.064

Wittmann, M. (1999). Time perception and temporal processing levels of the brain. *Chronobiology International, 16*(1), 17–32. https://doi.org/10.3109/07420529908998709

Wittmann, M., & Lehnhoff, S. (2005). Age effects in perception of time. *Psychological Reports, 97*(3), 921–935. https://doi.org/10.2466/pr0.97.3.921-935

Xu, R., & Church, R. M. (2017). Age-related changes in human and nonhuman timing. *Timing & Time Perception, 5*(3–4), 261–279. https://doi.org/10.1163/22134468-00002092

9

TIME PERCEPTION IN NEUROPSYCHIATRY

Anne Giersch
French Institute of Health and Medical Research, Strasbourg, France

Introduction

The diagnosis of schizophrenia relies on symptoms like hallucinations, delusions, social withdrawal, and a disorganization of thought (American Psychiatric Association, 2013). Nothing is explicitly mentioned about time, or the sense of bodily self. Nonetheless, a growing body of research, both in phenomenology and in experimental psychology, suggests it might play an important role in the development of psychotic symptoms. In the following we will focus on the relationship between timing and the sense of self (Fuchs, 2007; Martin et al., 2014; Vogeley & Kupke 2007). It should be emphasized from the onset that this relationship is not necessarily specific to schizophrenia. A disruption of the sense of bodily self is observed in other pathologies than schizophrenia, like the depersonalization-derealization disorder, or the borderline personality disorder. We will focus on patients with schizophrenia because it is in this pathology that we have most experimental results as well as clinical reports. The following, albeit focusing on schizophrenia, is mainly a development on the relationship between timing and the sense of self, as seen from an experimental psychology perspective.

The idea of a relationship between time and the sense of self is as old as phenomenology (Husserl, 1996; Minkowski, 2013). This relationship may not immediately appear as being self-evident. Over time, we grow from being a baby to an old, sometimes crippled being, we can change our viewpoints or our goals in life, we sometimes lose sight or audition. We learn new things, and we can change due to life events. If asked to define ourselves, this definition is likely to change over time. Despite all these changes, we usually have

DOI: 10.4324/9781003449546-9

no doubt that we are the same unique being from the beginning to the end. Even if we lose a sense, it is the same I who interacts with our surroundings, it is from the same first-person perspective that we envision the present moment. This ego is not defined verbally, it is neither an abstract concept nor a self that is separated, isolated from its surroundings. On the contrary, it is the very concrete ego that is felt as being at the origin of our apprehension of the world and of our actions. It is the ego that feels as being there (in our head and in our body) in the present moment, which leads to the concept of "Dasein" (being there, literally). Why we do experience this self as being the same over time is a mystery that is intimately related to the question of our sense of time continuity. As a matter of fact, when we say that we experience time as being continuous, we do not refer to an abstract representation of time, but to our everyday, every moment experience of the world, where time never seems to stop, jump, or disappear. In fact, we refer to the very same experience that is lived from our first-person perspective, i.e., our self. Time is thus intimately related to our experience of oneself. Conversely, the disorders of the sense of self in schizophrenia have been related to a disordered experience of time. We will first define the disorders of the self in schizophrenia. We will show how it is explored experimentally and will discuss to which extent the results relate with time disorders.

Disorders of the self in schizophrenia

The sense of self is manifold and entails at least the narrative self and the sense of bodily self (for extensive reviews and various views on this topic see Damasio, 2003; Gallagher, 2000). The narrative self is altered in schizophrenia, and regarding time, it is striking during clinical interviews that patients have difficulty to convey a clear chronology of their life events in time (Allé et al., 2016). Whether and how such alterations are related to more elementary disturbances in timing remains to be explored, and in this chapter, we will rather focus on the sense of bodily self. The sense of bodily self includes various nonverbal experiences (Dary et al., 2023). Patients with schizophrenia have a blurred representation of their body (Costantini et al., 2020), but several other components of the bodily self are impaired in patients. The first perspective has already been mentioned, and, as emphasized, plays an important role in the relation with timing (Fuchs, 2007; Martin et al., 2014; Vogeley & Kupke 2007). Additional components are body location, ownership, and agency. Body location allows us to implicitly know where we are and where our legs and arms are (Blanke & Metzinger, 2009). Body ownership refers to our feeling that we own our body (Tsakiris et al., 2007), whereas the experience of being at the origin of our actions defines agency. Initially the disorders of the self in schizophrenia have been mainly explored at the light of the concept of agency, i.e., the feeling of being the source of our actions and thoughts (Ebisch & Aleman, 2016; Franck et al., 2001). As a matter of fact, patients with schizophrenia sometimes believe that their own actions or thoughts are controlled by an external force. They thus misattribute their thoughts and actions to an external force, and this misattribution is interpreted as an impairment of agency, i.e., an impaired ability to feel as being the agent of one's own thoughts and actions.

Agency and "internal models"

Delusions of influence have been proposed to relate to defective predictions of the sensory consequences of the action (Franck et al., 2001; Frith, 2005; Jeannerod, 2009). Frith's hypotheses were based on the theory of "internal models" (Wolpert et al., 1995). In the case of a voluntary action, a motor program is built by means of an "inverse" model. When the motor program is finalized and sent to the motor apparatus, it generates an "efference copy," which is used by another internal model, i.e., the forward model, to predict the sensory outcome of the action (Von Holst and Mittelstaedt, 1950). The comparison between the expected outcome and the actual sensory feedback resulting from the action is used to adjust the action online. It has also been proposed to play a role in the sense of agency (Frith, 2005).

According to Frith's theory (2005), the sense of being the author of one's own actions would be reinforced if the expected outcome matches the actual one. Frith's most influential hypothesis states that the sense of agency is disturbed in patients with schizophrenia due to an alteration at the level of the efference copy (Franck et al., 2001; Frith, 2005; Jeannerod, 2009; Voss et al., 2010). A discussion of this hypothesis is beyond the scope of this chapter (see Giersch et al., 2016), and we will only highlight some results related to time issues. One of the paradigms most directly related to time is intentional binding (or temporal binding, Gutzeit et al., 2023). During this task, participants press a key actively or see a key being pressed, and this active or passive action is followed by a signal, e.g., a sound. The participants are asked to judge the time between the keypress and the sound, and the binding phenomenon results in this duration being judged as shorter in the case of an active, rather than a passive, action. Haggard et al. (2003) reported this effect to be increased in patients with schizophrenia. These effects have been replicated (Graham-Schmidt et al., 2016), although recently, Roth et al. (2023) have described decreased intentional (or temporal) binding in delusional patients, who have disturbed self-agency and have a difficulty to perceive the causal relationships between actions and their sensory consequences. In contrast, patients without those symptoms display a difficulty at estimating durations at baseline. The authors emphasize the potential link with timing.

Hughes et al. (2013) go further by suggesting that a voluntary action entails a timing prediction that might alone explain the results, independent of the action itself. Whether this explains the disorders observed in schizophrenia remains to be determined, though. In the case of intentional (or temporal) binding, we encounter a chicken and egg difficulty, which is frequent in a pathology such as schizophrenia, with diffuse impacts on cognition. More specifically, intentional (or temporal) binding results in a change of perceived duration, but it might also be the case that difficulties at estimating durations might affect the results in the binding task. Different cognitive impairments potentially interact in schizophrenia, and this difficulty should be kept in mind when interpreting the results. Nevertheless, the intentional (or temporal) binding task suggests the possibility of a link between time and voluntary action, which will be developed again down below.

What is also important regarding this chapter is the cancellation of predicted sensory signals, occurring as a direct consequence of voluntary actions (Brooks & Cullen, 2013; Izawa et al., 2008; Shergill et al., 2003). Since the sensory consequences of the actions are predicted, they do not represent pertinent information if they correspond to the expected one. This cancellation would help to avoid processing and reacting to self-generated signals inappropriately, i.e., as if they were externally generated. Motor programming would thus help to discard self-related signals in order to focus on more pertinent signals issued from the environment. Detecting unexpected information is indeed necessary to correct actions if necessary, and more generally to detect any change in the environment. Canceling non pertinent information helps to increase the signal-to-noise ratio of the signals conveying new information, or changes in our environment, thus helping us to better adapt to our surroundings. As we will see later in this chapter, cancellation may also concern spurious delays in the sensory consequences of actions. However, initial studies mainly suggested that patients with schizophrenia have difficulty to inhibit the sensory signals that result from voluntary actions (Blakemore et al., 1998; Shergill et al., 2005).

Minimal self-disorders

Other aspects of the sense of self may also be impaired, however (Parnas et al., 2003). Symptoms of depersonalization, a sense of emptiness, or a sense of derealization (when the world seems unreal), are part of what is called "minimal self-disorders." Derealization and depersonalization are widely shared across pathologies or personality disorder. They are part of the definition of borderline personality disorders (American Psychiatric Association, 2013), and even some patients with a vestibular syndrome display those symptoms (Lopez & Elzière, 2018). In patients with schizophrenia, they are associated with a range of additional, more or less strange symptoms, like the feeling that body limits include other individuals, a feeling of body dissolution or a brief feeling of not being human. More generally, patients with schizophrenia display a blurred demarcation between oneself and the world, a feeling of body disintegration, or a diminished "presence," as if the patient would not be entirely present in the world.

Such "minimal self-disorders," especially those more specifically related to schizophrenia, have been reported to predict the emergence of psychosis in individuals who are in a prodromal stage (Spark et al., 2021), i.e., who present attenuated symptoms of psychosis but do not answer to the criteria of schizophrenia. As such, they have gained importance in the literature: it has become critical to detect them as early as possible and to prevent the conversion to psychosis. Those disorders can occur during or independently of a voluntary action (Gallagher, 2000). For example, out of body experiences can occur at rest. We will summarize the findings evidencing bodily self-disorders that are independent of agency impairments.

Costantini et al. (2020) have shown that individuals with schizophrenia have difficulties to identify their fingers, which suggests that patients' representation of their body is altered. Other researchers have evidenced body ownership impairments. They have used body illusions: e.g., an avatar is seen as being stimulated tactilely while the subject is him- or herself stimulated tactilely in a synchronous way. This can lead all subjects, patients or not, to experience the avatar as being their own body. The most frequent approach is the rubber hand illusion, in which the illusion leads participants to self-attribute a rubber hand. The rubber hand is placed parallel to the participant's own hand, which is hidden from view. The experimenter strokes the rubber hand and the true hand of the participants simultaneously or asynchronously. The illusion of owning the rubber hand usually arises after some time, but only when the strokes are simultaneous. It has been described that patients with schizophrenia endorse this type of illusion more easily and faster than controls, thus revealing the fragility of their own boundaries (Graham et al., 2014). Interestingly, Graham et al. (2014) and Graham-Schmidt et al. (2018) showed that patients with first-rank "passivity" symptoms (i.e., with hallucinations and delusions, including agency disorders, as well as delusional perceptions) did not show the usual decrease of the rubber hand illusion when there is an asynchrony of 500 ms between the visual and tactile strokes. They suggested timing impairments as source of confusion between self- and other sources of information.

The rubber hand illusion is used frequently, as it works well. Whole body illusions are more difficult to evidence, however. Shaqiri et al. (2018) used the full-body illusion in patients with schizophrenia and were not successful in replicating the effects observed with the rubber hand illusion. Many other useful experimental approaches have confirmed disorders of the sense of bodily self, though, and especially body frontiers. For example, Lee et al. (2021) and Paredes et al. (2022) reported that the peri-personal space was closer to patients' body. As a whole the literature confirms the existence of self-disorders, whether affecting agency or bodily self-disorders (Braun et al., 2018; Hur et al., 2014; Rossetti et al. 2020). However, the link with timing disorders is until now at best anecdotal. This link is more at the heart of the following work. As a matter of fact, it has been proposed that the sense of self emerges as a consequence of the integration of various self-related signals in the same time window. We will first attempt at clarifying the concept of multisensory integration as regards the sense of self, before focusing on time integration itself.

Bodily self and multisensory integration

We own one unique bodily self. However, our sense of bodily self relies on information from different sensory modalities, i.e., proprioceptive, tactile, kinesthetic, interoceptive, vestibular. All sensory modalities convey information about the localization of our body and body limbs. This observation leads to the idea that our bodily experience involves the integration of information of all senses. Conversely, one popular mechanistic hypothesis for bodily self-disorders in schizophrenia is a deficit in multisensory processing (Northoff & Stanghellini, 2016). What this exactly entails is somewhat unclear, however (Di Cosmo et al., 2021; Noel et al., 2018). In fact, multisensory integration itself is not always altered in patients with schizophrenia (Martin et al., 2013; Noel et al., 2018). For example, when evaluating the ability to distinguish events in time, impairments in patients with schizophrenia are not disproportionally increased for multisensory relative to unisensory signals (Di Cosmo et al., 2021; Foucher et al., 2007; Martin et al., 2013).

As a matter of fact, the hypothesis of "multisensory integration" as a mechanism for the emergence of the sense of bodily self might be slightly misleading. The sensory modalities involved in the sense of bodily self are not integrated to the point of replacing multiple sensations by one unique perception, like what happens in the case of the McGurk effect for audiovisual information. In the McGurk effect (McGurk & MacDonald, 1976), a face is presented, and the visual information (moving lips) "GA" is combined with the sound "BA," leading to the perception of "DA." "DA" corresponds neither to visual nor to auditory information, which is no longer accessible consciously. This is not what happens for the sense of bodily self since access to the different information sources remains possible. It is in fact necessary to define what is meant by "integration," i.e., bottom-up sensory integration, vs. integration in a preexisting concept of bodily self (that affects sensory processing in a top-down manner), vs. something in-between (Braun et al., 2018). The latter would mean a late-occurring integration (after the processing of sensory information) allowing for diverse information sources (enteroception, body position) to be integrated in a representation of the body while preserving access to the information sources (Metzinger, 2007).

This point is important because it guides how research on time is conducted, when the aim is to understand the pathophysiology of the disorders of the sense of self. If multisensory integration is thought to be fundamental for the sense of self in the strong sense, i.e., the multisensory integration that leads to new percepts, then this should lead to investigate to which extent asynchronies between, for example, audio and visual information are tolerated and allow for multisensory integration. Since this is not how we understood the role of multisensory integration in the sense of self, we rather privileged a unisensory approach, which we detail in the following.

Time perception in patients with schizophrenia

Patients with schizophrenia report a fragmentation of thought and time experience (Fuchs, 2007; Stanghellini et al., 2016; Vogeley & Kupke, 2007), as if unable to follow information in a continuous way. Trying to understand what such a fragmentation means leads to question the sense of time continuity and the continuity of oneself in time. We intuitively think that our perception faithfully reflects the outer world, that it is as continuous as time seems to be, and that we ourselves are continuously in contact with our environment. Yet both the sense of time and of self are impaired in individuals with schizophrenia. Patients themselves report a link between those disorders, as illustrated by the following patient's citation:

> Time splits up and doesn't run forward anymore. There arise uncountable disparate now, now, now, all crazy and without rule or order. It is the same with myself. From moment to moment, various 'selves' arise and disappear entirely at random. There is no connection between my present ego and the one before.
>
> *(Kimura, cited in Fuchs, 2007)*

Such reports have led phenomenologists to hypothesize a link between the sense of self and timing (Fuchs, 2007; Fuchs & van Duppen, 2017; Stanghellini et al., 2016).

The first studies directly addressing the link between timing and schizophrenia investigated the duration estimation. This was justified by the fact that it is the most intuitive way to think about time, that it represents easy tasks for patients, and by the existence of a model of duration estimation, i.e., the clock model. The most popular version of the clock model includes a pacemaker, a switch, and an accumulator. When participants pay attention to time, a switch allows the pacemaker to send pulses in the accumulator, and the number of accumulated pulses reflects duration (Gibbon 1977; Gibbon et al., 1984, Treisman, 1963). Since this seminal version, a number of alternative models have been developed (e.g., Killeen & Grondin, 2022; Paton & Buonomano, 2018), but those are beyond the scope of this paper. Rammsayer (1990) had been one of the first to relate time duration impairments with dopamine impairments in schizophrenia. Andreasen et al. (1998) proposed to relate time estimation difficulties with what she called a dysmetria of thoughts, i.e., a disorganization of thought and behavior, and a number of studies have used various tasks to investigate time duration evaluation in schizophrenia (Bolbecker et al., 2014; Carroll et al., 2008, 2009; Davalos et al., 2003, 2011; Elevåg et al., 2003, 2004 ; Lee et al. 2009 ; Roy et al., 2012; reviewed in Ciullo et al., 2016). However, keeping a duration more than 2 s in mind involves working memory, and several studies have suggested that duration perception may be altered in schizophrenia due to working memory difficulties (Elevåg et al., 2004; Lee et al., 2009; Roy et al., 2012). Besides, alterations of duration perception in patients are inconsistent from study to study or even from moment to moment for a given patient (Thoenes & Oberfeld, 2017; Vogel et al., 2019; both papers review also in more detail the papers on duration perception in schizophrenia, and the phenomenological experience of patients, respectively). It is perfectly possible that alterations in duration estimation reflect some clinical impairments in patients (Bonnot et al., 2011), but duration estimation tasks, which mainly lead to interpretations of time accelerations or slowing down, may not be the best-adapted ones to capture time fragmentation. The roots of a time fragmentation seem to be best found in what allows time to have a structure, i.e., succession, order, and simultaneity (Vogel et al., 2019). In other words, it may be best to investigate time fragmentation by looking at the time of perception rather than at the perception of time. This distinction is even more justified that several findings suggest the perception of duration and asynchrony/order to rest on distinct mechanisms (Coull & Giersch, 2022). Processing order and distinguishing between simultaneous stimuli all help to structure events in time, and we will first review evidence that there is a limit to these abilities.

What are the limits of perception of healthy volunteers and in patients with schizophrenia?

The temporal structure of mental activity is tightly connected to the limits of our perception, i.e., to how often we can refresh our perception. A first limit in time perception is related to our ability to distinguish information in time (Elliott & Giersch, 2015; van Wassenhove, 2009; Wittmann, 2011). When two visual figures appear in different locations, neurotypicals need up to 30–50 ms to distinguish their onset in time (50 ms when stimuli are more difficult to detect, e.g., when their luminance changes only dimly, simultaneously or asynchronously). We showed that patients with schizophrenia have difficulties to detect stimulus onset asynchronies (Foucher et al., 2007; Giersch et al., 2009), i.e., to detect that two stimuli appear one after the other, rather than at the same time. In difficult conditions (e.g., dim changes in luminance), patients need up to several hundreds of milliseconds to detect the asynchrony between stimuli. Importantly we showed this difficulty to be independent of nonspecific deficits like altered conscious access, or inter-hemispheric transfer (Giersch et al., 2009; Lalanne et al., 2012a, 2012b). It is as if for patients everything can stop in time for several tens of milliseconds, which is worse than the time resolution of early movies in the 1930s. Those results have been replicated by other teams many times, in vision but also with audiovisual or audio-tactile stimuli (Di Cosmo et al., 2021; Noel et al., 2018; Schmidt et al., 2011; Stevenson et al., 2017; Wallace & Stevenson, 2014). Those are important results, as they show that patients' difficulties are not dependent on the visual modality. It is still unclear if the use of multi-sensory stimulations leads to a selective difficulty related to multisensory integration. Distinguishing two stimuli from different modalities is a more difficult task than in a unimodal task (van Wassenhove, 2009), which could itself explain that the amplitude of the difficulty is larger in patients, independent of the task at hand (Di Cosmo et al., 2021; Foucher et al., 2007). It is to be noted that large difficulties are also observed when participants are asked to order stimuli, by saying which one was the first or the second (Arrouet et al., 2022; Capa et al., 2014). Patients' difficulty appears to be larger for order than for simultaneity/asynchrony discrimination (Capa et al., 2014), although this comparison is always tricky (García-Pérez & Alcalá-Quintana, 2015; but see Coull & Giersch, 2022 for a review). Taken as a whole, these results show how difficult the structuration of information in time at a conscious level is. At least for order, this may originate from difficulties at the conscious level, as automatic processing of order appears at best limited (Chassignolle et al., 2021; Polgári et al., 2023). Several studies have suggested that access to consciousness is impaired in patients with schizophrenia (Berkovitch et al., 2018; Del Cul et al., 2006). However, we reasoned that difficult conscious ordering or asynchrony detection were unlikely telling the whole story. The relative preservation of everyday functioning of patients with schizophrenia suggests that at least some automatic processing in time remains preserved. This reasoning led us to question how patients deal with a bad time resolution at the conscious level, and what happens at a smaller timescale, i.e., subconsciously.

How do healthy volunteers compensate for the limits of perception?

For neurotypicals, Herzog et al. (2016) proposed that at a subconscious level, processing may be "quasi-continuous," to compensate for the limited time resolution of conscious perception. Consistent with this idea, we had shown repeatedly that healthy volunteers process stimulus onset asynchronies between two visual stimuli even when they do not detect them (Lalanne et al., 2012a, 2012b; Poncelet & Giersch, 2015). However, processing information at a high temporal scale does not rely on a millisecond-by-millisecond processing. As shown by Herzog et al. (2016), visual information requires time to be processed in detail: it requires feature extraction, integration of information over space and time to detect forms and distinguish them from one another, and comparison with form, semantic and lexical representations. This all can take up to more than 300 ms when detailed processing is required. Detailed information cannot be achieved at the millisecond level, and our own data show that sequences of successive information are anticipated and processed, not single events. We indeed showed that the processing of two visual stimuli displayed in different locations with an undetectable onset asynchrony leads attention to be moved in advance in the location of the second stimulus of the sequence (Poncelet & Giersch, 2015). Moreover, we showed that the processing of sequences of two successive visual stimuli is anticipated on the basis of the previous sequence, as if this previous sequence was replayed (Marques-Carneiro et al., 2020). These mechanisms would help to compensate for the low temporal resolution of conscious perception and would be crucial in reaching a sense of time continuity. Our model is illustrated in Figure 9.1. Instead of lagging behind information, trains of visual information are predicted in advance, thus allowing for a sense of continuity in perception.

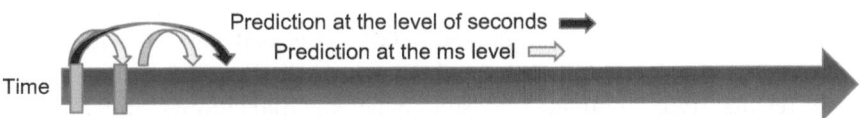

FIGURE 9.1   At a conscious level (in black) the ability to distinguish events in time and space is limited. Those limits are compensated for by the prediction of sequences of events at the level of milliseconds (gray arrows). Even though the two rectangle events are consciously judged as being simultaneous, they are followed over time thanks to predictions at the ms level.

Millisecond-level predictions in individuals with schizophrenia

We showed repeatedly that patients with schizophrenia are abnormally sensitive to short onset asynchronies between visual stimuli, even though they have difficulties to consciously report large asynchronies (Lalanne et al., 2012a, 2012b). In those studies, "short asynchronies" means asynchronies too short to be detectable consciously more than 50% of the times. We recently confirmed this abnormal sensitivity to short asynchronies in visual perception with EEG recordings (Marques-Carneiro et al., 2021). Participants decided whether two visual stimuli were shown asynchronously or simultaneously on a computer screen while EEG was recorded continuously. The results showed that the evoked EEG responses to 24 ms vs. perfect simultaneity differed in patients with schizophrenia but not in controls, consistent with the idea that patients are more sensitive to short asynchronies than controls. This sensitivity is reflected in EEG signals and appears to remain at a subconscious level, since EEG signals are not accompanied by a conscious detection of the 24 ms asynchronies.

Interestingly the analyses also showed decreased oscillations in the alpha range in controls, in the period preceding the trials, but only after a trial during which an asynchrony had been detected. Such a signal fits with the sequential effects (Marques-Carneiro et al., 2020), suggesting an attentional preparation from trial to trial. This decrease in alpha oscillations was lacking in patients with schizophrenia, suggesting they do not predict asynchronies efficiently, or at least that they do not mobilize attention to be fully prepared to process an asynchrony. The fact that there was still a response to short asynchronies fits with the usual contrast we found at the behavioral level between a large difficulty to consciously report an asynchrony, vs. an abnormal sensitivity to short asynchronies in implicit responses (i.e., that do not require an explicit judgment). Our interpretation was that if patients do not predict asynchronies correctly, such asynchronies would not be integrated in the time flow and would stand out.

It is here that we can make a parallel with the difficulty at discarding spurious signals described earlier in this chapter (the paragraph on agency, Blakemore et al., 1998; Shergill et al., 2005). It is as if patients with schizophrenia are unable to discard small asynchronies. It might be argued that our results were obtained in vision, whereas Blakemore's and Shergill's results were obtained in the motor domain. However, we found a similar result in the motor control domain, in a task during which participants performed a voluntary pointing action. During this task, the moment of the haptic feedback (haptic means kinesthetic and tactile) was manipulated by means of a virtual reality setup (Weibel et al., 2015). The haptic feedback was delayed by 15 ms, which was undetectable, or by 65 ms, which was easily perceptible. Participants were asked about their feeling of controlling the equipment, and their motor trajectory was continuously recorded. The motor results showed that patients with schizophrenia (but not with bipolar disorders) reacted abnormally to undetectable and unpredictable delays in the sensory feedback (Foerster et al., 2021). Moreover, we showed that the delays also affected patients' feeling of control: when there was an undetectable delay in the sensory consequence of the action, patients felt they did not control their action. These results are consistent with the hypothesis that patients with schizophrenia are unable to discard small delays from further processing.

Any neurobiological substrates for timing impairments in schizophrenia?

Patients' difficulty to discard small delays or asynchronies has been replicated several times within our team, although a replication in other laboratories is warranted. Nevertheless, the results are strong enough to indicate that for patients with schizophrenia, the onset of successive stimuli is not confused in time even for small delays, at least at an automatic level and prior to reaching consciousness. This observation constrains neurobiological explanations for the impairments observed in patients. As a matter of fact, the simplest explanation for the difficulties in judging order and asynchrony would have been noisy neural transmission, with a low signal-to-noise ratio (Winterer & Weinberger, 2004). Noisy processing of sensory signals might blur the time onset of a stimulus, thus making the comparison of two time onsets difficult.

There are a number of results suggesting noisy processing in patients with schizophrenia: variability in motor behavior from trial to trial (Carroll et al., 2009; Moussa-Tooks et al., 2019), in response times (Chidharom et al., 2021a), and in duration processing (Thoenes & Oberfeld, 2017). Widespread impairments in connectivity in schizophrenia (Uhlhaas, 2013) also fit with the idea that there might be noise at the level of neural processing. Several studies further confirm variability in the latency of evoked potentials, by showing impaired inter-trial coherence in EEG signals recorded from 100 to 300 ms after the signal (Chidharom et al., 2021b; Karanikolaou et al., 2022). Moreover, this variability in the EEG signal is related to variability in response times (Wolff et al., 2022) and in attention performance (Chidharom et al., 2021b). It makes sense to hypothesize that such noisy or imprecise signal coding accounts for at least patients' difficulty to detect asynchronies and judge order.

The question is whether neural variability already occurs at the most elementary level, when information is first processed, or at a secondary level, when information is integrated, compared, and when it is focused on. It can be noted that the variability in the inter-trial coherence has been mainly observed for signals arising 100–300 ms after the target occurrence, at a time when attention feedback is required. It is also observed in relation with attention disorders (Chidharom et al., 2021a, 2021b). Given the discrepancy between the implicit responses at very short asynchronies, and the difficulty at explicitly detecting order and asynchronies in general, it makes sense that the latter is related to a difficulty at mobilizing the large networks required to explicitly process time information. This may be especially true when the task requires comparing onsets of stimuli processed in different hemispheres. As a matter of fact, time information processing is based on a wide and interrelated network of subcortical and cortical areas (Coull & Giersch, 2022; Paton & Buonomano, 2018). Although there is less data on simultaneity/asynchrony discrimination tasks than on duration processing, at least left parietal and temporal cortices seem consistently involved in fMRI (Coull & Giersch, 2022). Regarding EEG, the simultaneity/asynchrony task has seldom been used in unisensory modalities, even for neurotypicals. Lange et al. (2012) have used electrical stimulations on the fingers and have shown that performance is predicted by the power of alpha and beta oscillations prior to the stimuli. Alpha oscillations have also been described as related to the individual abilities to detect that the auditory and visual stimulus are not simultaneous (Grabot et al., 2017), or to the time interval required between two successive flashes displayed in the same location, for those flashes to be distinguished in time (Samaha & Postle, 2015). Given the role of alpha oscillations in attention-related mechanisms, and given the widespread networks involved in the task, it is plausible that noisy processing may affect the considered tasks. However, once again we are still missing neurobiological data in patients.

We can nonetheless speculate on the neurobiological impairments in patients with schizophrenia. Even if noisy neural coding may account for impaired order and simultaneity/asynchrony explicit judgments, such a lack of temporal precision hardly accounts for the fact that patients are still sensitive, and even hypersensitive to short asynchronies. Northoff and Zilio (2022) have suggested that oscillatory activities on short timescales are not properly integrated in oscillatory activities on longer timescales. However, the considered timescales are not shorter than 100 ms, which limits the parallel with the results observed at the level of milliseconds. Another interesting hypothesis is the relation between the ability to predict the sensory consequences of the action precisely in time (Ford et al., 2014) and inter-trial variability in neuronal responses (Karanikolaou et al., 2022). Although generally consistent with the hypothesis illustrated in Figure 9.1, the question is the timescale of time prediction impairments, and in which conditions impairments occur. Two lines of research will be reminded here: the variable foreperiod task and the Smith predictors, which will be used to discuss how far we can go when trying to speculate on the neurobiological impairments of patients in relation with timing disorders.

Variable foreperiod

The variable foreperiod paradigm is used to assess implicit temporal preparation, which is what happens when waiting for a red traffic light to turn green. In the laboratory, a fixation point is displayed first, followed by a target occurring at a variable interval (or "foreperiod") after the initial fixation point. Participants are instructed to press on a key as soon as the target is displayed on the screen. When no other temporal information is available, the unidirectional flow of time automatically provides a degree of predictive power: since time flows inexorably forward ("time's arrow") an event that we expect to occur, but has not yet occurred, must do so at some time in the future. The "hazard function" measures the increasing conditional probability (and, hence, increasing sense of temporal expectation) over time that an event will occur, given that it has not already occurred (Elithorn & Lawrence, 1955; Luce, 1986). The increasing temporal certainty of target presentation over time allows the response plan to be updated online, i.e., as the trial unfolds, which translates into faster RTs at longer intervals (Niemi & Näätänen, 1981).

At the group level, patients with schizophrenia benefit from the passage of time to the same amount as controls (Ciullo et al., 2018; Martin et al., 2017). However, whenever there is an uncertainty regarding the occurrence of the target, the benefit of the passage of time disappears, as if patients become unable to prepare for the target (Ciullo et al., 2018; Martin et al., 2017). This is also observed in those patients who display disorders of the minimal self, as evaluated clinically. Those patients are impaired at benefiting from the passage of time independent of any uncertainty (Martin et al., 2017). Those results were interpreted as a fragility in the implicit prediction of time, which is certainly consistent with the hypothesis of time prediction impairments, even though those impairments occur independently from a voluntary action (see Hughues et al., 2013, for a discussion on the relative role of time prediction and agency). Other results are also consistent with the hypothesis of time prediction impairments. EEG studies show that a change in delay is coded even if the subject does not have to report it: when stimuli change duration on rare trials, i.e., oddballs, MMN (Mismatch Negativity) evoked potentials are observed (Näätänen et al., 1989). This occurs even for a change of a few tens of milliseconds (Michie et al., 2000; Suga et al., 2016). The amplitude of the MMN is typically reduced in individuals with schizophrenia, and interestingly, this effect is much larger when the deviant feature is duration, rather than pitch (Michie et al., 2000; Suga et al., 2016). Up until this point, all results might be explained by noisy processing.

However, several results suggest that at least some aspects of the time prediction are preserved or altered in unexpected ways. These contradictory findings may help to refine the hypothesis. As a matter of fact, some time prediction mechanisms have to be preserved, since they are required to do everyday actions as easy as walking, grasping an object, or speaking. The variable foreperiod task helps to evidence automatic sequential effects that are useful when training a gesture. Such sequential effects rely on the delays experienced in the previous trials, which leads subjects to expect similar delays on the current trial than on the trial before (Los and van den Heuvel, 2001; Vallesi et al., 2014). As a consequence, if the foreperiod on trial N is shorter than that on trial N-1, participants are not prepared, and reaction times are slowed. Such sequential effects were shown unimpaired in recent studies (Ciullo et al., 2018; Martin et al., 2017). If any, Zahn et al. (1963) have shown increased sequential effects in patients with schizophrenia relative to controls. Interestingly, in the Zahn et al. study (1963), there were only 12 patients, but none was treated. It might be argued that such sequential effects may work without requiring a high level of temporal precision. However, the results of Foerster et al. (2021), based on voluntary manual pointing actions (see above), suggest that patients are sensitive to small delays in the sensory consequences of the action: their feeling of control drops, and they adapt their trajectory when the haptic feedback is delayed by only 15 ms. This excessive sensitivity to unexpected delays contrasts with the hypothesis of a difficulty to predict sensory information precisely in time. Once again, this discrepancy with the results observed 100–300 ms after stimuli may originate in the nature of the explored mechanisms that require different levels of attention. In patients, the sensitivity to short delays is observed for mechanisms that are largely automatic, whereas inter-trial variability is observed for EEG signals or responses occurring at a time at which attentional control can affect neuronal and behavioral responses. The existence of very automatic mechanisms of temporal prediction is suggested by the model of Smith predictors.

Smith predictors

It should be reminded that it is still debated how time is predicted at the most elementary level. The internal models predict the sensory consequences of the action, but it is unclear whether this entails time prediction. It should be stressed too that the variable foreperiod task includes more than automatic temporal prediction. The variable foreperiod task includes a sensory detection task, and time prediction associated with the hazard function includes cortical regions, e.g., the parietal or prefrontal cortex, that are not necessarily mobilized for a simple voluntary action (Coull et al., 2016; Triviño et al., 2016). The idea of very basic time prediction comes from research on motor planning. Miall et al. (1993) worked on motor-related predictions and proposed that the cerebellum may act as a Smith predictor. This predictor would take into account the time constraints associated to the individual transmission delays, in order to increase the time accuracy of motor predictions. This led to the idea that the cerebellum could adapt to delays thanks to the Smith predictor. However, subsequent results suggested that such adaptations did not always work (Miall & Jackson, 2006), although this model is still discussed (Tolu et al., 2020). It is thus unclear if what we described as preserved time prediction in schizophrenia corresponds to the Smith predictor. We might need more knowledge on time prediction mechanisms in order to fully understand the impairments in patients with schizophrenia and their neural underpinnings. In particular, even if we assume that time prediction closely associated to motor prediction, like the Smith predictors, is preserved in patients, it remains to be understood to which extent this "motor" prediction is dissociated from the time prediction evidenced in the variable foreperiod tasks. It remains also to be seen whether and how different types of prediction interact with each other.

Pending questions and conclusions

If there is such a thing as elementary time prediction that is preserved in patients with schizophrenia, it is likely that such prediction can be modulated and adapted to the context. If this is true, diverse types of information might need to be integrated to account for time prediction adaptation (Markanday et al., 2023; Schubert & Zee, 2010). If one takes the example of small, spurious delays that need to be discarded, this may depend on the context (Weibel et al., 2015). For example, if an experimental block includes both small and large delays, small delays can be ignored, whereas they are processed if all delays are small in the experimental block (Weibel et al., 2015). Also, the frequency of delays may vary, like when we walk on a flat surface, vs. an irregular one, and this may also change how we take delays into account or not (this idea has been developed also for autism; see van de Cruys et al., 2014). In order to understand how and when time prediction is impaired in schizophrenia, and which neuronal networks are involved in those difficulties, we need to know how much, how, and which type of, time prediction is adapted to the task at hand. This question is all the more difficult for delays at the level of milliseconds, because short delays are processed subconsciously and cannot be corrected online, contrary to predictions at the level of hundreds of milliseconds. This means that usual internal models cannot work in closed loops as usual, and that adaptation needs to be different. At the level of milliseconds, delays have to be corrected from trial to trial and thus require additional mechanisms on top of forward models. This may require a change in usual models of predictive coding to integrate prediction at the level of milliseconds. In turn, integrating prediction at the level of milliseconds in models of predictive coding may shed light on the results in patients.

The results suggesting that patients with schizophrenia are abnormally sensitive to short delays and that they can be excessively influenced by delays experienced in prior trials suggest that the processing of time errors, and thus time prediction, is precise enough. Even the prediction of sequences at the level of millisecond may be preserved (Giersch et al., 2020). However, it is clear also from the results showing an abnormal sensitivity to short delays that the processing at the level of milliseconds is not totally preserved: how patients adjust to errors at the level of milliseconds, and integrate those in the time flow, seems to be affected. Given the small delays, such errors may have to be ignored or not, depending on the context (Foerster et al., 2021). In simultaneity/asynchrony discrimination tasks, for example, asynchronies vary in size during the experiment. In that case it is possible to adjust the span of to-be-detected asynchronies, to balance effort across the task (Marques-Carneiro et al., 2020; Weibel et al., 2015). Rather than a global impairment at predicting sequences of information, patients might rather have difficulties to ignore small asynchronies, and to adapt the prediction of sequences to the task. Whether such difficulties are due to impairments at larger time scales, induce them, or both, remains to be seen. Again, this requires to better understand the relationships between timing at different timescales, and the top-down influences on prediction at the level of milliseconds. Resolving those questions is important for therapeutics. Predicting sequences at the level of milliseconds involves the cerebellum, which would lead to targeting the cerebellum with non-invasive stimulation like transcranial magnetic stimulation. To improve the clinical state of patients, one has to take into consideration interactions with the cortical level, as already proposed in the literature (Andreasen et al., 1998; Brady et al., 2019; Parker et al., 2017). Understanding the cerebello-thalamo-cortical network as a way to integrate predictions at different timescales might help to devise more efficient approaches, e.g., by using stimulations during prediction tasks.

As is often the case, studies in schizophrenia lead to more questions than answers. Yet we hope that this short journey through time and the sense of self will have convinced the reader that trying to understand the specificities of the cognitive functioning in patients with schizophrenia helps to uncover some of the mysteries at work in our brains.

Acknowledgments

This research was constantly supported by the French National Institute for Health and Medical Research (INSERM) and the Centre Hospitalier Régional Universitaire of Strasbourg.

References

Allé, M. C., Gandolphe, M.-C., Doba, K., Köber, C., Potheegadoo, J., Coutelle, R., … Berna, F. (2016). Grasping the mechanisms of narratives' incoherence in schizophrenia: An analysis of the temporal structure of patients' life story. *Comprehensive Psychiatry*, *69*, 20–29. https://doi.org/10.1016/j.comppsych.2016.04.015

American Psychiatric Association. (2013). *Diagnostic and statistical manual of mental disorders* (5th ed.). https://doi.org/10.1176/appi.books.9780890425787

Andreasen, N. C., Paradiso, S., & O'Leary, D. S. (1998). "Cognitive dysmetria" as an integrative theory of schizophrenia: A dysfunction in cortical-subcortical-cerebellar circuitry? *Schizophrenia Bulletin*, *24*(2), 203–218. https://doi.org/10.1093/oxfordjournals.schbul.a033321

Arrouet, A., Polgári, P., Giersch, A., & Joos, E. (2022). Temporal order judgments in schizophrenia and bipolar disorders – Explicit and implicit measures. *Timing & Time Perception*, *11*(1-4), 362–385. https://doi.org/10.1163/22134468-bja10071

Berkovitch, L., Del Cul, A., Maheu, M., & Dehaene, S. (2018). Impaired conscious access and abnormal attentional amplification in schizophrenia. *NeuroImage. Clinical*, *18*, 835–848. https://doi.org/10.1016/j.nicl.2018.03.010

Blakemore, S.-J., Wolpert, D. M., & Frith, C. D. (1998). Central cancellation of self-produced tickle sensation. *Nature Neuroscience*, *1*(7), 635–640. https://doi.org/10.1038/2870

Blanke, O., & Metzinger, T. (2009). Full-body illusions and minimal phenomenal selfhood. *Trends in Cognitive Sciences*, *13*(1), 7–13. https://doi.org/10.1016/j.tics.2008.10.003

Bolbecker, A. R., Westfall, D. R., Howell, J. M., Lackner, R. J., Carroll, C. A., O'Donnell, B. F., & Hetrick, W. P. (2014). Increased timing variability in schizophrenia and bipolar disorder. *PloS One*, *9*(5), e97964. https://doi.org/10.1371/journal.pone.0097964

Bonnot, O., de Montalembert, M., Kermarrec, S., Botbol, M., Walter, M., & Coulon, N. (2011). Are impairments of time perception in schizophrenia a neglected phenomenon? *Journal of Physiology, Paris*, *105*(4–6), 164–169. https://doi.org/10.1016/j.jphysparis.2011.07.006

Brady, R. O., Gonsalvez, I., Lee, I., Öngür, D., Seidman, L. J., Schmahmann, J. D., … Halko, M. A. (2019). Cerebellar-prefrontal network connectivity and negative symptoms in schizophrenia. *The American Journal of Psychiatry*, https://doi.org/10.1176/appi.ajp.2018.18040429

Braun, N., Debener, S., Spychala, N., Bongartz, E., Sörös, P., Müller, H. H. O., & Philipsen, A. (2018). The senses of agency and ownership: A review. *Frontiers in Psychology*, *9*. https://www.frontiersin.org/articles/10.3389/fpsyg.2018.00535

Brooks, J. X., & Cullen, K. E. (2013). The primate cerebellum selectively encodes unexpected self-motion. *Current Biology: CB*, *23*(11), 947–955. https://doi.org/10.1016/j.cub.2013.04.029

Capa, R. L., Duval, C. Z., Blaison, D., & Giersch, A. (2014). Patients with schizophrenia selectively impaired in temporal order judgments. *Schizophrenia Research*, *156*(1), 51–55. https://doi.org/10.1016/j.schres.2014.04.001

Carroll, C. A., Boggs, J., O'Donnell, B. F., Shekhar, A., & Hetrick, W. P. (2008). Temporal processing dysfunction in schizophrenia. *Brain and Cognition*, *67*(2), 150–161. https://doi.org/10.1016/j.bandc.2007.12.005

Carroll, C. A., O'Donnell, B. F., Shekhar, A., & Hetrick, W. P. (2009). Timing dysfunctions in schizophrenia span from millisecond to several-second durations. *Brain and Cognition*, *70*(2), 181–190. https://doi.org/10.1016/j.bandc.2009.02.001

Chassignolle, M., Giersch, A., & Coull, J. T. (2021). Evidence for visual temporal order processing below the threshold for conscious perception. *Cognition*, *207*, 104528. https://doi.org/10.1016/j.cognition.2020.104528

Chidharom, M., Krieg, J., & Bonnefond, A. (2021a). Impaired frontal midline theta during periods of high reaction time variability in schizophrenia. *Biological Psychiatry: Cognitive Neuroscience and Neuroimaging*, *6*(4), 429–438. https://doi.org/10.1016/j.bpsc.2020.10.005

Chidharom, M., Krieg, J., Pham, B.-T., & Bonnefond, A. (2021b). Conjoint fluctuations of PFC-mediated processes and behavior: An investigation of error-related neural mechanisms in relation to sustained attention. *Cortex*, *143*, 69–79. https://doi.org/10.1016/j.cortex.2021.07.009

Ciullo, V., Spalletta, G., Caltagirone, C., Jorge, R. E., & Piras, F. (2016). Explicit time deficit in schizophrenia: Systematic review and meta-analysis indicate it is primary and not domain specific. *Schizophrenia Bulletin*, *42*(2), 505–518. https://doi.org/10.1093/schbul/sbv104

Ciullo, V., Piras, F., Vecchio, D., Banaj, N., Coull, J. T., & Spalletta, G. (2018). Predictive timing disturbance is a precise marker of schizophrenia. *Schizophrenia Research. Cognition*, *12*, 42–49. https://doi.org/10.1016/j.scog.2018.04.001

Costantini, M., Salone, A., Martinotti, G., Fiori, F., Fotia, F., Di Giannantonio, M., & Ferri, F. (2020). Body representations and basic symptoms in schizophrenia. *Schizophrenia Research*, *222*, 267–273. https://doi.org/10.1016/j.schres.2020.05.038

Coull, J. T., Cotti, J., & Vidal, F. (2016). Differential roles for parietal and frontal cortices in fixed versus evolving temporal expectations: Dissociating prior from posterior temporal probabilities with fMRI. *NeuroImage*, *141*, 40–51. https://doi.org/10.1016/j.neuroimage.2016.07.036

Coull, J. T., & Giersch, A. (2022). The distinction between temporal order and duration processing, and implications for schizophrenia. *Nature Reviews Psychology*, 1–15. https://doi.org/10.1038/s44159-022-00038-y

Damasio, A. (2003). Mental self : The person within. *Nature*, *423*(6937), 227–227. https://doi.org/10.1038/423227a

Dary, Z., Lenggenhager, B., Lagarde, S., Medina Villalon, S., Bartolomei, F., & Lopez, C. (2023). Neural bases of the bodily self as revealed by electrical brain stimulation: A systematic review. *Human Brain Mapping*. https://doi.org/10.1002/hbm.26253

Davalos, D. B., Kisley, M. A., & Ross, R. G. (2003). Effects of interval duration on temporal processing in schizophrenia. *Brain and Cognition*, *52*(3), 295–301. https://doi.org/10.1016/S0278-2626(03)00157-X

Davalos, D. B., Rojas, D. C., & Tregellas, J. R. (2011). Temporal processing in schizophrenia: Effects of task-difficulty on behavioral discrimination and neuronal responses. *Schizophrenia Research*, *127*(1), 123–130. https://doi.org/10.1016/j.schres.2010.06.020

Del Cul, A., Dehaene, S., & Leboyer, M. (2006). Preserved subliminal processing and impaired conscious access in schizophrenia. *Archives of General Psychiatry*, *63*(12), 1313–1323. https://doi.org/10.1001/archpsyc.63.12.1313

Di Cosmo, G., Costantini, M., Ambrosini, E., Salone, A., Martinotti, G., Corbo, M., ... Ferri, F. (2021). Body-environment integration: Temporal processing of tactile and auditory inputs along the schizophrenia continuum. *Journal of Psychiatric Research, 134*, 208–214. https://doi.org/10.1016/j.jpsychires.2020.12.034

Ebisch, S. J. H., & Aleman, A. (2016). The fragmented self: Imbalance between intrinsic and extrinsic self-networks in psychotic disorders. *The Lancet. Psychiatry, 3*(8), 784–790. https://doi.org/10.1016/S2215-0366(16)00045-6

Elithorn, A., & Lawrence, C. (1955). Central inhibition-some refractory observations. *Quarterly Journal of Experimental Psychology, 7*(3), 116–127. https://doi.org/10.1080/17470215508416684

Elliott, M. A., & Giersch, A. (2015). What happens in a moment. *Frontiers in Psychology, 6*, 1905. https://doi.org/10.3389/fpsyg.2015.01905

Elvevåg, B., McCormack, T., Gilbert, A., Brown, G. D. A., Weinberger, D. R., & Goldberg, T. E. (2003). Duration judgements in patients with schizophrenia. *Psychological Medicine, 33*(7), 1249–1261. https://doi.org/10.1017/S0033291703008122

Elvevåg, B., Brown, G. D. A., McCormack, T., Vousden, J. I., & Goldberg, T. E. (2004). Identification of tone duration, line length, and letter position: An experimental approach to timing and working memory deficits in schizophrenia. *Journal of Abnormal Psychology, 113*(4), 509–521. https://doi.org/10.1037/0021-843X.113.4.509

Foerster, F. R., Weibel, S., Poncelet, P., Dufour, A., Delevoye-Turrell, Y. N., Capobianco, A., ... Giersch, A. (2021). Volatility of subliminal haptic feedback alters the feeling of control in schizophrenia. *Journal of Abnormal Psychology, 130*(7), 775–784. https://doi.org/10.1037/abn0000703

Ford, J. M., Palzes, V. A., Roach, B. J., & Mathalon, D. H. (2014). Did I do that? Abnormal predictive processes in schizophrenia when button pressing to deliver a tone. *Schizophrenia Bulletin, 40*(4), 804–812. https://doi.org/10.1093/schbul/sbt072

Foucher, J. R., Lacambre, M., Pham, B.-T., Giersch, A., & Elliott, M. A. (2007). Low time resolution in schizophrenia Lengthened windows of simultaneity for visual, auditory and bimodal stimuli. *Schizophrenia Research, 97*(1-3), 118–127. https://doi.org/10.1016/j.schres.2007.08.013

Franck, N., Farrer, C., Georgieff, N., Marie-Cardine, M., Daléry, J., d'Amato, T., & Jeannerod, M. (2001). Defective recognition of one's own actions in patients with schizophrenia. *The American Journal of Psychiatry, 158*(3), 454–459. https://doi.org/10.1176/appi.ajp.158.3.454

Frith, C. (2005). The neural basis of hallucinations and delusions. *Comptes Rendus Biologies, 328*(2), 169–175.

Fuchs, T. (2007). The temporal structure of intentionality and its disturbance in schizophrenia. *Psychopathology, 40*(4), 229–235. https://doi.org/10.1159/000101365

Fuchs, T., & Van Duppen, Z. (2017). Time and events: On the phenomenology of temporal experience in schizophrenia (ancillary article to EAWE domain 2). *Psychopathology, 50*(1), 68–74. https://doi.org/10.1159/000452768

Gallagher, S. (2000). Philosophical conceptions of the self : Implications for cognitive science. *Trends in Cognitive Sciences, 4*(1), 14–21. https://doi.org/10.1016/S1364-6613(99)01417-5

García-Pérez, M. A., & Alcalá-Quintana, R. (2015). Converging evidence that common timing processes underlie temporal-order and simultaneity judgments: A model-based analysis. *Attention, Perception & Psychophysics, 77*(5), 1750–1766. https://doi.org/10.3758/s13414-015-0869-6

Giersch, A., Lalanne, L., & Isope, P. (2016). Implicit timing as the missing link between neurobiological and self disorders in schizophrenia? *Frontiers in Human Neuroscience, 10*, 303. https://doi.org/10.3389/fnhum.2016.00303

Giersch, A., Martin, B., & van der Burg, E. (2020). Time prediction and sense of self: lack of flexibility in patients with schizophrenia. *Schizophrenia Bulletin, 46*(Suppl_1), S63–S64, https://doi.org/10.1093/schbul/sbaa031.144

Gibbon, J. (1977). Scalar expectancy theory and Weber's law in animal timing. *Psychological Review, 84*(3), 279–325. https://doi.org/10.1037/0033-295X.84.3.279

Gibbon, J., Church, R. M., & Meck, W. H. (1984). Scalar timing in memory. *Annals of the New York Academy of Sciences, 423*(1), 52–77.

Giersch, A., Lalanne, L., Corves, C., Seubert, J., Shi, Z., Foucher, J., & Elliott, M. A. (2009). Extended visual simultaneity thresholds in patients with schizophrenia. *Schizophrenia Bulletin, 35*(4), 816–825. https://doi.org/10.1093/schbul/sbn016

Grabot, L., Kösem, A., Azizi, L., & van Wassenhove, V. (2017). Prestimulus alpha oscillations and the temporal sequencing of audiovisual events. *Journal of Cognitive Neuroscience, 29*(9), 1566–1582. https://doi.org/10.1162/jocn_a_01145

Graham, K. T., Martin-Iverson, M. T., Holmes, N. P., Jablensky, A., & Waters, F. (2014). Deficits in agency in schizophrenia, and additional deficits in body image, body schema, and internal timing, in passivity symptoms. *Frontiers in Psychiatry, 5*. https://doi.org/10.3389/fpsyt.2014.00126

Graham-Schmidt, K. T., Martin-Iverson, M. T., Holmes, N. P., & Waters, F. A. V. (2016). When one's sense of agency goes wrong: Absent modulation of time perception by voluntary actions and reduction of perceived length of intervals in passivity symptoms in schizophrenia. *Consciousness and Cognition, 45*, 9–23. https://doi.org/10.1016/j.concog.2016.08.006

Graham-Schmidt, K. T., Martin-Iverson, M. T., & Waters, F. A. V. (2018). Self- and other-agency in people with passivity (first rank) symptoms in schizophrenia. *Schizophrenia Research, 192*, 75–81. https://doi.org/10.1016/j.schres.2017.04.024

Gutzeit, J., Weller, L., Kürten, J., Huestegge, L. (2023). Intentional binding: Merely a procedural confound? *Journal of Experimental Psychology: Human Perception and Performance, 49*(6):759–773. https://doi.org/10.1037/xhp0001110

Haggard, P., Martin, F., Taylor-Clarke, M., Jeannerod, M., & Franck, N. (2003). Awareness of action in schizophrenia. *NeuroReport, 14*(7), 1081–1085. https://doi.org/10.1097/01.wnr.0000073684.00308.c0

Herzog, M. H., Kammer, T., & Scharnowski, F. (2016). Time slices: What is the duration of a percept? *PLoS Biology, 14*(4), e1002433. https://doi.org/10.1371/journal.pbio.1002433

Hughes, G., Desantis, A., & Waszak, F. (2013). Mechanisms of intentional binding and sensory attenuation : The role of temporal prediction, temporal control, identity prediction, and motor prediction. *Psychological Bulletin, 139*(1), 133–151. https://doi.org/10.1037/a0028566

Hur, J.-W., Kwon, J. S., Lee, T. Y., & Park, S. (2014). The crisis of minimal self-awareness in schizophrenia : A meta-analytic review. *Schizophrenia Research, 152*(1), 58–64. https://doi.org/10.1016/j.schres.2013.08.042

Husserl, E. (1996). *Leçons pour une phénoménologie de la conscience intime du temps.* Presses Universitaires de France - PUF. First German edition 1917.

Izawa, J., Rane, T., Donchin, O., & Shadmehr, R. (2008). Motor adaptation as a process of reoptimization. *The Journal of Neuroscience: The Official Journal of the Society for Neuroscience, 28*(11), 2883–2891. https://doi.org/10.1523/JNEUROSCI.5359-07.2008

Jeannerod, M. (2009). The sense of agency and its disturbances in schizophrenia: A reappraisal. *Experimental Brain Research, 192*(3), 527–532. https://doi.org/10.1007/s00221-008-1533-3

Karanikolaou, M., Limanowski, J., & Northoff, G. (2022). Does temporal irregularity drive prediction failure in schizophrenia? Temporal modelling of ERPs. *Schizophrenia (Heidelberg, Germany)*, *8*(1), 23. https://doi.org/10.1038/s41537-022-00239-7

Killeen, P. R., & Grondin, S. (2022). A trace theory of time perception. *Psychological Review*, *129*(4), 603–639. https://doi.org/10.1037/rev0000308

Lalanne, L., van Assche, M., & Giersch, A. (2012a). When predictive mechanisms go wrong: Disordered visual synchrony thresholds in schizophrenia. *Schizophrenia Bulletin*, *38*(3), 506–513. https://doi.org/10.1093/schbul/sbq107

Lalanne, L., Van Assche, M., Wang, W., & Giersch, A. (2012b). Looking forward: An impaired ability in patients with schizophrenia? *Neuropsychologia*, *50*(12), 2736–2744. https://doi.org/10.1016/j.neuropsychologia.2012.07.023

Lange, J., Halacz, J., van Dijk, H., Kahlbrock, N., & Schnitzler, A. (2012). Fluctuations of prestimulus oscillatory power predict subjective perception of tactile simultaneity. *Cerebral Cortex (New York, N.Y.: 1991)*, *22*(11), 2564–2574. https://doi.org/10.1093/cercor/bhr329

Lee, H.-S., Hong, S.-J. J., Baxter, T., Scott, J., Shenoy, S., Buck, L., ... Park, S. (2021). Altered peripersonal space and the bodily self in schizophrenia: A virtual reality study. *Schizophrenia Bulletin*, *47*(4), 927–937. https://doi.org/10.1093/schbul/sbab024

Lee, K.-H., Bhaker, R. S., Mysore, A., Parks, R. W., Birkett, P. B. L., & Woodruff, P. W. R. (2009). Time perception and its neuropsychological correlates in patients with schizophrenia and in healthy volunteers. *Psychiatry Research*, *166*(2), 174–183. https://doi.org/10.1016/j.psychres.2008.03.004

Lopez, C., & Elzière, M. (2018). Out-of-body experience in vestibular disorders—A prospective study of 210 patients with dizziness. *A Journal Devoted to the Study of the Nervous System and Behavior*, *104*, 193–206. https://doi.org/10.1016/j.cortex.2017.05.026

Los, S. A., & Van Den Heuvel, C. E. (2001). Intentional and unintentional contributions to nonspecific preparation during reaction time foreperiods. *Journal of Experimental Psychology: Human Perception and Performance*, *27*(2), 370–386. https://doi.org/10.1037/0096-1523.27.2.370

Luce, R. D. (1986). *Response times: Their role in inferring elementary mental organization*. Oxford University Press.

Markanday, A., Hong, S., Inoue, J., De Schutter, E., & Thier, P. (2023). Multidimensional cerebellar computations for flexible kinematic control of movements. *Nature Communications*, *14*, 2548. https://doi.org/10.1038/s41467-023-37981-0

Marques-Carneiro, J. E., Polgári, P., Koning, E., Seyller, E., Martin, B., Van der Burg, E., & Giersch, A. (2020). Where and when to look: Sequential effects at the millisecond level. *Attention, Perception & Psychophysics*, *82*(6), 2821–2836. https://doi.org/10.3758/s13414-020-01995-3

Marques-Carneiro, J. E., Krieg, J., Duval, C. Z., Schwitzer, T., & Giersch, A. (2021). Paradoxical sensitivity to sub-threshold asynchronies in schizophrenia: A behavioural and EEG approach. *Schizophrenia Bulletin Open*, (sgab011). https://doi.org/10.1093/schizbullopen/sgab011

Martin, B., Giersch, A., Huron, C., & van Wassenhove, V. (2013). Temporal event structure and timing in schizophrenia: Preserved binding in a longer "now". *Neuropsychologia*, *51*(2), 358–371. https://doi.org/10.1016/j.neuropsychologia.2012.07.002

Martin, B., Wittmann, M., Franck, N., Cermolacce, M., Berna, F., & Giersch, A. (2014). Temporal structure of consciousness and minimal self in schizophrenia. *Frontiers in Psychology*, *5*, 1175. https://doi.org/10.3389/fpsyg.2014.01175

Martin, B., Franck, N., Cermolacce, M., Falco, A., Benair, A., Etienne, E., ... Giersch, A. (2017). Fragile temporal prediction in patients with schizophrenia is related to minimal self disorders. *Scientific Reports*, *7*(1), 8278. https://doi.org/10.1038/s41598-017-07987-y

McGurk, H., & MacDonald, J. (1976). Hearing lips and seeing voices. *Nature*, *264*(5588), 746–748. https://doi.org/10.1038/264746a0

Metzinger, T. (2007). Empirical perspectives from the self-model theory of subjectivity: A brief summary with examples. In *Progress in Brain Research* (Vol. 168, pp. 215–278). Elsevier. https://doi.org/10.1016/S0079-6123(07)68018-2

Miall, R. C., Weir, D. J., Wolpert, D. M., & Stein, J. F. (1993). Is the cerebellum a smith predictor? *Journal of Motor Behavior*, *25*(3), 203–216. https://doi.org/10.1080/00222895.1993.9942050

Miall, R. C., & Jackson, J. K. (2006). Adaptation to visual feedback delays in manual tracking: Evidence against the Smith Predictor model of human visually guided action. *Experimental Brain Research*, *172*(1), 77–84. https://doi.org/10.1007/s00221-005-0306-5

Michie, P. T., Budd, T. W., Todd, J., Rock, D., Wichmann, H., Box, J., & Jablensky, A. V. (2000). Duration and frequency mismatch negativity in schizophrenia. *Clinical Neurophysiology*, *111*(6), 1054–1065. https://doi.org/10.1016/S1388-2457(00)00275-3

Minkowski, E. (2013). *Le temps vécu: Études phénoménologiques et psychopathologiques* (3e édition). PUF. 1st ed. 1933.

Moussa-Tooks, A. B., Kim, D.-J., Bartolomeo, L. A., Purcell, J. R., Bolbecker, A. R., Newman, S. D., ... Hetrick, W. P. (2019). Impaired Effective connectivity during a cerebellar-mediated sensorimotor synchronization task in schizophrenia. *Schizophrenia Bulletin*, *45*(3), 531–541. https://doi.org/10.1093/schbul/sby064

Näätänen, R., Paavilainen, P., & Reinikainen, K. (1989). Do event-related potentials to infrequent decrements in duration of auditory stimuli demonstrate a memory trace in man? *Neuroscience Letters*, *107*(1-3), 347–352. https://doi.org/10.1016/0304-3940(89)90844-6

Niemi, P., & Näätänen, R. (1981). Foreperiod and simple reaction time. *Psychological Bulletin*, *89*(1), 133–162. https://doi.org/10.1037/0033-2909.89.1.133

Noel, J.-P., Stevenson, R. A., & Wallace, M. T. (2018). Atypical audiovisual temporal function in autism and schizophrenia: Similar phenotype, different cause. *The European Journal of Neuroscience*, *47*(10), 1230–1241 https://doi.org/10.1111/ejn.13911

Northoff, G., & Stanghellini, G. (2016). How to link brain and experience? Spatiotemporal psychopathology of the lived body. *Frontiers in Human Neuroscience*, *10*, 76. https://doi.org/10.3389/fnhum.2016.00172

Northoff, G., & Zilio, F. (2022). From shorter to longer timescales: Converging integrated information theory (IIT) with the temporo-spatial theory of consciousness (TTC). *Entropy (Basel, Switzerland)*, *24*(2), 270. https://doi.org/10.3390/e24020270

Paredes, R., Ferri, F., & Seriès, P. (2022). Influence of E/I balance and pruning in peri-personal space differences in schizophrenia: A computational approach. *Schizophrenia Research*, *248*, 368–377. https://doi.org/10.1016/j.schres.2021.06.026

Parker, K., Kim, Y., Kelley, R., Nessler, A., Chen, K.-H., Muller-Ewald, V., ... Narayanan, N. (2017). Delta-frequency stimulation of cerebellar projections can compensate for schizophrenia-related medial frontal dysfunction. *Molecular Psychiatry*, *22*(5), 647–655. https://doi.org/10.1038/mp.2017.50

Parnas, J., Handest, P., Saebye, D., & Jansson, L. (2003). Anomalies of subjective experience in schizophrenia and psychotic bipolar illness. *Acta Psychiatrica Scandinavica*, *108*(2), 126–133.

Paton, J. J., & Buonomano, D. V. (2018). The neural basis of timing: distributed mechanisms for diverse functions. *Neuron*, *98*(4), 687–705. https://doi.org/10.1016/j.neuron.2018.03.045

Polgári, P., Jovanovic, L., van Wassenhove, V., & Giersch, A. (2023). The processing of subthreshold visual temporal order is transitory and motivation-dependent. *Scientific Reports*, *13*(1), 7699. https://doi.org/10.1038/s41598-023-34392-5

Poncelet, P. E., & Giersch, A. (2015). Tracking visual events in time in the absence of time perception: Implicit processing at the ms level. *PloS One*, *10*(6), e0127106. https://doi.org/10.1371/journal.pone.0127106

Rammsayer, T. (1990). Temporal discrimination in schizophrenic and affective disorders: Evidence for a dopamine-dependent internal clock. *International Journal of Neuroscience*, *53*(2-4), 111–120. https://doi.org/10.3109/00207459008986593

Rossetti, I., Romano, D., Florio, V., Doria, S., Nisticò, V., Conca, A., ... Maravita, A. (2020). Defective embodiment of alien hand uncovers altered sensorimotor integration in schizophrenia. *Schizophrenia Bulletin*, *46*(2), 294–302. https://doi.org/10.1093/schbul/sbz050

Roth, M. J., Lindner, A., Hesse, K., Wildgruber, D., Wong, H. Y., & Buehner, M. J. (2023). Impaired perception of temporal contiguity between action and effect is associated with disorders of agency in schizophrenia. *Proceedings of the National Academy of Sciences of the United States of America*, *120*(21), e2214327120. https://doi.org/10.1073/pnas.2214327120

Roy, M., Grondin, S., & Roy, M.-A. (2012). Time perception disorders are related to working memory impairment in schizophrenia. *Psychiatry Research*, *200*(2-3), 159–166. https://doi.org/10.1016/j.psychres.2012.06.008

Samaha, J., & Postle, B. R. (2015). The speed of alpha-band oscillations predicts the temporal resolution of visual perception. *Current Biology*, *25*(22), 2985–2990. https://doi.org/10.1016/j.cub.2015.10.007

Schmidt, H., McFarland, J., Ahmed, M., McDonald, C., & Elliott, M. A. (2011). Low-level temporal coding impairments in psychosis: Preliminary findings and recommendations for further studies. *Journal of Abnormal Psychology*, *120*(2), 476–482. https://doi.org/10.1037/a0023387

Schubert, M. C., & Zee, D. S. (2010). Saccade and vestibular ocular motor adaptation. *Restorative Neurology and Neuroscience*, *28*(1), 9–18. https://doi.org/10.3233/RNN-2010-0523

Shaqiri, A., Roinishvili, M., Kaliuzhna, M., Favrod, O., Chkonia, E., Herzog, M. H., ... Salomon, R. (2018). Rethinking body ownership in schizophrenia: Experimental and meta-analytical approaches show no evidence for deficits. *Schizophrenia Bulletin*, *44*(3), 643–652. https://doi.org/10.1093/schbul/sbx098

Shergill, S. S., Bays, P. M., Frith, C. D., & Wolpert, D. M. (2003). Two eyes for an eye: The neuroscience of force escalation. *Science*, *301*(5630), 187. https://doi.org/10.1126/science.1085327

Shergill, S. S., Samson, G., Bays, P. M., Frith, C. D., & Wolpert, D. M. (2005). Evidence for sensory prediction deficits in schizophrenia. *The American Journal of Psychiatry*, *162*(12), 2384–2386. https://doi.org/10.1176/appi.ajp.162.12.2384

Spark, J., Gawęda, Ł., Allott, K., Hartmann, J. A., Jack, B. N., Koren, D., ... Nelson, B. (2021). Distinguishing schizophrenia spectrum from non-spectrum disorders among young patients with first episode psychosis and at high clinical risk: The role of basic self-disturbance and neurocognition. *Schizophrenia Research*, *228*, 19–28. https://doi.org/10.1016/j.schres.2020.11.061

Stanghellini, G., Ballerini, M., Presenza, S., Mancini, M., Raballo, A., Blasi, S., & Cutting, J. (2016). Psychopathology of lived time: Abnormal time experience in persons with schizophrenia. *Schizophrenia Bulletin*, *42*(1), 45–55. https://doi.org/10.1093/schbul/sbv052

Stevenson, R. A., Park, S., Cochran, C., McIntosh, L. G., Noel, J.-P., Barense, M. D., ... Wallace, M. T. (2017). The associations between multisensory temporal processing and symptoms of schizophrenia. *Schizophrenia Research*, *179*, 97–103. https://doi.org/10.1016/j.schres.2016.09.035

Suga, M., Nishimura, Y., Kawakubo, Y., Yumoto, M., & Kasai, K. (2016). Magnetoencephalographic recording of auditory mismatch negativity in response to duration and frequency deviants in a single session in patients with schizophrenia. *Psychiatry and Clinical Neurosciences*, *70*(7), 295–302. https://doi.org/10.1111/pcn.12397

Thoenes, S., & Oberfeld, D. (2017). Meta-analysis of time perception and temporal processing in schizophrenia : Differential effects on precision and accuracy. *Clinical Psychology Review*, *54*, 44–64. https://doi.org/10.1016/j.cpr.2017.03.007

Tolu, S., Capolei, M. C., Vannucci, L., Laschi, C., Falotico, E., & Hernández, M. V. (2020). A cerebellum-inspired learning approach for adaptive and anticipatory control. *International Journal of Neural Systems*, *30*(1), 1950028. https://doi.org/10.1142/S012906571950028X

Treisman, M. (1963). Temporal discrimination and the indifference interval: Implications for a model of the "internal clock". *Psychological Monographs: General and Applied*, *77*(13), 1–31. https://doi.org/10.1037/h0093864

Triviño, M., Correa, Á., Lupiáñez, J., Funes, M. J., Catena, A., He, X., & Humphreys, G. W. (2016). Brain networks of temporal preparation: A multiple regression analysis of neuropsychological data. *NeuroImage*, *142*, 489–497. https://doi.org/10.1016/j.neuroimage.2016.08.017

Tsakiris, M., Schütz-Bosbach, S., & Gallagher, S. (2007). On agency and body-ownership: Phenomenological and neurocognitive reflections. *Consciousness and Cognition*, *16*(3), 645–660. https://doi.org/10.1016/j.concog.2007.05.012

Uhlhaas, P. J. (2013). Dysconnectivity, large-scale networks and neuronal dynamics in schizophrenia. *Current Opinion in Neurobiology*, *23*(2), 283–290. https://doi.org/10.1016/j.conb.2012.11.004

Vallesi, A., Arbula, S., & Bernardis, P. (2014). Functional dissociations in temporal preparation: Evidence from dual-task performance. *Cognition*, *130*(2), 141–151. https://doi.org/10.1016/j.cognition.2013.10.006

Van de Cruys, S., Evers, K., Van der Hallen, R., Van Eylen, L., Boets, B., de-Wit, L., & Wagemans, J. (2014). Precise minds in uncertain worlds: Predictive coding in autism. *Psychological Review*, *121*(4), 649–675. https://doi.org/10.1037/a0037665

van Wassenhove, V. (2009). Minding time in an amodal representational space. *Philosophical Transactions of the Royal Society of London. Series B, Biological Sciences*, *364*(1525), 1815–1830. https://doi.org/10.1098/rstb.2009.0023

Vogel, D. H. V., Beeker, T., Haidl, T., Kupke, C., Heinze, M., & Vogeley, K. (2019). Disturbed time experience during and after psychosis. *Schizophrenia Research. Cognition*, *17*, 100136. https://doi.org/10.1016/j.scog.2019.100136

Vogeley, K., & Kupke, C. (2007). Disturbances of time consciousness from a phenomenological and a neuroscientific perspective. *Schizophrenia Bulletin*, *33*(1), 157–165. https://doi.org/10.1093/schbul/sbl056

von Holst, E., & Mittelstaedt, H. (1950). Das Reafferenzprinzip. *Naturwissenschaften*, *37*(20), 464–476. https://doi.org/10.1007/BF00622503

Voss, M., Moore, J., Hauser, M., Gallinat, J., Heinz, A., & Haggard, P. (2010). Altered awareness of action in schizophrenia: A specific deficit in predicting action consequences. *Brain: A Journal of Neurology*, *133*(10), 3104–3112. https://doi.org/10.1093/brain/awq152

Wallace, M. T., & Stevenson, R. A. (2014). The construct of the multisensory temporal binding window and its dysregulation in developmental disabilities. *Neuropsychologia, 64*, 105–123. https://doi.org/10.1016/j.neuropsychologia.2014.08.005

Weibel, S., Poncelet, P. E., Delevoye-Turrell, Y., Capobianco, A., Dufour, A., Brochard, R., ... Giersch, A. (2015). Feeling of control of an action after supra and subliminal haptic distortions. *Consciousness and Cognition, 35*, 16–29. https://doi.org/10.1016/j.concog.2015.04.011

Winterer, G., & Weinberger, D. R. (2004). Genes, dopamine and cortical signal-to-noise ratio in schizophrenia. *Trends in Neurosciences, 27*(11), 683–690. https://doi.org/10.1016/j.tins.2004.08.002

Wittmann, M. (2011). Moments in time. *Frontiers in Integrative Neuroscience, 5*, 66. https://doi.org/10.3389/fnint.2011.00066

Wolff, A., Gomez-Pilar, J., Zhang, J., Choueiry, J., de la Salle, S., Knott, V., & Northoff, G. (2022). It's in the timing: Reduced temporal precision in neural activity of schizophrenia. *Cerebral Cortex, 32*(16), 3441–3456. https://doi.org/10.1093/cercor/bhab425

Wolpert, D. M., Ghahramani, Z., & Jordan, M. I. (1995). An internal model for sensorimotor integration. *Science, 269*(5232), 1880–1882. https://doi.org/10.1126/science.7569931

Zahn, T. P., Rosenthal, D., & Shakow, D. (1963). Effects of irregular preparatory intervals on reaction time in schizophrenia. *Journal of Abnormal and Social Psychology, 67*, 44–52. https://doi.org/10.1037/h0049269

10

INDIVIDUAL DIFFERENCES IN THE STUDY OF TIME PERCEPTION

Joseph Glicksohn
Bar-Ilan University, Israel

Introduction

In 1796, a 24-year-old assistant was dismissed by his boss, the Astronomer Royal, "on the grounds that [he] differed from him by 800 ms in judging stellar transits—that is, in estimating the moment a given star passed the meridian wire in the Greenwich telescope" (Mollon & Perkins, 1996, p. 101). This was a serious error, "for upon such observations depended the calibration of the clocks, and upon the clock depended all other observations of place and time" (Boring, 1950, p. 135); indeed, "on the clock depended estimates of longitude. And on longitude depended the British Empire" (Mollon & Perkins, 1996, p. 101). It was this difference in "measured transit times recorded by observers in the same situation" (Schaffer, 1988, p. 115) that was given the name the *personal equation* in astronomy (Schaffer, 1988). The personal equation is also the title of Chapter 8 in Boring's *A History of Experimental Psychology*, for the personal equation paved the way for the "reaction experiment of the new scientific psychology" and was viewed as an early example of "the effect of predisposing attitude on perception and reaction" (Boring, 1950, p. 149). Of course, as Schaffer (1988, p. 125) informs us, for the astronomer Edward Walter Maunder, "The problem of personality was an aspect of human character, but it was *therefore* manageable by astronomical discipline." Indeed, a distinction could be readily drawn between, in one astronomer's words, "the eager, quick, impulsive man who habitually anticipates, as it were, the instant when he sees star and wire together" and the "phlegmatic, slow-and-sure man who carefully waits till he is quite sure

DOI: 10.4324/9781003449546-10

that the contact has taken place, and then deliberately and firmly records it" (Schaffer, 1988, p. 125). In fact, as Gregory (1993, p. 1260) writes,

> Experimental psychology, to my mind ... started off on the wrong foot, by failing to see the significance of the astronomers' personal equation.... From observations with an artificial star, it turned out that some observers pressed the key before the star's image crossed the line. So, it was clear—and should have been clear to psychologists!—that observers do not respond to stimuli—they predict when the star will cross the line.

The personal equation thus already signifies the importance of individual differences in time perception—the topic of the present chapter. For, as Doob (1971, pp. 104–105), in his comprehensive study of time perception, writes:

> Variations in predispositional factors, ones that cannot always be excluded even when psychophysicists do their best to control and purify their subjects, likewise may have marked effects upon the outcome: the individual's attention span; his attitude toward the sources of stimulation; his fatigue during a given experimental session (some sessions require hundreds of judgments, one right on top of the other); and idiosyncratic differences among subjects or, as these differences were called during the early part of this century, the personal equation.

Individual differences and the testing of models: Implications for the internal clock

In Treisman's (1963) model of an internal clock for time perception, an arousal-dependent pacemaker produces a sequence of pulses, which are subsequently stored, counted, and ultimately transformed into an estimate of a time interval. Scalar Expectancy Theory (SET), which is a dominant paradigm in current research on time perception (Grondin, 2001; Matthews & Meck, 2016; Wearden, 2016), presents one instantiation of this internal clock model. The influential "attentional-gate" model (Zakay & Block, 1997) is also an offshoot of Treisman's model, as will be seen in comparing the two. The counter now is termed a "cognitive counter," and short-term memory is now viewed as working memory, but essentially the cognitive architecture is preserved—though with the notable addition of an attentional gate to the switch, which was originally incorporated by Gibbon et al. (1984, p. 53). The transition from a counter to a "cognitive counter" implicates a major shift in thinking, because instead of pulses being counted, these are "subjective time units" that are accumulated (Zakay, 1989, p. 368), and this accumulation is influenced by attention (Block & Zakay, 1996, pp. 182–183). As McClain (1983, p. 185) suggests, these units are based on "the average mental content per unit of duration." Produced time, for example, is "based on a predetermined number of subjective time units which are associated in one's mind with the required objective time" (Zakay, 1993, p. 93). Nevertheless, the terminology used is not strict. Some authors remain with the notion of pulses (e.g., Lake et al., 2016), even though the nature of these pulses "remains quite speculative" (Macar et al., 1994, p. 674). Other authors prefer the term "subjective time unit," which allows for a consideration of the possibility that these units can change in size (Glicksohn, 2001, 2022; Kent et al., 2019).

Irrespective of which particular type of internal clock model you subscribe to, by positing the existence of such an internal clock, one gains three things. First, as Allan (1998, p. 114) noted, "Psychophysicists have always found that the concept of an internal-clock provides a convenient metaphor to guide thinking and research." This is still true, as indicated by several prominent researchers over the past 20 years (e.g., Eisler, 2003; Rammsayer & Troche, 2014). Second, by positing such a model, one can consider which major components of the internal clock are implicated by differences in time perception, when using the various methods for assessing time perception. Among these methods, one is time production, wherein the participant is asked to produce a time interval (see, e.g., Glicksohn & Hadad, 2012). Another is time reproduction, wherein the participant is asked to reproduce a time interval by, for example, delimiting the appearance of a colored circle, such that it appears on screen for exactly the same time as it had done previously (see, e.g., Glicksohn et al., 2006). As Pouthas and Perbal (2004, p. 372) suggest, "the reproduction task should provide information about the memory component and the comparison of durations in memory whereas the production task would reveal effects of changes in the speed of the internal clock." In fact, this has been shown in a number of studies (e.g., Baudouin et al., 2006). Third, one can consider what type of personality-environment interactions should these same components reveal. For example, as Block et al. (1999, p. 206) suggest,

Compared to adolescents and young adults, children may have more difficulty waiting for something to happen… While a target duration is elapsing, children may be relatively impatient, focusing on the time at which the duration will end. This may increase the amount of attention they allocate to time, thereby lengthening their prospective duration judgments.…

Working memory

One line of thought here is that adults presenting with attention deficit hyperactivity disorder (ADHD) should exhibit impaired performance on a task of time reproduction, because of their deficient working memory. As Barkley et al. (2001, p. 357) suggest, what is impaired is their "capacity to retain a temporal interval in mind, online through working memory, so as to subsequently use it to guide the temporal duration of the response in the reproduction task." Note that in this type of study, one can either *assume* an impaired working memory impacting on time reproduction, or, on uncovering impaired performance on a time reproduction task, *together* with no impairment on a time production task, *infer* an impaired working memory. A complementary line of thought is that individuals possessing high working memory capacity should perform better on a task of time reproduction (i.e., being more accurate and/or less variable) than would those having low working memory capacity (Broadway & Engle, 2011). It is also important to note the usefulness of considering such individual differences in working memory in line with what Woehrle and Magliano (2012, p. 315) refer to as the *working memory capacity hypothesis*. This hypothesis is that individuals possessing high working memory capacity will be better able to perform a demanding task and, concomitantly, will be less accurate in concurrent time estimation, using reproduction or, in fact, any other method of time estimation. Of course, one should first agree that the method of reproduction can reliably draw out such individual differences. While Blankenship and Anderson (1976, p. 168) expressed a nonconsensual view that "The method of reproduction is untrustworthy because it necessarily confounds factors affecting the subjective durations of the initial, presentation event and the subsequent, reproduction event; differences in attentional and set characteristics of these two events cannot be controlled and may introduce time-order errors," Zakay (1990, p. 64) supported the consensual view that "reproduction is considered to be more accurate and reliable than production and verbal estimation." In fact, for some researchers, the method of reproduction is *the* method to employ when studying individual differences in time perception (e.g., Eisler & Eisler, 1992).

Attention

One line of thought here is that with increasing demands on one's attention to an ongoing task, less attention is paid to the passage of time (Zakay & Block, 1997), hence one should exhibit impaired performance on both a task of time production and of time reproduction (i.e., being less accurate and/or more variable in both tasks). Thus, for example, based on their meta-analytic review of sex differences in time perception, Block et al. (2000, p. 1339) concluded that female participants made both shorter reproductions and shorter productions of target time intervals (and larger verbal estimations), relative to male participants. Based on this, they suggested (p. 1341) that "females focus attention on time more than males do, with the result that they accumulate subjective temporal units at a faster rate...." Note that in the "attentional-gate" version of the internal clock, "attention" is instantiated as an attentional gate, a gate that enables the pulses or "subjective time units" to pass through (Zakay & Block, 1997). One should, however, stress that both the "subjective time unit" and the nature of this "attention to time" that are referred to are fuzzy notions (Sawyer et al., 1994, p. 656). While Zakay (1993) refers to "subjective time units," Block and Zakay (1996, pp. 182–183) subsequently revert to the notion of pulses. In this type of study, one can either *assume* an open gate impacting on time production, resulting in shorter time productions, or, on uncovering impaired performance on a time production task, *infer* an impaired, disabled, more fluctuating, or rapidly closing attentional gate. It is important to distinguish here between the switch of the internal clock, which can be either on (closed) or off (open), and this attentional gate, whose functioning is on a gradient. As opposed to the view expressed by Fortin et al. (1993, p. 536) that the gate seems to act exactly like the switch, namely is also in an on/off state, Zakay et al. (1999, p. 563) clarify that the gate can open wider when attention is allocated to monitoring time. As Wearden et al. (2010, p. 541) conclude, this allows for a "graded effect on performance." The complementary line of thought here is that when task requirements are minimal and not engaging, then more attention is paid to the passage of time (i.e., the attentional gate stays open wide for much longer than usual), resulting in longer productions of time intervals.

Arousal

One line of thought here is that with increasing arousal level, the faster will be the functioning of the pacemaker component of the internal clock, namely the emission rate of pulses (Mioni et al., 2016, p. 536), hence the more pulses that are generated, with subsequent shorter time production (Penton-Voak et al., 1996). In contrast, when the arousal level decreases, fewer pulses should be generated, resulting in longer time production. A difference in pacemaker emission rate of pulses has been suggested to result in the sex difference for time production—women having a relatively faster internal clock (Glicksohn & Hadad, 2012; see also Bartholomew et al., 2015; Hancock & Block, 2016). In addition, the more focused one is on ongoing and unfolding subjective experience, the slower the emission rate of pulses of the pacemaker, which again results in longer time production (Glicksohn, 2001; Glicksohn & Berkovich-Ohana, 2012). In discussing such a change in the pacemaker emission rate of pulses (or, the "speed" of the internal clock), Allman et al. (2014) refer to this as a "first-order" principle of subjective time and suggest that exposure to flicker stimulation at different frequencies would be a ready way for assessing such changes. This has been shown recently, using the method of production (Glicksohn & Weisinger, 2022).

Using the "internal clock" as a working hypothesis thus enables a consideration of individual differences in the study of time perception. This, however, must be combined with a variety of methods of time perception in order to converge on the locus of such individual differences. Thus, a difference between target groups in performance on a task of time reproduction, coupled with a lack of difference between these groups on a task of time production, would pinpoint the locus of the effect on the component of working memory—and such seems to be the case for participants presenting with ADHD. Are sex differences in time perception to be understood in terms of the working of attention (or the attentional gate), or rather in terms of the arousal-dependent rate of functioning of the pacemaker component of the internal clock? Bartholomew et al. (2015) "substantiate a modest sex difference on time production" (p. 1), but what they mean is that "female participants are more likely to underestimate in a prospective time production task when an explicit counting strategy is not employed" (p. 12), and that "it should be noted that the effect of sex was not significant during the counting condition of our time production task" (p. 14). On their part, Block et al. (1999, p. 199) concluded that "the effect of sex also appears to be attributable to method; when we looked only at reproduction data, there was no sex difference." In fact, they have also suggested that "the reproduction method may not reveal anything interesting concerning sex differences in duration judgments" (Block et al., 2000, p. 1335)—though other researchers looking at sex differences using the method of reproduction draw a different conclusion (e.g., A. D. Eisler & Eisler, 1994; H. Eisler & Eisler, 1992). To further complicate matters, it also depends on which particular measure one employs to analyze the data. Thus, Glicksohn and Hadad (2012) reported a sex difference in time production, with female participants having a relatively faster internal clock, making shorter time productions, and having a smaller exponent of the psychophysical function. This, however, was not the case in their data when analyzed using measures of absolute discrepancy and absolute error (for a recent discussion of sex differences in time perception, see Grondin, 2020).

Three ways of looking at individual differences

In a task of time production, you produce a target duration (P) by signaling when that duration (T) is thought to have elapsed. For example, if the required duration is 16 s, individual **A** might produce a duration of 16 s, individual **B** one of 18 s, and individual **C** one of 8 s. While for all three individuals, produced duration (P) is subjectively viewed as lasting 16 s (T), individual **A** exhibits veridical time perception (i.e., $P = T$; 1 subjective s = 1 chronometric s; the internal clock is synchronized with actual, chronometric, time), individual **B** would be viewed as having a *slower* internal clock ($P > T$), and individual **C** would be viewed as having a *faster* internal clock ($P < T$). The power function relating P to T (Eisler, 1976) is given by $P = aT^\beta$, linearized as $\log(P) = \log(a) + \beta\log(T) = \alpha + \beta\log(T)$, α being the intercept, and β being the slope. Extending the example to T values of 4, 8, 16, and 32 s, if individual **A** exhibits respective P values of 4, 8, 16, and 32 s, the psychophysical function here would be characterized by $\alpha = 0$, and $\beta = 1$. If individual **B** is consistent, exhibiting P values of 6, 10, 18, and 34 s, then the psychophysical function would have $\alpha \neq 0$, indicating a consistent bias (+2) in producing durations. When $\beta \neq 1$, then the untransformed data are not consistent with a linear function. This psychophysical function is not in itself dependent on any particular instantiation of an "internal clock." Nevertheless, as discussed in the previous section, one can, for example, look at the degree to which produced duration is discrepant from target duration, given ongoing exposure to flicker stimulation, which should be changing the pacemaker emission rate of pulses (Glicksohn & Weisinger, 2022), or, for example, whether male and female participants have different psychophysical functions (Glicksohn & Hadad, 2012).

Individual differences in the parameter values of a psychophysical model

Individuals **A**, **B**, and **C**, above, exemplify the value of looking at individual differences in the parameter values. For example, while individuals **A** and **B** would both have the same slope ($\beta = 1$), they would differ in intercept ($\alpha = 0$ vs. $\alpha \neq 0$). Agostino et al. (2017) found that, for 14 of their 35 adult participants, $\beta = 1$, while for 12 of these, $\beta > 1$, and for the remaining 9 participants, $\beta < 1$. The existence of such individual differences in the exponent of the psychophysical function suggests that one should not pool their data to compute a group average or profile. As Glicksohn et al. (2007, p. 203) suggested with regard to such growth functions, "Our 'take-home' message: look at the individual data, pool when you can, and only then contrast groups." The individual parameter values can be subsequently correlated both with various personality traits and with performance measures on other tasks. For example, in using two different tasks of time reproduction (one termed "immediate," in that the participant provides an estimation of elapsed duration with practically no delay from offset of the interval; the second termed "remote," in that the participants have to wait for a certain interval before providing this time estimation), individual parameter values for each task could be computed and then correlated with both trait impulsivity and sensation seeking, as well as cognitive impulsivity (Glicksohn et al., 2006). We reported (among other things) that trait impulsivity was negatively correlated ($r = -.20$) with the slope value for immediate time reproduction, while sensation seeking was positively correlated with ($r = .21$) with the slope value for remote time reproduction.

Individual differences and deviations from linearity

In the linearized psychophysical function, presented above, wherein α is the intercept and β is the slope, these intercept and slope values are themselves negatively correlated. Any data point deviating from this linearity will be readily noticed, and if this point is quite removed from the regression line, then one can identify the data point as being an outlier. This notion was first outlined in connection with the relationship between impulsivity and time perception (Glicksohn, 2002), and there it was suggested that such outliers could be subsequently investigated for both trait and cognitive impulsivity. Using the data subsequently published by Glicksohn et al. (2006) to exemplify this point, it was shown how two such outliers were both easily identified and were clearly indicative of either trait impulsivity or cognitive impulsivity. As Glicksohn (2002, p. 15) wrote, "The first outlier ... scored 89 (maximum for this sample) on [trait impulsivity] ... and her slope measure was at the minimum (0.27). The second outlier scored 86 on [trait impulsivity] ... and her slope measure was 0.37." This idea, however, is a general one for identifying outliers in the data. In addition, a deviation from the linearity that is expected from the linearized psychophysical function is of potential interest. As Glicksohn et al. (2017, p. 1) reported regarding time perception and the experience of time when immersed in an altered sensory environment, "for those participants reporting a marked change in time experience, such as 'the sense of time disappeared' and 'time became slower', their time-production data could not be linearized using a log-log plot." Who those individuals are, and what their personality profile might be, are surely of interest to those studying individual differences in time perception.

Differences between individuals in time perception

Perhaps the most common form of research design adopted by those who are interested in individual differences in time perception is that discussed in this section, namely the comparison of a target group (e.g., ADHD, or meditators) who should differ from a suitable control group in both personality and time perception. Here are three examples of such research. When a group of individuals presenting with ADHD are compared to a control group, they invariably perform differently on a task of time reproduction (making shorter reproductions), even when controlling for differences in IQ (Barkley et al., 2001; Smith et al., 2002). This is attributed to their impulsivity (but also to their impaired working memory, as noted above). Meditators, in turn, may also score high on trait impulsivity (Weiner et al., 2016; Wittmann et al., 2015), and sometimes are reported to differ from controls, or from their own pre-meditation baseline, on a task of time perception (Droit-Volet et al., 2015), including that of time production (making longer time productions; Berkovich-Ohana et al., 2012). This, however, has been attributed to their more focused, internally directed attention, and the subsequent slower rate of functioning of their internal clock, coupled with larger subjective time units (Glicksohn, 2001). The third example is that of delinquents, who should be impulsive and hence should have a faster rate of emission of pulses, resulting in shorter time productions (Barratt, 1981).

Using the psychophysical function relating subjective time to actual (chronometric) time is a good way of investigating individual differences in time perception. Not only does one investigate how a participant estimates a number of target durations (informative in itself), but one can also quantify this using such measures as the slope and the constant of the (linearized) function, perhaps do a more refined analysis using nonlinear regression (Kornbrot, 2016), and also look for discontinuities and aberrant performance in the data. The literature, however, is full of examples wherein either single target durations are requested, or where even with multiple target durations, a psychophysical approach is not adopted. As Eisler (1996, p. 67) comments,

> Presenting data as ratios of responses over standard durations, for instance, reproduction to standards, runs counter to the concept of a psychophysical function. Such ratios obscure the course of growth of subjective duration with clock duration. However, ratios of this kind may prove useful in some contexts.

In the next section, the psychobiology of personality and time perception is addressed, which both continues this line of group comparisons and enables correlative studies with individual parameter values of the psychophysical function.

The psychobiology of personality and time perception

In positing the existence of an internal clock whose pacemaker component is arousal-dependent, it is only natural to turn to those individual differences that are identified in personality traits as having a distinct psychobiological underpinning (Glicksohn, 2002; Zuckerman, 1989, 1998; Zuckerman & Glicksohn, 2016). These are the Big Three (Extraversion, Neuroticism, and Psychoticism), identified with the personality space defined by Eysenck (1990), Sensation Seeking, which is identified with the personality space defined by Zuckerman (2005), and Impulsivity (see, e.g., Glicksohn et al., 2016; Leshem & Glicksohn, 2007). A number of these have a direct bearing on the constructs of arousal and arousability (Zuckerman, 1987; Zuckerman & Glicksohn, 2016), and while both *arousal* and *attention* have been viewed with suspicion in the past (Allport, 1980; Claridge, 1981), their role in the architecture of the internal clock does lead to testable predictions regarding the plausibility of individual differences in time perception. With this in mind, individual differences in time perception are addressed, first in line with Eysenckian personality space, and then in terms of the sister traits of Sensation Seeking and Impulsivity (Zuckerman, 1993).

Eysenckian personality space and individual differences in time perception

In Eysenck's (1990) theory of personality, Extraversion is associated with cortical arousal (extraverts being chronically hypoaroused, best assessed using an EEG alpha index—extraverts should exhibit higher alpha power); Neuroticism is associated with autonomic arousal (those scoring high should exhibit higher variability in arousal, such as greater lability in electrodermal activity); and Psychoticism, which might well be "partly determined by cortical arousal" (Eysenck, 1994, p. 168), though this is an open question, as is the study of Psychoticism in general (for an overview, see Eysenck, 1992). What are the implications for time perception? When Rammsayer (2002) investigated to what degree these three dimensions were related to time perception (time discrimination and verbal estimation), he found no relationship with either Extraversion or Neuroticism, but rather one with Psychoticism—those scoring high on Psychoticism performing better on the verbal estimation task. Rammsayer (2002, p. 835) suggested that:

> in the light of the attentional-gate model, superior performance on temporal processing of intervals in the range of seconds may reflect high-P[Psychoticism] individuals' greater ability to direct attention more efficiently to temporal information when performing a time-estimation task.... High-P subjects' better allocation of attention to temporal information may also account for the more veridical temporal reproductions of high-P compared with low-P subjects established by means of the temporal reproduction method in a previous study.
>
> *(Rammsayer, [1997])*

Thus, Rammsayer's interpretation of the relationship found between Psychoticism (P) and both verbal estimation (Rammsayer, 2002) and time reproduction (Rammsayer, 1997) is in terms of the better attention of high-P participants. And yet, he continues to suggest (Rammsayer, 2002, p. 835) that "Such an interpretation would be functionally equivalent with the idea of high-P subjects having faster internal clocks than low-P subjects...." To my mind, these are two different hypotheses: high-P is related to either "better" attention (presumably, this would be instantiated in the attentional-gate model as the attentional gate being left open for longer) or as a faster pacemaker rate (i.e., more pulses being produced). High-P individuals have been shown to be more responsive to flicker stimulation, and "If one equates responsivity with arousability ... then P+ [high-P] participants can be characterized as being more arousable than P- participants. This is not quite the Eysenckian hypothesis (Eysenck, 1994), but is certainly of sufficient interest to warrant further research" (Glicksohn & Naftuliev, 2005, p. 1088).

Impulsivity and sensation seeking

This relationship between Psychoticism and time perception subsequently leads one to consider whether one of its subtraits and specifically that of Impulsivity (Eysenck, 1992) would bear a stronger relationship with time perception. In fact, Rammsayer (2002, p. 834) himself made this suggestion, and he is not alone here in suggesting such an important relationship. Barratt's (1993) conceptualization of Impulsivity has a strong emphasis on temporality and time perception. He suggested that "High-impulsive subjects have fast cognitive tempos like manic patients" (Barratt, 1993, p. 52) and that "temporal information processing underlies these two aspects of impulsiveness [motor and cognitive dimensions of impulsivity]" (Barratt & Stanford, 1995, p. 112). In research conducted within Barratt's model of Impulsivity, it has been reported that "High impulsive individuals ... showed significantly less accuracy than low impulsive on the time estimation tasks [time production], consistently overestimating the passage of time to a greater degree than the controls" (Lawrence & Stanford, 1999, p. 203). Other researchers have also included tasks of time perception in their battery of tests assessing cognitive impulsivity. These include Lynam (1997) and Lilienfeld et al. (1996), who employed both time production and verbal estimation in their studies. Our own working hypothesis here is that impulsive individuals will have a faster pacemaker rate (Glicksohn et al., 2006, p. 262), though we note that Sharma and Khan (2018, p. 828), in contrast, argue that "a faster internal clock or a faster cognitive processing speed favors a more deliberate processing of the choice at hand which results in more patient decisions." They suggest, further, that

The proposal that a faster internal clock may lead to impulsive intertemporal choices (Wittmann & Paulus, 2008) was based on the observation that impulsive clinical populations who tended to exhibit impatience in intertemporal choice also tended to have altered time perception. But there had been a lack of research that studied the association between interval timing and intertemporal choice among healthy samples.

(p. 828)

At this point in time, this discrepancy brings to mind four findings in the literature that deserve further attention. The first is that the relationship between impulsivity and time perception (specifically, time reproduction) might only be seen in male participants (Rammsayer & Rammstedt, 2000). The second is that impulsivity is not only a trait but also a state. As Wingrove and Bond (1997, p. 333) suggest, "people who become aware that they are likely to behave impulsively compensate by slowing down their responses." This would surely impact on their performance on a task of time perception. The third is that cognitive impulsivity and trait impulsivity do not overlap, such that participants scoring high on one of these do not necessarily score high on the other (Glicksohn et al., 2016). The fourth is that if participants are instructed not to count when performing a task of time perception (Rattat & Droit-Volet, 2012), this "may itself engage executive inhibitory processes" (Brown et al., 2013, p. 951).

Turning now to the trait of Sensation Seeking, while little research has been conducted relating this to time perception, what we do know is that Sensation Seeking will most probably interact with the sensory environment to which one is exposed in determining time perception. Glicksohn (1996) reported such an interactive effect that was specific to the intercept value of the linearized psychophysical function for time production. Glicksohn et al. (2006) further uncovered a relationship between Sensation Seeking and time reproduction. Given that time reproduction data cannot be indicative of the pacemaker emission rate of pulses (Block et al., 2000), it is plausible that it is the attentional gate which is implicated here. Brocke et al. (1999, p. 1106) remind us that "Zuckerman predicts stronger attention mechanisms for high than for low sensation seekers. Specifically, he postulates that high sensation seekers are good at focused (selective) and divided attention."

Using these psychobiological traits of personality as predictors is a good way of investigating individual differences in time perception. Three traits stand out: Psychoticism, Impulsivity, and Sensation Seeking. As a working hypothesis for future studies, it would seem that Impulsivity is related to the pacemaker emission rate of pulses (i.e., arousal level), while Sensation Seeking might be related to the attentional gate (i.e., attention). Nevertheless, Sensation Seeking has also been related to arousability (Zuckerman &

Glicksohn, 2016). It is important to stress that Impulsivity and Sensation Seeking are separate traits (Zuckerman & Glicksohn, 2016) and that arousal and arousability are also independent factors (Hagemann & Naumann, 2009). Perhaps a neurocognitive study of time perception using electrophysiological measures of arousal (e.g., baseline, eyes-closed EEG alpha power) and arousability (e.g., change in EEG alpha power from baseline to that recorded while performing the task of time perception) would be instructive. Indeed, both functional and structural brain imaging should be considered because arousal or physiological activation does refer not only to general cortical activity but also to sub-cortical activity, and further to activity in certain regions of interest. In the next section, I turn to the biological foundations of time perception.

Cognitive neuroscience and time perception

Not only does the internal clock model serve to guide research regarding its major components which are implicated by individual differences in time perception, but it also suggests which areas of the brain might be relevant for this understanding.

One line of thought here is that the insula helps to form "our internal feelings moment after moment, the development of which would be at the basis of our inner experience of duration" (Nani et al., 2019, p. 1818). A recent review has highlighted both the *left* and the *right* insula's role in time perception (assessed using fMRI, Chapter 2), in various contexts and modalities (Naghibi et al., 2023), supporting the notion that time perception (and, timing in general) is embodied. Furthermore, a study that carefully dissociated the insular activity of time perception from that related to a control task of color discrimination pinpointed activity in the *anterior* insula (Livesey et al., 2007). When the participant employs chronometric counting during time production, it is the *left* insula that is more activated (Wiener et al., 2010). Indeed, there is an inverse relationship between time reproduction and activation in the insula (Wittmann et al., 2011), indicating that impulsivity is coupled with both shorter time reproductions and greater insular activation. Insular activation (together with activity in the DLPFC— see below) also reflects the "degree of inter-subject variability" (Tregellas et al., 2006, p. 310) in time perception.

Both the accumulator (the memory store in which the pulses produced by the pacemaker are retained) and the working memory components of the internal clock could find their expression in the prefrontal cortex, especially in the dorsolateral prefrontal cortex (DLPFC). DLPFC activity (assessed primarily by fMRI and/or EEG; see Chapters 2 and 3), while reflecting the degree of inter-subject variability in time perception, also indicates its major role in time perception—especially the *right* DLPFC (Smith et al., 2003). The major task employed in these studies is that of time discrimination. Rubia (2006) proposes that the DLPFC could act as the accumulator of the internal clock, but continues to suggest that the DLPFC would then be the working memory component of the internal clock (see also Rubia & Smith, 2004, p. 333). Recall, however, that in the internal clock models, the accumulator and the working memory store are viewed as separate components (e.g., Gibbon et al., 1984; Zakay & Block, 1997). Thus, individual differences in time perception should be readily seen in differential prefrontal activity. Indeed, Rubia and Smith stress that an impairment in prefrontal activity has been found in ADHD (Smith et al., 2002).

The supplementary motor area (SMA) is also a prime candidate for representing the accumulator of the internal clock, as Macar et al. (1999) have suggested—thus proposing an additional cortical area to that of the DLPFC. Activity in the SMA can be assessed using both fMRI and EEG. A recent review has suggested that the rostral or pre-SMA (hence, closer to the DLPFC) is more engaged when longer temporal durations are involved, while the caudal aspect of the SMA is more engaged with subsecond durations (Nani et al., 2019). From their study, Macar et al. (1999) concluded that longer time productions were indexed by larger negativity of the amplitude of slow cortical activity recorded from the SMA during the task. Based on this, they suggested that SMA level of activation could well index the "pulse accumulation" process. Individual differences in time perception have been reported to correlate with such SMA activation, as Teghil et al. (2020) have noted. As Teghil et al. (2020) conclude, "interval timing relies on a broad neural network, involving the basal ganglia, the motor and supplementary motor cortex, the posterior supramarginal and posterior temporal gyrus, the middle frontal, inferior frontal, and insular cortex" (p. 255).

As Ivry and Spencer (2004, p. 1) propose, "the cerebellum can be characterized as an internal timing system. In brief, this hypothesis states that the cerebellum provides representations of the precise timing of salient events, the onset and offset of movements or the duration of a stimulus." Can one be more specific here? The cerebellum (whose activity can be monitored using fMRI) could be a pacemaker. Thus, for example, O'Leary et al. (2003) reported that smoking marijuana appears to accelerate a cerebellar clock altering self-paced behaviors. These included self-paced tapping, and the rate of counting repeatedly from one to three. On their part, Tracy et al. (2000) reported that while their participants were engaged in a task of time production, activation of the cerebellum was recorded. Thus, while Ivry and Spencer (2004) suggest their own hypothesis that "the cerebellum is best conceptualized as forming a system of multiple timing elements rather than a single amodal 'clock' .... This hypothesis assumes that specific timing elements within the cerebellum are recruited in a task-specific manner" (p. 2), given the two studies reported here, the notion that the cerebellum might be a pacemaker seems quite plausible. As such, individual differences in the rate of functioning of the pacemaker might well be linked to individual differences in cerebellar activation. Indeed, the cerebellum seems to be gaining in prominence in recent years for what is termed "predicting time" (Bareš et al., 2019).

While support for the hypothesis linking internal-clock models and the EEG alpha rhythm has waxed and waned over time (Glicksohn et al., 2009), it is still the case that the baseline rate of functioning of the pacemaker might be linked to EEG activity. Whether this be the theta band (4 to 8 Hz), given the recent report that for time production while exposed to flicker stimulation, veridical perception was achieved at a flicker rate of 6 Hz (Glicksohn & Weisinger, 2022); the beta band (14 to 18 Hz), given the recent report of a positive correlation between verbal time estimation and beta power (da Silva et al., 2021); or the alpha band (8 to 13 Hz), given the recent report providing "the first evidence that we are aware of for a causal link between individual alpha levels and the rate of the internal clock in time estimation" (Mioni et al., 2020, p. 7). The novel finding that we had reported (Glicksohn et al., 2009) was that it was not alpha *power*, but rather peak alpha frequency (PAF)—and specifically the left-right asymmetry in PAF—*during* time production (with eyes closed), which was positively correlated with time production. Van Wassenhove et al. (2019) amplify on this, writing (p. 14):

> In the internal clock proposed by Treisman (1963), one α cycle was assumed to represent one tick. Following this reasoning, the peak of the α rhythm should be linked to the subjective speed at which time is felt. The higher the individual's α peak frequency, the smaller its period and the more ticks would be accumulated in the same amount of time, thus predicting a lengthening of subjective time; conversely, the lower the α peak (the larger the period), the less ticks accumulated in the internal clock, thereby predicting a shortening of subjective time.

Whether the positive correlation that we reported between hemispheric asymmetry in EEG peak alpha frequency and time production (Glicksohn et al., 2009) is "mediated" by chronometric counting—as recently argued by Venskus and Hughes (2021)—is worthy of further research attention.

Block and Zakay (2001), in their turn-of-the-century review of time perception, wrote (p. 169): "New directions for interdisciplinary connections may enrich the study of time. One possible thread that may unify several disciplines is the answer to the old question of individual differences in the experience and judgment of time." In concluding this chapter, what will be the answer to Block and Zakay's (2001) call for studying individual differences in time perception? The study of individual differences in time perception will be highly informative regarding the actual functioning of the internal clock, as of other cognitive processes (e.g., Vogel & Awh, 2008). Indeed, the study of individual differences in performing a task of time reproduction was very much on the agenda of Hannes and Anna Eisler, as they repeatedly noted (e.g., A. D. Eisler & Eisler, 1994; A. D. Eisler et al., 1997). And as Hancock (2020, p. 6) recently stressed, "Distilling the reasons for such wide individual differences remains a great challenge to the psychology and neurophysiology of time...." This chapter has hopefully provided the reader with plenty of food for thought regarding individual differences in the study of time perception.

References

Agostino, C. S., Caetano, M. S., Balcı, F., Claessens, P. M., & Zana, Y. (2017). Individual differences in long-range time representation. *Attention, Perception, & Psychophysics, 79*(3), 833–840. https://doi.org/10.3758/s13414-017-1286-9

Allan, L. G. (1998). The influence of the scalar timing model on human timing research. *Behavioural Processes, 44*(2), 101–117. https://doi.org/10.1016/S0376-6357(98)00043-6

Allman, M. J., Teki, S., Griffiths, T. D., & Meck, W. H. (2014). Properties of the internal clock: First- and second-order principles of subjective time. *Annual Review of Psychology, 65*, 743–771. https://doi.org/10.1146/annurev-psych-010213-115117

Allport, D. A. (1980). Attention and performance. In G. Claxton (Ed.), *Cognitive psychology: New directions* (pp. 112–153). Routledge and Kegan Paul.

Bareš, M., Apps, R., Avanzino, L., Breska, A., D'Angelo, E., Filip, P., ... Petter, E. A. (2019). Consensus paper: Decoding the contributions of the cerebellum as a time machine. From neurons to clinical applications. *The Cerebellum, 18*, 266–286. https://doi.org/10.1007/s12311-018-0979-5

Barkley, R. A., Murphy, K. R., & Bush, T. (2001). Time perception and reproduction in young adults with attention deficit hyperactivity disorder. *Neuropsychology, 15*(3), 351–360. https://doi.org/10.1037/0894-4105.15.3.351

Barratt, E. S. (1981). Time perception, cortical evoked potentials, and impulsiveness among three groups of adolescents. In J. R. Hays, T. K. Roberts, & K. S. Solway (Eds.), *Violence and the violent individual* (pp. 87–95). Spectrum Publications, Inc.

Barratt, E. S. (1993). Impulsivity: Integrating cognitive, behavioral, biological, and environmental data. In W. G. McCown, J. L. Johnson, & M. B. Shure (Eds.), *The impulsive client: Theory, research, and treatment* (pp. 39–56). American Psychological Association.

Barratt, E. S., & Stanford, M. S. (1995). Impulsiveness. In C. G. Costello (Ed.), *Personality characteristics of the personality disordered* (pp. 91–119). John Wiley & Sons.

Bartholomew, A. J., Meck, W. H., & Cirulli, E. T. (2015). Analysis of genetic and non-genetic factors influencing timing and time perception. *PLoS ONE, 10*, e0143873. https://doi.org/10.1371/journal.pone.0143873

Baudouin, A., Vanneste, S., Isingrini, M., & Pouthas, V. (2006). Differential involvement of internal clock and working memory in the production and reproduction of duration: A study on older adults. *Acta Psychologica, 121*(3), 285–296. https://doi.org/10.1016/j.actpsy.2005.07.004

Berkovich-Ohana, A., Glicksohn, J., & Goldstein, A. (2012). Mindfulness-induced changes in gamma band activity—Implications for the default mode network, self-reference and attention. *Clinical Neurophysiology, 123*, 700–710. https://doi.org/10.1016/j.clinph.2011.07.048

Blankenship, D. A., & Anderson, N. H. (1976). Subjective duration: A functional measurement analysis. *Perception & Psychophysics, 20*(3), 168–172. https://doi.org/10.3758/BF03198596

Block, R. A., Hancock, P. A., & Zakay, D. (2000). Sex differences in duration judgments: A meta-analytic review. *Memory & Cognition, 28*(8), 1333–1346. https://doi.org/10.3758/BF03211834

Block, R. A., & Zakay, D. (1996). Models of psychological time revisited. In H. Helfrich (Ed.), *Time and mind* (pp. 171–195). Hogrefe & Huber.

Block, R. A., & Zakay, D. (2001). Psychological time at the millennium: Some past, present, future, and interdisciplinary issues. In M. P. Soulsby & J. T. Fraser (Eds.), *Time: Perspectives at the millennium (The study of time X)* (pp. 157–173). Bergin & Garvey.

Block, R. A., Zakay, D., & Hancock, P. A. (1999). Developmental changes in human duration judgments: A meta-analytic review. *Developmental Review, 19*(1), 183–211. https://doi.org/10.1006/drev.1998.0475

Boring, E. G. (1950). *A history of experimental psychology* (2nd ed.). Appleton-Century-Crofts.

Broadway, J. M., & Engle, R. W. (2011). Lapsed attention to elapsed time? Individual differences in working memory capacity and temporal reproduction. *Acta Psychologica, 137*(1), 115–126. https://doi.org/10.1016/j.actpsy.2011.03.008

Brocke, B., Beauducel, A., & Tasche, K. G. (1999). Biopsychological bases and behavioral correlates of sensation seeking: Contributions to a multilevel validation. *Personality and Individual Differences, 26*(6), 1103–1123. https://doi.org/10.1016/S0191-8869(98)00215-3

Brown, S. W., Collier, S. A., & Night, J. C. (2013). Timing and executive resources: Dual-task interference patterns between temporal production and shifting, updating, and inhibition tasks. *Journal of Experimental Psychology: Human Perception and Performance, 39*(4), 947–963. https://doi.org/10.1037/a0030484

Claridge, G. (1981). Arousal. In G. Underwood & R. Stevens (Eds.), *Aspects of consciousness, Vol. 2: Structural issues* (pp. 119–147). Academic Press.

da Silva, K., Curvina, M., Araújo, S., Rocha, K., Marinho, F. V., Magalhães, F. E., … Silva-Júnior, F. (2021). Male practitioners of physical activity present lower absolute power of beta band in time perception test. *Neuroscience Letters, 764*. https://doi.org/10.1016/j.neulet.2021.136210

Doob, L. W. (1971). *Patterning of time.* Yale University Press.

Droit-Volet, S., Fanget, M., & Dambrun, M. (2015). Mindfulness meditation and relaxation training increases time sensitivity. *Consciousness and Cognition, 31*, 86–97. https://doi.org/10.1016/j.concog.2014.1010.1007

Eisler, A. D., & Eisler, H. (1994). Subjective time scaling: Influence of age, gender, and type A and type B behavior. *Chronobiologia, 21*, 185–200.

Eisler, A. D., Eisler, H., & Derwinger, A. (1997). Time perception in extravert and introvert personalities. In M. Guirao (Ed.), *Procesos Sensoriales y Cognitivos* (pp. 371–387). Ediciones Dunken.

Eisler, H. (1976). Experiments on subjective duration 1868-1975: A collection of power function exponents. *Psychological Bulletin, 83*(6), 1154–1171. https://doi.org/1110.1037/0033-2909.1183.1156.1154

Eisler, H. (1996). Time perception from a psychophysicist's perspective. In H. Helfrich (Ed.), *Time and mind* (pp. 65–86). Hogrefe & Huber.

Eisler, H. (2003). The parallel-clock model: A tool for quantification of experienced duration. In R. Buccheri, M. Saniga, & W. M. Stuckey (Eds.), *The nature of time: Geometry, physics and perception* (pp. 19–26). Kluwer Academic.

Eisler, H., & Eisler, A. D. (1992). Time perception: Effects of sex and sound intensity on scales of subjective duration. *Scandinavian Journal of Psychology, 33*(4), 339–358. https://doi.org/10.1111/j.1467-9450.1992.tb00923.x

Eysenck, H. J. (1990). Biological dimensions of personality. In L. A. Pervin (Ed.), *Handbook of personality: Theory and research* (1st ed., pp. 244–276). Guilford Press.

Eysenck, H. J. (1992). The definition and measurement of psychoticism. *Personality and Individual Differences, 13*(7), 757–785. https://doi.org/10.1016/0191-8869(92)90050-Y

Eysenck, H. J. (1994). Personality: Biological foundations. In P. A. Vernon (Ed.), *The neuropsychology of individual differences* (pp. 151–207). Academic Press.

Fortin, C., Rousseau, R., Bourque, P., & Kirouac, E. (1993). Time estimation and concurrent nontemporal processing: Specific interference from short-term-memory demands. *Perception & Psychophysics, 53*(5), 536–548. https://doi.org/10.3758/BF03205202

Gibbon, J., Church, R. M., & Meck, W. H. (1984). Scalar timing in memory. In J. Gibbon & L. Allan (Eds.), *Timing and time perception. Annals of the New York Academy of Sciences* (Vol. 423, pp. 52–77). The New York Academy of Sciences.

Glicksohn, J. (1996). Entering trait and context into a cognitive-timer model for time estimation. *Journal of Environmental Psychology, 16*(4), 361–370. https://doi.org/10.1006/jevp.1996.0030

Glicksohn, J. (2001). Temporal cognition and the phenomenology of time: A multiplicative function for apparent duration. *Consciousness and Cognition, 10*(1), 1–25. https://doi.org/10.1006/ccog.2000.0468

Glicksohn, J. (2002). Criminality, personality and cognitive neuroscience. In J. Glicksohn (Ed.), *The neurobiology of criminal behavior* (pp. 3–24). Kluwer Academic.

Glicksohn, J. (2022). From illusion to reality and back in time perception. *Frontiers in Psychology, 13*. https://doi.org/10.3389/fpsyg.2022.1031564

Glicksohn, J., & Berkovich-Ohana, A. (2012). Absorption, immersion, and consciousness. In J. Gackenbach (Ed.), *Video game play and consciousness* (pp. 83–99). Nova Science Publishers, Inc.

Glicksohn, J., Berkovich-Ohana, A., Balaban Dotan, T., Goldstein, A., & Donchin, O. (2009). Time production and EEG alpha revisited. *NeuroQuantology, 7*(1), 138–151. https://doi.org/10.14704/nq.2009.7.1.215

Glicksohn, J., Berkovich-Ohana, A., Mauro, F., & Ben-Soussan, T. D. (2017). Time perception and the experience of time when immersed in an altered sensory environment. *Frontiers in Human Neuroscience, 11*, 487. https://doi.org/10.3389/fnhum.2017.00487

Glicksohn, J., & Hadad, Y. (2012). Sex differences in time production revisited. *Journal of Individual Differences, 33*(1), 35–42. https://doi.org/10.1027/1614-0001/a000059

Glicksohn, J., Hadad, Y., & Ben-Yaacov, T. (2016). "Now you see me, now you don't": The assessment of impulsivity. *Cogent Psychology*, *3*, 1242682. https://doi.org/10.1080/23311908.2016.1242682

Glicksohn, J., Leshem, R., & Aharoni, R. (2006). Impulsivity and time estimation: Casting a net to catch a fish. *Personality and Individual Differences*, *40*(2), 261–271. https://doi.org/10.1016/j.paid.2005.07.003

Glicksohn, J., & Naftuliev, Y. (2005). In search of an electrophysiological index for psychoticism. *Personality and Individual Differences*, *39*, 1083–1092. https://doi.org/10.1016/j.paid.2005.07.013

Glicksohn, J., Naor-Ziv, R., & Leshem, R. (2007). Impulsive decision making: Learning to gamble wisely? *Cognition*, *105*(1), 195–205. https://doi.org/10.1016/j.cognition.2006.08.003

Glicksohn, J., & Weisinger, B. (2022). Time production intensively studied in one observer. *Journal for Person-Oriented Research*, *8*(1), 24–36. https://doi.org/10.17505/jpor.2022.24219

Gregory, R. (1993). Seeing in time. *Perception*, *22*(11), 1257–1260. https://doi.org/10.1068/p221257

Grondin, S. (2001). From physical time to the first and second moments of psychological time. *Psychological Bulletin*, *127*(1), 22–44. https://doi.org/10.1037/0033-2909.1127.1031.1022

Grondin, S. (2020). *The perception of time: Your questions answered*. Routledge.

Hagemann, D., & Naumann, E. (2009). States vs. traits: An integrated model for the test of Eysenck's arousal/arousability hypothesis. *Journal of Individual Differences*, *30*(2), 87–99. https://doi.org/10.1027/1614-0001.30.2.87

Hancock, P. A. (2020). How and why the brain evolves time. *Behavioural Brain Research*, *373*. https://doi.org/10.1016/j.bbr.2019.112071

Hancock, P. A., & Block, R. A. (2016). A new law for time perception. *American Journal of Psychology*, *129*(2), 111–124. https://doi.org/10.5406/amerjpsyc.129.2.0111

Ivry, R. B., & Spencer, R. M. (2004). Evaluating the role of the cerebellum in temporal processing: Beware of the null hypothesis. *Brain*, *127*(8), e13. https://doi.org/10.1093/brain/awh226

Kent, L., Van Doorn, G., & Klein, B. (2019). Time dilation and acceleration in depression. *Acta Psychologica*, *194*, 77–86. https://doi.org/10.1016/j.actpsy.2019.02.003

Kornbrot, D. E. (2016). Human psychophysical functions, an update: Methods for identifying their form; estimating their parameters; and evaluating the effects of important predictors. *Psychometrika*, *81*(1), 201–216. https://doi.org/10.1007/s11336-014-9418-9

Lake, J. I., LaBar, K. S., & Meck, W. H. (2016). Emotional modulation of interval timing and time perception. *Neuroscience and Biobehavioral Reviews*, *64*, 403–420. https://doi.org/10.1016/j.neubiorev.2016.03.003

Lawrence, J. B., & Stanford, M. S. (1999). Impulsivity and time of day: Effects on performance and cognitive tempo. *Personality and Individual Differences*, *26*(2), 199–207. https://doi.org/10.1016/S0191-8869(98)00022-1

Leshem, R., & Glicksohn, J. (2007). The construct of impulsivity revisited. *Personality and Individual Differences*, *43*(4), 681–691. https://doi.org/10.1016/j.paid.2007.01.015

Lilienfeld, S. O., Hess, T., & Rowland, C. (1996). Psychopathic personality traits and temporal perspective: A test of the short time horizon hypothesis. *Journal of Psychopathology and Behavioral Assessment*, *18*(3), 285–314. https://doi.org/10.1007/BF02229050

Livesey, A. C., Wall, M. B., & Smith, A. T. (2007). Time perception: Manipulation of task difficulty dissociates clock functions from other cognitive demands. *Neuropsychologia*, *45*(2), 321–331. https://doi.org/10.1016/j.neuropsychologia.2006.06.033

Lynam, D. R. (1997). Pursuing the psychopath: Capturing the fledgling psychopath in a nomological net. *Journal of Abnormal Psychology*, *106*(3), 425–438. https://doi.org/10.1037/0021-843X.106.3.425

Macar, F., Grondin, S., & Casini, L. (1994). Controlled attention sharing influences time estimation. *Memory & Cognition*, *22*(6), 673–686. https://doi.org/10.3758/BF03209252

Macar, F., Vidal, F., & Casini, L. (1999). The supplementary motor area in motor and sensory timing: Evidence from slow brain potential changes. *Experimental Brain Research*, *125*(3), 271–280. https://doi.org/10.1007/s002210050683

Matthews, W. J., & Meck, W. H. (2016). Temporal cognition: Connecting subjective time to perception, attention, and memory. *Psychological Bulletin*, *142*(8), 865–907. https://doi.org/10.1037/bul0000045

McClain, L. (1983). Interval estimation: Effect of processing demands on prospective and retrospective reports. *Perception & Psychophysics*, *34*(2), 185–189. https://doi.org/10.3758/BF03211347

Mioni, G., Shelp, A., Stanfield-Wiswell, C. T., Gladhill, K. A., Bader, F., & Wiener, M. (2020). Modulation of individual alpha frequency with tACS shifts time perception. *Cerebral Cortex Communications*, *1*(1), 1–9. https://doi.org/10.1093/texcom/tgaa064

Mioni, G., Stablum, F., Prunetti, E., & Grondin, S. (2016). Time perception in anxious and depressed patients: A comparison between time reproduction and time production task. *Journal of Affective Disorders*, *196*, 154–163. https://doi.org/10.1016/j.jad.2016.02.047

Mollon, J. D., & Perkins, A. J. (1996). Errors of judgement at Greenwich in 1796. *Nature*, 380, 101–102. https://doi.org/10.1038/380101a0

Naghibi, N., Jahangiri, N., Khosrowabadi, R., Eickhoff, C. R., Eickhoff, S. B., Coull, J. T., & Tahmasian, M. (2023). Embodying time in the brain: A multi-dimensional neuroimaging meta-analysis of 95 duration processing studies. *Neuropsychology Review*, 1–22. https://doi.org/10.1007/s11065-023-09588-1

Nani, A., Manuello, J., Liloia, D., Duca, S., Costa, T., & Cauda, F. (2019). The neural correlates of time: A meta-analysis of neuroimaging studies. *Journal of Cognitive Neuroscience*, *31*(12), 1796–1826. https://doi.org/10.1162/jocn_a_01459

O'Leary, D. S., Block, R. I., Turner, B. M., Koeppel, J., Magnotta, V. A., Ponto, L. B., . . . Andreasen, N. C. (2003). Marijuana alters the human cerebellar clock. *NeuroReport*, *14*(8), 1145–1151. https://doi.org/10.1097/00001756-200306110-00009

Penton-Voak, I. S., Edwards, H., Percival, A., & Wearden, J. H. (1996). Speeding up an internal clock in humans? Effects of click trains on subjective duration. *Journal of Experimental Psychology: Animal Behavior Processes*, *22*(3), 307–320. https://doi.org/10.1037/0097-7403.22.3.307

Pouthas, V., & Perbal, S. (2004). Time perception depends on accurate clock mechanisms as well as unimpaired attention and memory processes. *Acta Neurobiologiae Experimentalis*, *64*(3), 367–385.

Rammsayer, T. H. (1997). On the relationship between personality and time estimation. *Personality and Individual Differences*, *23*(5), 739–744. https://doi.org/10.1016/S0191-8869(97)00117-7

Rammsayer, T. H. (2002). Temporal information processing and basic dimensions of personality: Differential effects of psychoticism. *Personality and Individual Differences*, *32*(5), 827–838. https://doi.org/10.1016/S0191-8869(01)00089-7

Rammsayer, T. H., & Rammstedt, B. (2000). Sex-related differences in time estimation: The role of personality. *Personality and Individual Differences*, *29*(2), 301–312. https://doi.org/10.1016/S0191-8869(99)00194-4

Rammsayer, T. H., & Troche, S. (2014). Elucidating the internal structure of psychophysical timing performance in the sub-second and second range by utilizing confirmatory factor analysis. In H. Merchant & V. de Lafuente (Eds.), *Neurobiology of interval Timing* (pp. 33–47). Springer.

Rattat, A.-C., & Droit-Volet, S. (2012). What is the best and easiest method of preventing counting in different temporal tasks? *Behavior Research Methods*, *44*(1), 67–80. https://doi.org/10.3758/s13428-011-0135-3

Rubia, K. (2006). The neural correlates of timing functions. In J. Glicksohn & M. S. Myslobodsky (Eds.), *Timing the future: The case for a time-based prospective memory* (pp. 213–238). World Scientific.

Rubia, K., & Smith, A. (2004). The neural correlates of cognitive time management: A review. *Acta Neurobiologiae Experimentalis*, *64*(3), 329–340.

Sawyer, T. F., Meyers, P. J., & Huser, S. J. (1994). Contrasting task demands alter the perceived duration of brief time intervals. *Perception & Psychophysics*, *56*(6), 649–657. https://doi.org/10.3758/BF03208358

Schaffer, S. (1988). Astronomers mark time: Discipline and the personal equation. *Science in Context*, *2*(1), 115–145. https://doi.org/10.1017/S026988970000051X

Sharma, S. N., & Khan, A. (2018). Interval timing predicts impulsivity in intertemporal choice: combined behavioral and drift-diffusion model evidence. *Journal of Cognitive Psychology*, *30*(8), 816–831. https://doi.org/10.1080/20445911.2018.1539002

Smith, A., Taylor, E., Lidzba, K., & Rubia, K. (2003). A right hemispheric frontocerebellar network for time discrimination of several hundreds of milliseconds. *NeuroImage*, *20*(1), 344–350. https://doi.org/10.1016/S1053-8119(03)00337-9

Smith, A., Taylor, E., Rogers, J. W., Newman, S., & Rubia, K. (2002). Evidence for a pure time perception deficit in children with ADHD. *Journal of Child Psychology and Psychiatry and Allied Disciplines*, *43*(4), 529–542. https://doi.org/10.1111/1469-7610.00043

Teghil, A., Di Vita, A., D'Antonio, F., & Boccia, M. (2020). Inter-individual differences in resting-state functional connectivity are linked to interval timing in irregular contexts. *Cortex*, *128*, 254–269. https://doi.org/10.1016/j.cortex.2020.03.021

Tracy, J. I., Faro, S. H., Mohamed, F. B., Pinsk, M., & Pinus, A. (2000). Functional localization of a "time keeper" function separate from attentional resources and task strategy. *NeuroImage*, *11*(3), 228–242. https://doi.org/10.1006/nimg.2000.0535

Tregellas, J. R., Davalos, D. B., & Rojas, D. C. (2006). Effect of task difficulty on the functional anatomy of temporal processing. *NeuroImage*, *32*(1), 307–315. https://doi.org/10.1016/j.neuroimage.2006.02.036

Treisman, M. (1963). Temporal discrimination and the indifference interval: Implications for a model of the "internal clock". *Psychological Monographs: General and Applied*, *77*(13), 576. https://doi.org/10.1037/h0093864

van Wassenhove, V., Herbst, S., & Kononowicz, T. W. (2019). Timing the brain to time the mind: Critical contributions of time-resolved neuroimaging for temporal cognition. In S. Supek & C. J. Aine (Eds.), *Magnetoencephalography* (pp. 1–50). Springer.

Venskus, A., & Hughes, G. (2021). Individual differences in alpha frequency are associated with the time window of multisensory integration, but not time perception. *Neuropsychologia*. https://doi.org/10.1016/j.neuropsychologia.2021.107919

Vogel, E. K., & Awh, E. (2008). How to exploit diversity for scientific gain: Using individual differences to constrain cognitive theory. *Current Directions in Psychological Science, 17*(2), 171–176. https://doi.org/10.1111/j.1467-8721. 2008.00569.x

Wearden, J. (2016). *The psychology of time perception.* Palgrave Macmillan.

Wearden, J. H., O'Rourke, S. C., Matchwick, C., Min, Z., & Maeers, S. (2010). Task switching and subjective duration. *The Quarterly Journal of Experimental Psychology, 63*(3), 531–543. https://doi.org/10.1080/17470210903024768

Weiner, L., Wittmann, M., Bertschy, G., & Giersch, A. (2016). Dispositional mindfulness and subjective time in healthy individuals. *Frontiers in Psychology, 7,* 786. https://doi.org/10.3389/fpsyg.2016.00786

Wiener, M., Turkeltaub, P., & Coslett, H. B. (2010). The image of time: A voxelwise meta-analysis. *NeuroImage, 49*(2), 1728–1740. https://doi.org/10.1016/j. neuroimage.2009.09.064

Wingrove, J., & Bond, A. J. (1997). Impulsivity: A state as well as trait variable. Does mood awareness explain low correlations between trait and behavioural measures of impulsivity? *Personality and Individual Differences, 22*(3), 333–339. https://doi. org/10.1016/S0191-8869(96)00222-X

Wittmann, M., Otten, S., Schötz, E., Sarikaya, A., Lehnen, H., Jo, H.-G., . . . Meissner, K. (2015). Subjective expansion of extended time-spans in experienced meditators. *Frontiers in Psychology, 5,* 1586. https://doi.org/10.3389/fpsyg. 2014.01586

Wittmann, M., & Paulus, M. P. (2008). Decision making, impulsivity and time perception. *Trends in Cognitive Sciences, 12*(1), 7–12. https://doi.org/10.1016/j.tics. 2007.10.004

Wittmann, M., Simmons, A. N., Flagan, T., Lane, S. D., Wackermann, J., & Paulus, M. P. (2011). Neural substrates of time perception and impulsivity. *Brain Research, 1406,* 43–58. https://doi.org/10.1016/j.brainres.2011.06.048

Woehrle, J. L., & Magliano, J. P. (2012). Time flies faster if a person has a high working-memory capacity. *Acta Psychologica, 139*(2), 314–319. https://doi.org/ 10.1016/j.actpsy.2011.12.006

Zakay, D. (1989). Subjective time and attentional resource allocation: An integrated model of time estimation. In I. Levin, & D. Zakay (Eds.), *Time and human cognition: A life-span perspective* (pp. 365–397). Elsevier Science Publishers.

Zakay, D. (1990). The evasive art of subjective time measurement: Some methodological dilemmas. In R. A. Block (Ed.), *Cognitive models of psychological time* (pp. 59–84). Lawrence Erlbaum Associates, Inc.

Zakay, D. (1993). Time estimation methods—Do they influence prospective duration estimates? *Perception, 22*(1), 91–101. https://doi.org/10.1068/p220091

Zakay, D., & Block, R. A. (1997). Temporal cognition. *Current Directions in Psychological Science, 6*(1), 12–16. https://doi.org/10.1111/1467-8721.ep115 12604

Zakay, D., Block, R. A., & Tsal, Y. (1999). Prospective duration estimation and performance. In D. Gopher, & A. Koriat (Eds.), *Attention and performance XVII. Cognitive regulation of performance: Interaction of theory and application* (Vol. 17, pp. 557–580). MIT Press.

Zuckerman, M. (1987). A critical look at three arousal constructs in personality theories: Optimal levels of arousal, strength of the nervous system, and sensitivities to signals of reward and punishment. In J. Strelau, & H. J. Eysenck (Eds.), *Personality dimensions and arousal* (pp. 217–231). Plenum.

Zuckerman, M. (1989). Personality in the third dimension: A psychobiological approach. *Personality and Individual Differences, 10*(4), 391–418. https://doi.org/10.1016/0191-8869(89)90004-4

Zuckerman, M. (1993). Sensation seeking and impulsivity: A marriage of traits made in biology? In W. G. McCown, J. L. Johnson, & M. B. Shure (Eds.), *The impulsive client: Theory, research, and treatment* (pp. 71–91). American Psychological Association.

Zuckerman, M. (1998). Psychobiological theories of personality. In D. F. Barone, M. Hersen, & V. B. Van Hassett (Eds.), *Advanced personality* (pp. 123–154). Plenum.

Zuckerman, M. (2005). *Psychobiology of personality* (2nd ed., revised and updated). Cambridge University Press.

Zuckerman, M., & Glicksohn, J. (2016). Hans Eysenck's personality model and the constructs of sensation seeking and impulsivity. *Personality and Individual Differences, 103*, 48–52. https://doi.org/10.1016/j.paid.2016.04.003

INDEX